partial differential equation methods in control and shape analysis

LECTURE NOTES IN PURE AND APPLIED MATHEMATICS

1. *N. Jacobson,* Exceptional Lie Algebras
2. *L.-Å. Lindahl and F. Poulsen,* Thin Sets in Harmonic Analysis
3. *I. Satake,* Classification Theory of Semi-Simple Algebraic Groups
4. *F. Hirzebruch, W. D. Newmann, and S. S. Koh, Differentiable Manifolds and Quadratic Forms*
5. *I. Chavel,* Riemannian Symmetric Spaces of Rank One
6. *R. B. Burckel,* Characterization of C(X) Among Its Subalgebras
7. *B. R. McDonald, A. R. Magid, and K. C. Smith,* Ring Theory: Proceedings of the Oklahoma Conference
8. *Y.-T. Siu,* Techniques of Extension on Analytic Objects
9. *S. R. Caradus, W. E. Pfaffenberger, and B. Yood,* Calkin Algebras and Algebras of Operators on Banach Spaces
10. *E. O. Roxin, P.-T. Liu, and R. L. Sternberg,* Differential Games and Control Theory
11. *M. Orzech and C. Small,* The Brauer Group of Commutative Rings
12. *S. Thomier,* Topology and Its Applications
13. *J. M. Lopez and K. A. Ross,* Sidon Sets
14. *W. W. Comfort and S. Negrepontis,* Continuous Pseudometrics
15. *K. McKennon and J. M. Robertson,* Locally Convex Spaces
16. *M. Carmeli and S. Malin,* Representations of the Rotation and Lorentz Groups: An Introduction
17. *G. B. Seligman,* Rational Methods in Lie Algebras
18. *D. G. de Figueiredo,* Functional Analysis: Proceedings of the Brazilian Mathematical Society Symposium
19. *L. Cesari, R. Kannan, and J. D. Schuur,* Nonlinear Functional Analysis and Differential Equations: Proceedings of the Michigan State University Conference
20. *J. J. Schäffer,* Geometry of Spheres in Normed Spaces
21. *K. Yano and M. Kon,* Anti-Invariant Submanifolds
22. *W. V. Vasconcelos,* The Rings of Dimension Two
23. *R. E. Chandler,* Hausdorff Compactifications
24. *S. P. Franklin and B. V. S. Thomas,* Topology: Proceedings of the Memphis State University Conference
25. *S. K. Jain,* Ring Theory: Proceedings of the Ohio University Conference
26. *B. R. McDonald and R. A. Morris,* Ring Theory II: Proceedings of the Second Oklahoma Conference
27. *R. B. Mura and A. Rhemtulla,* Orderable Groups
28. *J. R. Graef,* Stability of Dynamical Systems: Theory and Applications
29. *H.-C. Wang,* Homogeneous Branch Algebras
30. *E. O. Roxin, P.-T. Liu, and R. L. Sternberg,* Differential Games and Control Theory II
31. *R. D. Porter,* Introduction to Fibre Bundles
32. *M. Altman,* Contractors and Contractor Directions Theory and Applications
33. *J. S. Golan,* Decomposition and Dimension in Module Categories
34. *G. Fairweather,* Finite Element Galerkin Methods for Differential Equations
35. *J. D. Sally,* Numbers of Generators of Ideals in Local Rings
36. *S. S. Miller,* Complex Analysis: Proceedings of the S.U.N.Y. Brockport Conference
37. *R. Gordon,* Representation Theory of Algebras: Proceedings of the Philadelphia Conference
38. *M. Goto and F. D. Grosshans,* Semisimple Lie Algebras
39. *A. I. Arruda, N. C. A. da Costa, and R. Chuaqui,* Mathematical Logic: Proceedings of the First Brazilian Conference
40. *F. Van Oystaeyen,* Ring Theory: Proceedings of the 1977 Antwerp Conference
41. *F. Van Oystaeyen and A. Verschoren,* Reflectors and Localization: Application to Sheaf Theory
42. *M. Satyanarayana,* Positively Ordered Semigroups
43. *D. L Russell,* Mathematics of Finite-Dimensional Control Systems
44. *P.-T. Liu and E. Roxin,* Differential Games and Control Theory III: Proceedings of the Third Kingston Conference, Part A
45. *A. Geramita and J. Seberry,* Orthogonal Designs: Quadratic Forms and Hadamard Matrices
46. *J. Cigler, V. Losert, and P. Michor,* Banach Modules and Functors on Categories of Banach Spaces
47. *P.-T. Liu and J. G. Sutinen,* Control Theory in Mathematical Economics: Proceedings of the Third Kingston Conference, Part B

48. *C. Byrnes,* Partial Differential Equations and Geometry
49. *G. Klambauer,* Problems and Propositions in Analysis
50. *J. Knopfmacher,* Analytic Arithmetic of Algebraic Function Fields
51. *F. Van Oystaeyen,* Ring Theory: Proceedings of the 1978 Antwerp Conference
52. *B. Kadem,* Binary Time Series
53. *J. Barros-Neto and R. A. Artino,* Hypoelliptic Boundary-Value Problems
54. *R. L. Sternberg, A. J. Kalinowski, and J. S. Papadakis,* Nonlinear Partial Differential Equations in Engineering and Applied Science
55. *B. R. McDonald,* Ring Theory and Algebra III: Proceedings of the Third Oklahoma Conference
56. *J. S. Golan,* Structure Sheaves Over a Noncommutative Ring
57. *T. V. Narayana, J. G. Williams, and R. M. Mathsen,* Combinatorics, Representation Theory and Statistical Methods in Groups: YOUNG DAY Proceedings
58. *T. A. Burton,* Modeling and Differential Equations in Biology
59. *K. H. Kim and F. W. Roush,* Introduction to Mathematical Consensus Theory
60. *J. Banas and K. Goebel,* Measures of Noncompactness in Banach Spaces
61. *O. A. Nielson,* Direct Integral Theory
62. *J. E. Smith, G. O. Kenny, and R. N. Ball,* Ordered Groups: Proceedings of the Boise State Conference
63. *J. Cronin,* Mathematics of Cell Electrophysiology
64. *J. W. Brewer,* Power Series Over Commutative Rings
65. *P. K. Kamthan and M. Gupta,* Sequence Spaces and Series
66. *T. G. McLaughlin,* Regressive Sets and the Theory of Isols
67. *T. L. Herdman, S. M. Rankin III, and H. W. Stech,* Integral and Functional Differential Equations
68. *R. Draper,* Commutative Algebra: Analytic Methods
69. *W. G. McKay and J. Patera,* Tables of Dimensions, Indices, and Branching Rules for Representations of Simple Lie Algebras
70. *R. L. Devaney and Z. H. Nitecki,* Classical Mechanics and Dynamical Systems
71. *J. Van Geel,* Places and Valuations in Noncommutative Ring Theory
72. *C. Faith,* Injective Modules and Injective Quotient Rings
73. *A. Fiacco,* Mathematical Programming with Data Perturbations I
74. *P. Schultz, C. Praeger, and R. Sullivan,* Algebraic Structures and Applications: Proceedings of the First Western Australian Conference on Algebra
75. *L Bican, T. Kepka, and P. Nemec,* Rings, Modules, and Preradicals
76. *D. C. Kay and M. Breen,* Convexity and Related Combinatorial Geometry: Proceedings of the Second University of Oklahoma Conference
77. *P. Fletcher and W. F. Lindgren,* Quasi-Uniform Spaces
78. *C.-C. Yang,* Factorization Theory of Meromorphic Functions
79. *O. Taussky,* Ternary Quadratic Forms and Norms
80. *S. P. Singh and J. H. Burry,* Nonlinear Analysis and Applications
81. *K. B. Hannsgen, T. L. Herdman, H. W. Stech, and R. L. Wheeler,* Volterra and Functional Differential Equations
82. *N. L. Johnson, M. J. Kallaher, and C. T. Long,* Finite Geometries: Proceedings of a Conference in Honor of T. G. Ostrom
83. *G. I. Zapata,* Functional Analysis, Holomorphy, and Approximation Theory
84. *S. Greco and G. Valla,* Commutative Algebra: Proceedings of the Trento Conference
85. *A. V. Fiacco,* Mathematical Programming with Data Perturbations II
86. *J.-B. Hiriart-Urruty, W. Oettli, and J. Stoer,* Optimization: Theory and Algorithms
87. *A. Figa Talamanca and M. A. Picardello,* Harmonic Analysis on Free Groups
88. *M. Harada,* Factor Categories with Applications to Direct Decomposition of Modules
89. *V. I. Istrățescu,* Strict Convexity and Complex Strict Convexity
90. *V. Lakshmikantham,* Trends in Theory and Practice of Nonlinear Differential Equations
91. *H. L. Manocha and J. B. Srivastava,* Algebra and Its Applications
92. *D. V. Chudnovsky and G. V. Chudnovsky,* Classical and Quantum Models and Arithmetic Problems
93. *J. W. Longley,* Least Squares Computations Using Orthogonalization Methods
94. *L. P. de Alcantara,* Mathematical Logic and Formal Systems
95. *C. E. Aull,* Rings of Continuous Functions
96. *R. Chuaqui,* Analysis, Geometry, and Probability
97. *L. Fuchs and L. Salce,* Modules Over Valuation Domains
98. *P. Fischer and W. R. Smith,* Chaos, Fractals, and Dynamics
99. *W. B. Powell and C. Tsinakis,* Ordered Algebraic Structures

100. *G. M. Rassias and T. M. Rassias,* Differential Geometry, Calculus of Variations, and Their Applications
101. *R.-E. Hoffmann and K. H. Hofmann,* Continuous Lattices and Their Applications
102. *J. H. Lightbourne III and S. M. Rankin III,* Physical Mathematics and Nonlinear Partial Differential Equations
103. *C. A. Baker and L. M. Batten,* Finite Geometrics
104. *J. W. Brewer, J. W. Bunce, and F. S. Van Vleck,* Linear Systems Over Commutative Rings
105. *C. McCrory and T. Shifrin,* Geometry and Topology: Manifolds, Varieties, and Knots
106. *D. W. Kueker, E. G. K. Lopez-Escobar, and C. H. Smith,* Mathematical Logic and Theoretical Computer Science
107. *B.-L. Lin and S. Simons,* Nonlinear and Convex Analysis: Proceedings in Honor of Ky Fan
108. *S. J. Lee,* Operator Methods for Optimal Control Problems
109. *V. Lakshmikantham,* Nonlinear Analysis and Applications
110. *S. F. McCormick,* Multigrid Methods: Theory, Applications, and Supercomputing
111. *M. C. Tangora,* Computers in Algebra
112. *D. V. Chudnovsky and G. V. Chudnovsky,* Search Theory: Some Recent Developments
113. *D. V. Chudnovsky and R. D. Jenks,* Computer Algebra
114. *M. C. Tangora,* Computers in Geometry and Topology
115. *P. Nelson, V. Faber, T. A. Manteuffel, D. L. Seth, and A. B. White, Jr.,* Transport Theory, Invariant Imbedding, and Integral Equations: Proceedings in Honor of G. M. Wing's 65th Birthday
116. *P. Clément, S. Invernizzi, E. Mitidieri, and I. I. Vrabie,* Semigroup Theory and Applications
117. *J. Vinuesa,* Orthogonal Polynomials and Their Applications: Proceedings of the International Congress
118. *C. M. Dafermos, G. Ladas, and G. Papanicolaou,* Differential Equations: Proceedings of the EQUADIFF Conference
119. *E. O. Roxin,* Modern Optimal Control: A Conference in Honor of Solomon Lefschetz and Joseph P. Lasalle
120. *J. C. Díaz,* Mathematics for Large Scale Computing
121. *P. S. Milojevič,* Nonlinear Functional Analysis
122. *C. Sadosky,* Analysis and Partial Differential Equations: A Collection of Papers Dedicated to Mischa Cotlar
123. *R. M. Shortt,* General Topology and Applications: Proceedings of the 1988 Northeast Conference
124. *R. Wong,* Asymptotic and Computational Analysis: Conference in Honor of Frank W. J. Olver's 65th Birthday
125. *D. V. Chudnovsky and R. D. Jenks,* Computers in Mathematics
126. *W. D. Wallis, H. Shen, W. Wei, and L. Zhu,* Combinatorial Designs and Applications
127. *S. Elaydi,* Differential Equations: Stability and Control
128. *G. Chen, E. B. Lee, W. Littman, and L. Markus,* Distributed Parameter Control Systems: New Trends and Applications
129. *W. N. Everitt,* Inequalities: Fifty Years On from Hardy, Littlewood and Pólya
130. *H. G. Kaper and M. Garbey,* Asymptotic Analysis and the Numerical Solution of Partial Differential Equations
131. *O. Arino, D. E. Axelrod, and M. Kimmel,* Mathematical Population Dynamics: Proceedings of the Second International Conference
132. *S. Coen,* Geometry and Complex Variables
133. *J. A. Goldstein, F. Kappel, and W. Schappacher,* Differential Equations with Applications in Biology, Physics, and Engineering
134. *S. J. Andima, R. Kopperman, P. R. Misra, J. Z. Reichman, and A. R. Todd,* General Topology and Applications
135. *P Clément, E. Mitidieri, B. de Pagter,* Semigroup Theory and Evolution Equations: The Second International Conference
136. *K. Jarosz,* Function Spaces
137. *J. M. Bayod, N. De Grande-De Kimpe, and J. Martínez-Maurica,* *p*-adic Functional Analysis
138. *G. A. Anastassiou,* Approximation Theory: Proceedings of the Sixth Southeastern Approximation Theorists Annual Conference
139. *R. S. Rees,* Graphs, Matrices, and Designs: Festschrift in Honor of Norman J. Pullman
140. *G. Abrams, J. Haefner, and K. M. Rangaswamy,* Methods in Module Theory
141. *G. L. Mullen and P. J.-S. Shiue,* Finite Fields, Coding Theory, and Advances in Communications and Computing
142. *M. C. Joshi and A. V. Balakrishnan,* Mathematical Theory of Control: Proceedings of the International Conference

143. *G. Komatsu and Y. Sakane*, Complex Geometry: Proceedings of the Osaka International Conference
144. *I. J. Bakelman*, Geometric Analysis and Nonlinear Partial Differential Equations
145. *T. Mabuchi and S. Mukai*, Einstein Metrics and Yang–Mills Connections: Proceedings of the 27th Taniguchi International Symposium
146. *L. Fuchs and R. Göbel*, Abelian Groups: Proceedings of the 1991 Curaçao Conference
147. *A. D. Pollington and W. Moran*, Number Theory with an Emphasis on the Markoff Spectrum
148. *G. Dore, A. Favini, E. Obrecht, and A. Venni*, Differential Equations in Banach Spaces
149. *T. West*, Continuum Theory and Dynamical Systems
150. *K. D. Bierstedt, A. Pietsch, W. Ruess, and D. Vogt*, Functional Analysis
151. *K. G. Fischer, P. Loustaunau, J. Shapiro, E. L. Green, and D. Farkas*, Computational Algebra
152. *K. D. Elworthy, W. N. Everitt, and E. B. Lee*, Differential Equations, Dynamical Systems, and Control Science
153. *P.-J. Cahen, D. L. Costa, M. Fontana, and S.-E. Kabbaj*, Commutative Ring Theory
154. *S. C. Cooper and W. J. Thron*, Continued Fractions and Orthogonal Functions: Theory and Applications
155. *P. Clément and G. Lumer*, Evolution Equations, Control Theory, and Biomathematics
156. *M. Gyllenberg and L. Persson*, Analysis, Algebra, and Computers in Mathematical Research: Proceedings of the Twenty-First Nordic Congress of Mathematicians
157. *W. O. Bray, P. S. Milojevič and Č. V. Stanojevič*, Fourier Analysis: Analytic and Geometric Aspects
158. *J. Bergen and S. Montgomery*, Advances in Hopf Algebras
159. *A. R. Magid*, Rings, Extensions, and Cohomology
160. *N. H. Pavel*, Optimal Control of Differential Equations
161. *M. Ikawa*, Spectral and Scattering Theory: Proceedings of the Taniguchi International Workshop
162. *X. Liu and D. Siegel*, Comparison Methods and Stability Theory
163. *J.-P. Zolésio*, Boundary Control and Variation
164. *M. Křížek, P. Neittaanmäki, and R. Stenberg*, Finite Element Methods: Fifty Years of the Courant Element
165. *G. Da Prato and L. Tubaro*, Control of Partial Differential Equations
166. *E. Ballico*, Projective Geometry with Applications
167. *M. Costabel, M. Dauge, and S. Nicaise*, Boundary Value Problems and Integral Equations in Nonsmooth Domains
168. *G. Ferreyra, G. R. Goldstein, and F. Neubrander*, Evolution Equations
169. *S. Huggett*, Twistor Theory
170. *H. Cook, W. T. Ingram, K. T. Kuperberg, A. Lelek, and P. Minc*, Continua: With the Houston Problem Book
171. *D. F. Anderson and D. E. Dobbs*, Zero-Dimensional Commutative Rings
172. *K. Jarosz*, Function Spaces: The Second Conference
173. *V. Ancona, E. Ballico, and A. Silva*, Complex Analysis and Geometry
174. *E. Casas*, Control of Partial Differential Equations and Applications
175. *N. Kalton, E. Saab, and S. Montgomery-Smith*, Interaction Between Functional Analysis, Harmonic Analysis, and Probability
176. *Z. Deng, Z. Liang, G. Lu, and S. Ruan*, Differential Equations and Control Theory
177. *P. Marcellini, G. Talenti, and E. Vesentini*, Partial Differential Equations and Applications: Collected Papers in Honor of Carlo Pucci
178. *A. Kartsatos*, Theory and Applications of Nonlinear Operators of Accretive and Monotone Type
179. *M. Maruyama*, Moduli of Vector Bundles
180. *A. Ursini and P. Aglianò*, Logic and Algebra
181. *X. H. Cao, S. X. Liu, K. P. Shum, and C. C. Yang*, Rings, Groups, and Algebras
182. *D. Arnold and R. M. Rangaswamy*, Abelian Groups and Modules
183. *S. R. Chakravarthy and A. S. Alfa*, Matrix-Analytic Methods in Stochastic Models
184. *J. E. Andersen, J. Dupont, H. Pedersen, and A. Swann*, Geometry and Physics
185. *P.-J. Cahen, M. Fontana, E. Houston, and S.-E. Kabbaj*, Commutative Ring Theory: Proceedings of the II International Conference
186. *J. A. Goldstein, N. E. Gretsky, and J. J. Uhl, Jr.*, Stochastic Processes and Functional Analysis
187. *A. Sorbi*, Complexity, Logic, and Recursion Theory
188. *G. Da Prato and J.-P. Zolésio*, Partial Differential Equation Methods in Control and Shape Analysis

Additional Volumes in Preparation

partial differential equation methods in control and shape analysis

edited by

Giuseppe Da Prato

Scuola Normale Superiore
Pisa, Italy

Jean-Paul Zolésio

Institut Non Linéaire de Nice
Centre National de la Recherche Scientifique
and Centre de Mathématiques Appliquées
École des Mines de Paris
Sophia Antipolis, France

CRC Press
Taylor & Francis Group
Boca Raton London New York

CRC Press is an imprint of the
Taylor & Francis Group, an **informa** business

CRC Press
Taylor & Francis Group
6000 Broken Sound Parkway NW, Suite 300
Boca Raton, FL 33487-2742

First issued in hardback 2017

ISBN 13: 978-1-138-41323-8 (hbk)
ISBN 13: 978-0-8247-9837-6 (pbk)

Visit the Taylor & Francis Web site at
http://www.taylorandfrancis.com

and the CRC Press Web site at
http://www.crcpress.com

Library of Congress Cataloging-in-Publication Data

Partial differential equation methods in control and shape analysis /
edited by G. Da Prato and J.-P. Zolésio.
p. cm. (Lecture notes in pure and applied mathematics ; v. 188)
ISBN 0-8247-9837-6 (alk. paper)
1. Shape theory (Topology)—Congresses. 2. Control theory—Congresses. 3. Differential equations, Partial—Congresses.
I. Da Prato, Giuseppe. II. Zolésio, J.-P. III. Series.
QA612.7.P37 1997
629.8'312—dc21

96-29978
CIP

Preface

The International Federation for Information Processing (IFIP) working group 7.2 Conference on Control and Shape Optimization was held at Scuola Normale Superiore di Pisa, Italy. The meeting was sponsored by Scuola Normale Superiore di Pisa and CNR Gruppo Nazionale di Analisi Funzionale. The purpose of the workshop was to exchange ideas between the group working on control theory and the group working on shape optimization. It was part of an ongoing collaboration between Scuola Normale Superiore di Pisa and the Centre de Recherche en Mathématiques Appliquées de l'Ecole des Mines de Paris.

Optimization and control theory are recurrent themes in the modeling of real-life systems from many areas: real-time systems, material sciences, lifting profiles, thermal testing, elastic shells, and biodynamics. The Hamilton-Jacobi approach is beginning to play a major role in solving concrete problems where active control is needed, while shape optimization is the tool of choice for passive control problems. The challenge is to bring these two approaches together (e.g., the optimal location of actuators/sensors for tracking improvement, the best shape of a plate for enhancing the stabilizing control). We hope this volume will stimulate further research.

We would like to thank all contributors and Mrs. Caterina D'Elia, at Scuola Normale, for their efforts on behalf of the conference.

Guiseppe Da Prato
Jean-Paul Zolésio

Contents

Contributors

Jean-Christophe Aguilar Ecole des Mines de Paris, Centre de Mathématiques Appliquées, Sophia Antipolis, France

S. Andrieux EDF-DER-MMN, Clamart, France

F. Bagagiolo Università di Trento, Povo (Trento), Italy

M. Bardi Università di Padova, Padova, Italy

A. Ben Abda ENIT, LAMAP, Tunis, and Institut Préparatoire aux Études Scientifiques et Techniques, La Marsa, Tunisia

Lucio Boccardo Universitá di Roma I, Rome, Italy

Francesca Bucci Università di Modena, Modena, Italy

Eduardo Casas ETSI Caminos - Universidad de Cantabria, Santander, Spain

Michel C. Delfour Université de Montréal, Montréal, Québec, Canada

F. R. Desaint CNRS-Institut Non-Linéaire de Nice, Sophia Antipolis, France

I. Capuzzo Dolcetta Università di Roma "La Sapienza," Rome, Italy

Raja Dziri École des Mines de Paris, Centre de Mathématiques Appliquées, Sophia Antipolis, France

Roberto Ferretti Seconda Università di Roma "Tor Vergata," Rome, Italy

Fausto Gozzi Università di Pisa, Pisa, Italy

Y. Guido CMA, ENSMP, Sophia Antipolis, France

M. Jaoua ENIT, LAMAP, Tunis, and Institut Préparatoire aux Études Scientifiques et Techniques, La Marsa, Tunisia

Irena Lasiecka University of Virginia, Charlottesville, Virginia

Jean-Paul Marmorat CMA-EMP, Sophia Antipolis, France

Zoubida Mghazli Université Ibn Tofail, Kénitra, Morocco

J. Pousin Swiss Federal Institute of Technology, Lausanne, Switzerland, and INSA LYON, Laboratoire Modélisation et Calcul Scientifique, URA-CNRS, Villeurbanne, France

Jean R. Roche Université de Nancy 1, URA -CNRS, Vandoeuvre-Les-Nancy, France

Giulia Sargenti Università di Roma "La Sapienza," Rome, Italy

Carlo Sinestrari Università di Roma "Tor Vergata," Rome, Italy

Jan Sokołowski Université de Nancy I, URA-CNRS, Vandoeuvre-Les Nancy, France, and Systems Research Institute of the Polish Academy of Sciences, Warsaw, Poland

Gianmario Tessitore Università degli Studi "G. Sansone," Florence, Italy

Maria Elisabetta Tessitore Università di Roma "La Sapienza," Rome, Italy

Roberto Triggiani University of Virginia, Charlottesville, Virginia

Vincenzo Vespri Università dell'Aquila, Coppito (AQ), Italy

P. Villaggio Università di Pisa, Pisa, Italy

Jean-Paul Zolésio Institut Non Linéare de Nice, CNRS, Sophia Antipolis, France, and Centre de Mathematiques Appliquées, INRIA-ENSMP, Sophia Antipolis, France

partial differential equation
methods in control and
shape analysis

Shape Control of a Hydrodynamic Wake

JEAN-CHRISTOPHE AGUILAR, ECOLE DES MINES DE PARIS, CENTRE DE MATHÉMATIQUES APPLIQUÉES, B.P.207, 06904 SOPHIA ANTIPOLIS, FRANCE

JEAN PAUL ZOLESIO, CNRS-INLN, 1361 ROUTE DES LUCIOLES, 06560 VALBONNE, FRANCE

ABSTRACT. This paper propounds a shape variational formulation of a hydrodynamic free interface which appears behind a three dimensional lifting profile. We prove the existence of an optimal wake under Density Perimeter constraints. We derive from this formulation the standard equilibrium condition in the classical case where this interface is a regular surface.

1. INTRODUCTION

We consider a "hydrodynamicaly well profiled" body B. B has a uniform stationary velocity U_∞. A thin viscous boundary layer is developed around B and in that study we neglect it, in the sense that we consider that the shape B coincide with the shape of the body augmented by its boundary layer, according with the classical boundary layer theory. Then we consider a sliding condition , $U.n = 0$ on $Q = \partial B$ (V being the stationary speed of the fluid). Nevertheless, we cannot completely neglect the vorticity in that flow in view of the modeling of the lifting effect. It is classical in engineering to consider the vorticity of the flow as being supported by a piece of surface S in addition to Q. S is called the wake. We assume the flow is governed by Euler's equations in $\Omega \setminus S$, Ω being the outer domain and S is said "in equilibrium" when the resulting jump of pressure $[\![p]\!]$ across S is zero. The objective of that paper is to solve that free boundary problem whose solution is the couple $(U = \nabla\phi, S)$ with $\phi \in H^1(\Omega \setminus S)$. We develop a new variational formulation on the variables U and S. We introduce an energy $J_\varepsilon(S)$ in the form

$$J_\varepsilon(S) = \min_{y \in H^1(\Omega\setminus S)} \int_{\Omega\setminus S} \left(\frac{\varepsilon}{2} y^2 + \frac{1}{2} \mid \nabla y \mid^2 + i.\nabla y\right) dx$$

and the analysis of the optimality condition for $J_\varepsilon(S)$ makes use of the shape analysis technics. In order to insure the existence of S, we introduce a surface tension $\sigma > 0$ via a surface energy for S which is represented by the use of the Density Perimeter which is the adapted perimeter concept for this kind of shape variational problem.

2. DEFINITIONS AND MAIN PROPERTIES

B is a bounded domain in $I\!R^N$ ($N \geq 2$) with boundary Q. The fluid will occupy the outer domain. More precisely, we consider a "large" bounded domain D with $\bar{B} \subset D$ and ∂D being lipschitzian. The fluid occupies the domain $\Omega = D \setminus \bar{B}$. The boundary of Ω is made of two connected components Q and ∂D.

The stationary speed field U of the fluid in the domain Ω is assumed irrotational in $\Omega \setminus S$ where S is a closed subset in Ω with zero measure and empty interior. Our modeling is assuming that the body Q is "well profiled" in such a way that the support of the $curl(U)$ will be in $S \cup Q$. In this fist paper, we neglect boundary layer effect in the neighborhood of $S \cup Q$. For each closed set S in Ω, we consider the Sobolev space $H^1(\Omega \setminus S)$. The open set $\Omega \setminus S$ is non smooth and $H^1(\Omega \setminus S)$ is defined as

$$H^1(\Omega \setminus S) = \{y \in L^2(\Omega \setminus S),\ \nabla y \in L^2(\Omega \setminus S; I\!R^N)\}$$

In the case where S is contained in a smooth orientable surface Σ, the traces of any element y are defined on both sides of Σ and may be different functions in $H^{\frac{1}{2}}(\Sigma)$.

In that case, we shall denote by $[\![y]\!]$ the jump of y through the surface Σ. Of course, $H^1(\Omega \setminus S)$ in not a subspace of $H^1(\Omega)$, but we have $H^1(\Omega) \subset H^1(\Omega \setminus S)$ for any closed set S in Ω. For any y in $H^1(\Omega)$, we have $[\![y]\!] = 0$ on S. From irrotationality assumption, we have $U\mid_{\Omega\setminus S} = \nabla\phi$ in $\Omega \setminus S$ for some scalar potential ϕ in $H^1(\Omega \setminus S)$.

That ϕ defines an element ϕ^o of $L^2(\Omega)$ as S has a zero measure. That element ϕ^o defines a distribution over Ω, $\phi^o \in \mathcal{D}'(\Omega)$, and we consider its gradient $\nabla\phi^o \in \mathcal{D}'(\Omega)$. In fact, as ϕ^o is uniquely associated to ϕ, the restriction of the distribution $\nabla\phi^o$, element of $\mathcal{D}'(\Omega; I\!R^N)$, to the open set $\Omega \setminus S$ is $\nabla\phi$ and $\nabla(\phi^o) = (\nabla\phi)^o + \mu$ where $\mu = \gamma_S^*([\![\phi]\!]\vec{n})$ is a measure, $\mu \in \mathcal{D}^{o\prime}(\Omega; I\!R^N)$, supported by S. We take $U = (\nabla\phi)^0 = \nabla(\phi^o) - \mu$. In such a situation, we get $curl(U) = curl((\nabla\phi)^0) - curl(\mu)$. The distribution $curl\,\mu$ is supported by S (as was μ), the restriction to $\Omega \setminus S$ of $curl((\nabla\phi)^o)$ is zero (as $curl\nabla = 0$) so that $curl((\nabla\phi)^o) = \mu$ is a distribution of order one supported by S. Finally, we get $curl\,U = \gamma_S^*(\vec{n} \wedge \nabla_\Gamma[\![\phi]\!])$ and $divU = \gamma_S^*([\![\frac{\partial\phi}{\partial n}]\!])$

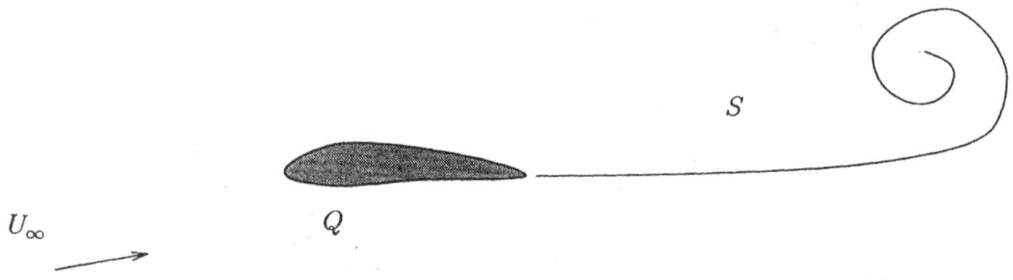

FIGURE 1. Fluid domain

3. Reduction to a bounded domain contained in D

We introduce the perturbation velocity potential φ so that $U = u_\infty(i + (\nabla\varphi)^o)$ in Ω, $U_\infty = u_\infty i$; $\phi_M = u_\infty(x_M + \varphi_M)$ with $x_M = \langle OM, i \rangle_{I\!R^3}$ and $\varphi \in H^1(\Omega \setminus S)$

More precisely, we consider a "large" bounded domain D with $\bar{B} \subset D$ and ∂D being lipschitzian.
So, when D is large enough the perturbation speed will be zero out of D.
The fluid occupies the domain $\Omega = D \setminus \bar{B}$. The boundary of Ω is made of two connected components Q and ∂D.

4. Weakly compressible flow

In order to insure the uniformity of the classical Poincaré constant in the non smooth domains $\Omega \setminus S$, we introduce a zero order term in the energy leading to a weakly compressible condition controlled by ε. Given $\varepsilon > 0$, we consider the energy functional

$$E^\varepsilon_{\Omega\setminus S}(y) = \int_{\Omega\setminus S} \left(\frac{\varepsilon}{2}y^2 + \frac{1}{2} \mid \nabla y \mid^2 + i.\nabla y\right) dx$$

The minimizer φ of that functional over $H^1(\Omega \setminus S)$ is the solution of the weak problem

$$\forall y \in H^1(\Omega \setminus S), \int_{\Omega\setminus S} (\varepsilon\varphi\, y + \nabla\varphi.\nabla y + i.\nabla y)\, dx = 0 \tag{4.1}$$

So that, performing by part, we can see that the problem takes the following form

$$\begin{cases} \Delta\varphi = \varepsilon\varphi & \text{in } \Omega \\ \dfrac{\partial\varphi}{\partial n} = -i.n & \text{on } Q \cup S^+ \cup S^- \\ \varphi = 0 & \text{on } \partial D \end{cases}$$

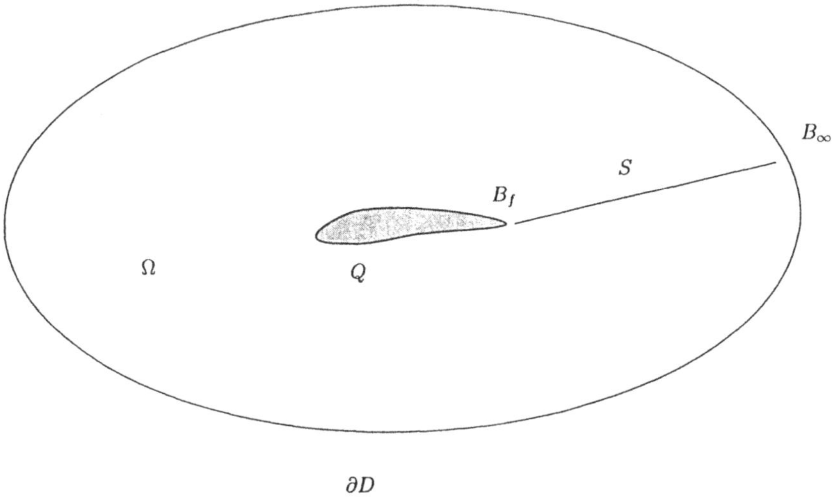

FIGURE 2. Bounded fluid domain

We have the following estimates

Lemma 1.

$$\|\nabla\varphi\|_{L^2(\Omega\setminus S;\mathbb{R}^N)} \leq |\,\Omega\,|^{\frac{1}{2}}$$

Proof. with $y = \varphi$ in 4.1,

$$\begin{aligned} \int_{\Omega\setminus S} |\,\nabla\varphi\,|^2\, dx &\leq \int_{\Omega\setminus S} |\,i\,|\;|\,\nabla\varphi\,|\, dx \\ &\leq \int_{\Omega\setminus S} |\,\nabla\varphi\,|\, dx \\ &\leq |\,\Omega\,|^{\frac{1}{2}} \Big(\int_{\Omega\setminus S} |\,\nabla\varphi\,|^2\, dx\Big)^{\frac{1}{2}} \end{aligned}$$

□

Lemma 2.

$$\sqrt{\varepsilon}\,\|\varphi\|_{L^2(\Omega)} \leq |\,\Omega\,|^{\frac{1}{2}}$$

Proof. with $y = \varphi$ in (4.1),

$$\begin{aligned}(\sqrt{\varepsilon}\|\varphi\|_{L^2(\Omega\setminus S)})^2 &= -\|\nabla\varphi\|^2_{L^2(\Omega\setminus S)} - \int_\Omega i.\nabla\varphi\, dx \\ &\le \int_{\Omega\setminus S} |\nabla\varphi|\, dx \\ &\le |\Omega|^{\frac{1}{2}} (\int_{\Omega\setminus S} |\nabla\varphi|^2\, dx)^{\frac{1}{2}} \\ &\le |\Omega|\end{aligned}$$

□

Lemma 3.

$$\|\Delta\varphi\|_{L^2(\Omega\setminus S)} \le \sqrt{\varepsilon}\ |\Omega|^{\frac{1}{2}}$$

Proof.

$$\begin{aligned}\|\Delta\varphi\|_{L^2(\Omega\setminus S)} &= \varepsilon\ \|\varphi\|_{L^2(\Omega\setminus S)} \\ &\le \sqrt{\varepsilon}\ |\Omega|^{\frac{1}{2}} \text{ with lemma 2}\end{aligned}$$

□

In view of that last estimate, we see that $div(U\ |_{\Omega\setminus S}) = u_\infty\ \Delta\varphi$ goes to zero with ε. Then, the flow is almost incompressible.

In the case where S is a smooth surface, we would get, denoting by $\vec{n}$ the normal field on S and performing by part on (4.1).

$$\frac{\partial\varphi_+}{\partial n} = \frac{\partial\varphi_-}{\partial n} = -i.n \text{ on } S$$

For each closed set S in Ω, $\varphi(\Omega\setminus S)$ denoting the solution of problem (4.1), we consider the energy functional, for given $\varepsilon > 0$,

$$J_\varepsilon(S) = \min_{y\in H^1(\Omega\setminus S)} E^\varepsilon_{\Omega\setminus S}(y)$$

Lemma 4.

$$J_\varepsilon(S) = -\frac{1}{2}\int_{\Omega\setminus S}(\varepsilon\varphi^2(\Omega\setminus S) + |\nabla\varphi(\Omega\setminus S)|^2)\, dx$$

Proof. with $y = \varphi$ in (4.1),

$$J_\varepsilon(S) = \frac{1}{2}\int_{\Omega\setminus S} < i, \nabla\varphi(\Omega\setminus S) > dx = \frac{1}{2} < i, \int_{\Omega\setminus S} \nabla\varphi(\Omega\setminus S) > dx >$$

□

Lemma 5.

$$0 \ge J_\varepsilon(S) \ge -|\Omega|$$

Proof.

$$J_\varepsilon(S) = -\frac{1}{2}(\varepsilon\|\varphi\|^2_{L^2(\Omega)} + \|\nabla\varphi\|^2_{L^2(\Omega)})$$

with lemma 1 and lemma 2 □

5. Deformations of the domains

For any $V \in C^0([0,\tau[;\mathbb{R}^N)$

$$V = 0 \text{ on } Q \cup \partial D$$

we consider the flow mapping $T_t(V) : X \longmapsto x(t,X)$
With $x(t,X)$ solution to the system of ordinary differential equations

$$\begin{cases} \frac{d}{dt}x(t,X) = V(t,x(t,X)) \\ x(0,X) = X \end{cases}$$

We know from [3] that T_t is a diffeomorphism from $D \setminus \bar{B}$ onto itself.

6. Optimal wake existence

We consider now the extremality of the functional J_ε. The energy associated to S is related to its length. We choose here the density perimeter $P_\gamma(S)$ for a given $\gamma > 0$ which could be related to a surface tension concept, see [1].

$$P_{\gamma,H}(S) \stackrel{def}{=} \sup_{\varepsilon \in (0,\gamma)} [\frac{m(S^\varepsilon)}{2\varepsilon} + H(\varepsilon)]$$

$$S^\varepsilon = \bigcup_{x \in S} B(x,\varepsilon)$$

The main properties of $P_{\gamma,H}$ are

Proposition 1.

$$\Omega_n \xrightarrow{H^c} \Omega \Rightarrow P_{\gamma,H}(\partial\Omega) \leq \liminf_{n \to +\infty} P_{\gamma,H}(\partial\Omega_n)$$

Proposition 2.

$$\Omega_n \xrightarrow{H^c} \Omega \Rightarrow \Omega_n \xrightarrow{char} \Omega$$

Proposition 3.

$$P_{\gamma,H}(\partial\Omega) < \infty \Rightarrow meas(\partial\Omega) = 0$$

H^c is the Hausdorff topology.

$$d_{H^d}(\Omega_1,\Omega_2) = \sup_{x \in \mathbb{R}^N} | d_{\Omega_1}(x) - d_{\Omega_2}(x) |$$

where $d_{\Omega_1}(x) = \inf_{y \in \Omega_1} \|x - y\|$

$$d_{H^c}(\Omega_1,\Omega_2) = d_{H^d}(\Omega_1^c,\Omega_2^c)$$

B_f (Resp. B_∞) is a closed set in Q (Resp. ∂D) with $n-1$ dimensional Hausdorff measure $| B_f |_{\mathcal{H}^{n-1}} = | B_\infty |_{\mathcal{H}^{n-1}} = 0$. The admissible family of closed sets S is chosen as

$$\mathcal{S}_0 = \{S = \bar{S},\ meas(S) = 0,\ \bar{S} \supset B_f \cup B_\infty,\ \#(S) = 1\}$$

Where $\bar{S}$ is the closure of S in $\mathbb{R}^N$ and $\#(S)$ is the number of connected components of S.

Proposition 4. $\forall M > 0$, $\mathcal{S}_0^M = \{S \in \mathcal{S}_0 \mid P_\gamma(S) \leq M\}$ equipped with the Hausdorff metric is a compact metric space.

Proof. From [1] we know that given a sequence S_n in $\mathcal{S}_0$ with $P_\gamma(S_n) \leq M$ there exists a subsequence still denoted by S_n such that $S_n \overset{H}{\longrightarrow} S$ in Hausdorff metric where S is a closed set in Ω. Moreover $\chi_{\Omega \backslash S_n} \longrightarrow \chi_{\Omega \backslash S}$ in $L^2(\Omega)$. So that $meas(S) = 0$. Also, we know ([2]) that $\#$ is lower semi continuous for the Hausdorff topology then $\#S \leq 1$ but as $S_n \supset B_f \cup B_\infty$ we get $S \supset B_f \cup B_\infty$ then S is non empty and then $\#(S) = 1$. Finally P_γ is lower semi continuous ([1]), then $P_\gamma(S) \leq M$ □

Given $\sigma > 0$, we consider the optimality problem

$$Min\ \{J_\varepsilon(S) + \sigma\ P_\gamma(S)\ \mid\ S \in \mathcal{S}_0\} \tag{6.1}$$

Theorem 1. *For each $\varepsilon > 0$, the problem (6.1) has optimal solutions in the family $\mathcal{S}_0$.*

Before showing this theorem, we need the following result

Lemma 6. J_ε is lower semi continuous on $\mathcal{S}_0^M$

Proof. Let $S_n \overset{H}{\longrightarrow} S$, let $\varphi_n = \varphi(\Omega \setminus S_n)$. From Lemma (1) and (2),

$$\begin{aligned} \|(\nabla\varphi_n)^o\|_{L^2(\Omega)} &\leq |\,\Omega\,|^{\frac{1}{2}} \\ \|(\varphi_n)^o\|_{L^2(\Omega)} &\leq \frac{1}{\sqrt{\varepsilon}}\,|\,\Omega\,| \end{aligned}$$

Then, after extraction of subsequences

$$\begin{aligned} (\nabla\varphi_n)^o \rightharpoonup f \quad &\text{weakly in} \quad L^2(\Omega; \mathbb{R}^N) \\ (\varphi_n)^o \rightharpoonup g \quad &\text{weakly in} \quad L^2(\Omega) \end{aligned}$$

From the Hausdorff convergence of S_n to S, we get: Let $\psi \in \mathcal{D}(\Omega \setminus S)$, $\exists n_\psi = n(d(S, K))$, where $K = supp\psi$. Such that $n \geq n_\psi$ implies $\psi \in \mathcal{D}(\Omega \setminus S_n)$ then, we see easily that $f\,|_{\Omega \backslash S} = \nabla(g\,|_{\Omega \backslash S})$ We set $\varphi = g\,|_{\Omega \backslash S}$ so that $\varphi \in H^1(\Omega \setminus S)$

On the other hand, we have

$$J_\varepsilon(S_n) = \frac{1}{2}\int_{\Omega \backslash S_n} < i, \nabla\varphi_n > dx = \frac{1}{2}\int_\Omega < i, (\nabla\varphi_n)^o > dx$$

Which converges, as $n \longrightarrow \infty$, to

$$\frac{1}{2}\int_\Omega < i, f > dx = \frac{1}{2}\int_{\Omega \backslash S} < i, f > dx = \frac{1}{2}\int_{\Omega \backslash S} < i, \nabla\varphi > dx \geq J_\varepsilon(S)$$

□

Proof. [of the theorem] Let S_n be a minimizing sequence for problem (6.1). We assume $J_\varepsilon(S_n) + \sigma\ P_\gamma(S_n)$ monotonically decreasing to the infimum as $n \longrightarrow 0$. Then

$$J_\varepsilon(S_n) + \sigma\ P_\gamma(S_n) \leq J_\varepsilon(S_1) + \sigma\ P_\gamma(S_1) = a$$

$\sigma P_\gamma(S_n) \leq a - J_\varepsilon(S_n)$ then, form lemma 5

$$P_\gamma(S_n) \leq M = \frac{1}{\sigma}(a + |\,\Omega\,|) \tag{6.2}$$

From proposition 4, we can assume that $S_n \longrightarrow S$ in Hausdorff metric with $S \in \mathcal{S}_0$. From [1] we know that P_γ is lower semi continuous on $\mathcal{S}_0$ and, as J_ε is semi continuous inferiorly, the result classically derives. □

7. Viscous wake

The perimeter $P_\gamma(S)$ can be considered as a viscous term associated to S. We can take in account a more general contribution of the viscous effect. In the flow in $\Omega \setminus S$, we can neglect the viscosity but it is not reasonable on S because of the jump of the speed flow through S. Then, we can not neglect a viscous effect and we chose a classical term in the following form:

$$J_\varepsilon(S) + \int_S (\sigma + \nu \| [\![\nabla_\Gamma \varphi(S)]\!] \|^2)\, d\Gamma$$

8. Necessary optimality condition

8.1. The smooth case. We assume that the optimal wake S is smooth enough. Then, using shape sensitivity analysis, we derive the shape gradient of the energy $J_\varepsilon(S)$. We perturb S using a one parameter family of transformation T_t mapping Ω on itself, $\partial\Omega$ onto $\partial\Omega$, with $T_t(B_f) = B_f$ and $T_t(B_\infty) = B_\infty$

$$J_\varepsilon(S) = \min_{y \in H^1(\Omega \setminus S)} E^\varepsilon_{\Omega \setminus S}(y)$$

so that the wake equilibrium problem (6.1) take the following shape variational form

$$\min_{S \in \mathcal{S}_0} \left(\min_{y \in H^1(\Omega \setminus S)} E^\varepsilon_{\Omega \setminus S}(y) + \sigma\, P_\gamma(S) \right) \tag{8.1}$$

We apply the results concerning the derivative of a Minimum with respect to a parameter s [4]. For a given $s > 0$, we set $S_s = T_s(V)(S)$

Lemma 7. The family $\mathcal{S}_0$ is stable under transformations $T_s(V)$:

$$\forall V,\ \forall s,\ S_s \in \mathcal{S}_0$$

Proof. $T_t : D \longrightarrow D$ is a smooth one to one transformation then $| S_s | = 0$, $\#(S_s) = \#S$, S_s is closed □

Lemma 8. The elements of the Sobolev space are transported by $T_s(V)$:

$$y \in H^1(\Omega \setminus S_s) \iff z = y \circ T_s(V) \in H^1(\Omega \setminus S)$$

Then, problem (8.1) lead to the extremization of the functional

$$J_\varepsilon(S_s) = \min_{z \in H^1(\Omega \setminus S)} E^\varepsilon_{\Omega \setminus S_s}(z \circ T_s(V)^{-1}) \tag{8.2}$$

We set

$$F(s, z) = E^\varepsilon_{\Omega \setminus S}(z \circ T_s(V)^{-1})$$
$$f(s) = J_\varepsilon(S_s)$$

and we make use of the

Theorem 2. *Let K be a compact set, $F : [0, \tau] \times K \to \mathbb{R}$ a differentiable mapping and let $f(t) = \min\{F(t, y) \mid y \in K\}$. Denote by K^* the subset of K of elements φ which realize minimum at t=0.*
f is side differentiable at $t = 0$ and

$$\begin{aligned} f'(0, +1) &= \lim_{t \downarrow 0} \frac{f(t) - f(0)}{t} \\ &= \min\{\frac{\partial}{\partial t} F(0, \varphi) \mid \varphi \in K^*\} \end{aligned}$$

In order to apply that result, we need to reduce the minimum to a compact set K. This derives from the coercivity of $E^{\varepsilon}_{\Omega\setminus S}$
Effectively, we have

$$\begin{aligned} E^{\varepsilon}_{\Omega\setminus S}(y) &\geq \frac{\varepsilon}{2}\|y\|^2_{L^2(\Omega)} + \frac{1}{2}\|\nabla y\|^2_{L^2(\Omega)} - \mid \Omega \mid^{\frac{1}{2}} \ \|\nabla y\|_{L^2(\Omega\setminus S)} \\ &\geq \frac{\varepsilon}{2}\|y\|^2_{H^1(\Omega\setminus S)} - \mid \Omega \mid^{\frac{1}{2}} \ \|\nabla y\|_{L^2(\Omega\setminus S)} \\ &\geq \frac{\varepsilon}{4}\|y\|^2_{H^1(\Omega\setminus S)} \end{aligned}$$

as soon as $\|y\|_{H^1(\Omega\setminus S)} \geq \frac{4}{\varepsilon} \mid \Omega \mid^{\frac{1}{2}} = M$
then $\|y\|_{H^1(\Omega\setminus S)} \geq M$ implies $E^{\varepsilon}_{\Omega\setminus S}(y) \geq \frac{4}{\varepsilon} \mid \Omega \mid \geq 0$

But the minimum $J_\varepsilon(S_s)$ being negative, in the minimization problem (8.2) $H^1(\Omega \setminus S)$ can be replaced by

$$K = \{y \in H^1(\Omega \setminus S) \mid \|y\| \leq M\}$$

K is weakly compact in $H^1(\Omega \setminus S)$

Lemma 9. The Eulerian derivative of the domain functional $J_\varepsilon(S)$ in the direction of the vector field V acting on S is

$$dJ_\varepsilon(S;V) = \int_S [\![\frac{1}{2}|\nabla\varphi|^2 + \nabla\varphi.i + \frac{\varepsilon}{2}\varphi^2]\!] V.n d\Gamma$$

Proof.

$$f(t) = \int_{\Omega_t} \frac{1}{2}|\nabla(\varphi o T_t^{-1})|^2 + \nabla(\varphi o T_t^{-1}).i \, dx$$

we make use of:

$$\begin{aligned} \frac{\partial}{\partial t}\left(\int_{\Omega_t} \frac{1}{2}|\nabla(\varphi o T_t^{-1})|^2 dx\right)_{|t=0} &= \int_\Omega \nabla\varphi.\nabla(-\nabla\varphi.V)\, dx + \int_S \frac{1}{2}[\![|\nabla\varphi|^2]\!]V.n \, d\Gamma \\ &= \int_\Omega \nabla\varphi.V\Delta\varphi \, dx - \int_S [\![\frac{\partial\varphi}{\partial n}\nabla\varphi]\!].V \, d\Gamma + \int_S \frac{1}{2}[\![|\nabla\varphi|^2]\!]V.n \, d\Gamma \\ &= \int_S [\![\nabla\varphi.V]\!]\, i.n \, d\Gamma + \int_S \frac{1}{2}[\![|\nabla\varphi|^2]\!]V.n \, d\Gamma \end{aligned}$$

and

$$\begin{aligned} \frac{\partial}{\partial t}\left(\int_{\Omega_t} \nabla(\varphi o T_t^{-1}).i \, dx\right)_{|t=0} &= \frac{\partial}{\partial t}\left(\int_{\Omega_t} div(\varphi o T_t^{-1} i) dx\right)_{|t=0} \\ &= \int_\Omega div(-\nabla\varphi.Vi)\, dx + \int_S [\![div(\varphi i)]\!]V.n \, d\Gamma \\ &= \int_S [\![-\nabla\varphi.V]\!] i.n \, d\Gamma + \int_S [\![\nabla\varphi.i]\!] \ V.n \, d\Gamma \end{aligned}$$

then

$$f'(0) = \int_S [\![\frac{1}{2}|\nabla\varphi|^2 + \nabla\varphi.i]\!] V.n d\Gamma$$

□

We get now the necessary optimality condition.

Proposition 5. Let S be a minimizer for the problem (6.1). Then the pressure is defined on both sides of S and is given by the Bernoulli's equation. Moreover, its jump across S is zero.

$$[\![p]\!] = 0 \text{ on } S$$

Proof. On the optimal wake,

$$\int_S [\![\frac{1}{2}|\nabla\varphi|^2 + \nabla\varphi.\vec{U}_\infty + \frac{\varepsilon}{2}\varphi^2]\!]\, V.n\, d\Gamma = 0, \quad \forall\, V$$

Considering the following Bernoulli's equation on both sides of S,

$$\frac{1}{2}(U^2 - u_\infty^2 + \varepsilon\phi^2) + \frac{p}{\rho} + gz = \frac{p_0}{\rho}$$

$(p_0 = p_{z=0}$is the atmospheric pressure)

in term of speed perturbation potential, this expression turns to be:

$$u_\infty^2(\frac{1}{2}|\nabla\varphi|^2 + \nabla\varphi.i + \frac{\varepsilon}{2}\varphi^2) + \frac{p}{\rho} + gz = \frac{p_0}{\rho}\text{on } S$$

which permits to conclude. □

8.2. The non smooth case. In general, S could be non smooth (as up to now we have derive no smoothness results on S). The same shape sensitivity analysis can be performed but avoiding any boundary integral on S. Then, taking volume integrals, we give now the necessary condition which will be a relaxed formulation of the previous one.

Proposition 6.

$$dJ_\varepsilon(S;V) = \int_{\Omega\setminus S} div\{(\frac{1}{2}\mid\nabla\varphi\mid^2 + \nabla\varphi.i + \frac{\varepsilon}{2}\varphi^2)\ V\}\, dx$$

Proof. Using the two following propositions 7 and 8. □

Proposition 7.

$$\begin{aligned}\frac{\partial}{\partial t}(\int_{\Omega\setminus S_t} \frac{1}{2}|\nabla(\varphi o T_t^{-1})|^2 dx)_{|t=0} = \ & \frac{1}{2}\int_{\Omega\setminus S} div(\mid\nabla\varphi\mid^2\ V)\, dx \\ & - \int_{\Omega\setminus S} div(V.\nabla\varphi\ \nabla\varphi)\, dx\end{aligned}$$

Before showing this proposition, we need the two following lemmas

Lemma 10.

$$\int_{\Omega\setminus S} < \epsilon(V)\nabla\varphi, \nabla\varphi > dx = \int_{\Omega\setminus S} div(V.\nabla\varphi\ \nabla\varphi) - < D^2\varphi\nabla\varphi, V > dx$$

Proof.

$$\begin{aligned}\int_{\Omega\setminus S} \partial_i V_j\, \partial_i\varphi\, \partial_j\varphi\, dx &= \int_{\Omega\setminus S} -V_j\, \partial_i(\partial_i\varphi\, \partial_j\varphi) + \partial_i\{V_j(\partial_i\varphi\partial_j\varphi)\}\, dx \\ &= \int_{\Omega\setminus S} - < V, \nabla\varphi > \Delta\varphi - < D^2\varphi\ \nabla\varphi, V > + div(V.\nabla\varphi\ \nabla\varphi)\, dx\end{aligned}$$

□

Lemma 11.

$$\int_{\Omega\setminus S} \frac{1}{2} divV\, |\nabla\varphi|^2 dx = \int_{\Omega\setminus S} - < D^2\varphi\nabla\varphi, V > + \frac{1}{2} div(|\nabla\varphi|^2 V)\, dx$$

Proof.

$$\int_{\Omega\setminus S} divV\,|\nabla\varphi|^2dx = \int_{\Omega\setminus S} -V.\nabla(|\nabla\varphi|^2)dx + div(|\nabla\varphi|^2)\ V\}dx$$

$$\text{Where}\quad V.\nabla(|\nabla\varphi|^2) =< 2D^2\varphi\nabla\varphi, V >$$

□

Proof. [of the proposition 7]

$$\begin{aligned}\frac{\partial}{\partial t}\Big(\int_{\Omega\setminus S_t}\frac{1}{2}|\nabla(\varphi o T_t^{-1})|^2dx\Big)_{|t=0} &= \frac{\partial}{\partial t}\Big(\int_{\Omega\setminus S}\frac{1}{2}<^*DT_t^{-1}\nabla\varphi,^*DT_t^{-1}\nabla\varphi > det(DT_t)\,dx\Big)_{|t=0}\\ &= \int_{\Omega\setminus S} < \{\frac{1}{2}I_d\ divV(0) - \epsilon(V(0))\}\nabla\varphi, \nabla\varphi > dx\end{aligned}$$

And concluding with lemmas 10 and 11. □

Proposition 8.

$$\frac{\partial}{\partial t}\Big(\int_{\Omega_t}\nabla(\varphi o T_t^{-1}).i\,dx\Big)_{|t=0} = \int_{\Omega\setminus S} div(\nabla\varphi.i\ V) + \int_{\Omega\setminus S} div(V.\nabla\varphi\ \nabla\varphi)\,dx$$

Proof.

$$\begin{aligned}\frac{\partial}{\partial t}\Big(\int_{\Omega_t}\nabla(\varphi o T_t^{-1}).i\,dx\Big)_{|t=0} &= \int_{\Omega\setminus S} < -DV^*(0)\nabla\varphi, i > + < \nabla\varphi, i > divV(0)\,dx\\ &= \int_{\Omega\setminus S} div(\nabla\varphi.i\ V) - < D^2\varphi\, i, V > dx\\ &\quad - \int_{\Omega\setminus S} div(V.\nabla\varphi\ i) + < D^2\varphi\ V, i > dx\\ &= \int_{\Omega\setminus S} div(\nabla\varphi.i\ V) + \int_{\Omega\setminus S} div(V.\nabla\varphi\ \nabla\varphi)\,dx\end{aligned}$$

□

References

1. D. BUCUR and J.P. ZOLESIO, *Pseudo courbure dans l'optimisation de forme*, C.R. Acad. Sci. Paris tome 320, Serie I, 1995.
2. M. DELFOUR and J.P. ZOLESIO, *Shape Analysis via Oriented Distance Functions*, Journal of Functional Analysis, 1994.
3. J. SOKOLOWSKI and J.P. ZOLESIO, *Introduction to Shape Optimisation: Shape sensitivity analysis*, Computational Mathematics, vol. 16, Springer-Verlag, New York, Berlin, Heidelberg, 1992.
4. J.P. ZOLESIO, *The material derivative (or speed) method for shape optimization*, E.J. Haug and J.Cea, Sijthoff and Noordhoff, Alphen aan den Rijn, 1981, pp 1089-1151.

JEAN-CHRISTOPHE AGUILAR, ECOLE DES MINES DE PARIS, CENTRE DE MATHÉMATIQUES APPLIQUÉES, B.P.207, 06904 SOPHIA ANTIPOLIS, FRANCE

JEAN PAUL ZOLESIO, CNRS-INLN, 1361 ROUTE DES LUCIOLES, 06560 VALBONNE, FRANCE

On Some Inverse Geometrical Problems

S. Andrieux EDF-DER-MMN, Clamart, France

A. Ben Abda M. Jaou ENIT, LAMAP, Tunis, and Institut Préparatoire aux Études Scientifiques et Techniques, La Marsa, Tunisia

I Introduction

Recent efforts have focused on an industrial process : Nondestructive thermal testing of materials. They are generated by a growing interest in the detection and location of structural internal flaws. These methods give rise to a class of identification problems : Inverse geometrical problems defined by overspecified data.

These kind of problems are posed as follows : Consider a material occupying a domain Ω in IR^n, $n \geq 2$. and let Γ be the unknown geometry.

One wishes to determine Γ by injecting a heat flux Φ (or a current flux in the case of electrical testing) across $\partial\Omega$ and measuring the temperature f (or the voltage) on an open subset of $\partial\Omega$: M.

The temperature field u satisfies the steady state heat conduction problem :

$$\begin{cases} -\Delta u = 0 \text{ in } \Omega_\Gamma \\ \dfrac{\partial u}{\partial n} = \Phi \text{ on } \partial\Omega_\Gamma \end{cases}$$

$$\left(\int_{\partial\Omega_\Gamma} u = 0 \text{ and } \int_{\partial\Omega_\Gamma} \Phi = 0 \right)$$

Thus the problem is to know if Γ can be determined by one choice of the heat flux Φ (and the corresponding measurement f).

The determination of Γ consists in seeking the solution of three questions.

1) The uniqueness : Does Φ (and f) uniquely determine the unknown Γ.

2) The stability : Because of the error in measurements and in view of numerical treatments one has to study the variation of the geometry with respect to a variation of the measurement.

3) The inversion process : The goal of the problem is the determination of the unknown geometry by finding an inversion process which can be explicit or iterative.

The uniqueness question has been widely studied by many authors for different kind of geometrical flaws. In the case of inclusions [13] proved that when the inclusion D is a priori known to be a convex polyhedron , the shape and the location of D are determined by one measurement only. In the case where the unknown Γ is a part of the outer part of $\partial\Omega$, one heat flux with its correspondant measurement, suffices to determine Γ [5].

In the case of a buried insulated crack [14] showed that two specific current fluxes together with correspondant voltage suffice to determine the crack. Furthermore, they proved that this result is the best one possible. This result was extended in the case of a family of n cracks [9], it was proved that a family of n+1 fluxes with their corresponding voltage suffices to establish the uniqueness result. Recently [2] improved this result showing that two specific fluxes suffice (and are necessary) to establish the identifiability.

Notice that in all these works the crucial step towards the identifiability result rests on the knowledge of the shape of the level lines and therefore one can point out the bidimensional character of the proofs. It is shown here that in the case of a crack with a known emerging point on the boundary, one specific heat flux (or current flux) with the corresponding temperature field (or voltage) suffice to determine the crack .

The second question (the stability) can be viewed as the continuity of the mapping that associates the geometry to the data.

For that purpose, and in the cases of unknown boundaries as well as in the case of segment cracks with an emerging point on the boundary, compact metric spaces of admissible geometries are constructed. The stability result is derived from the uniqueness theorems and the continuity of the direct problems.The method given in[8] for inclusions is followed to reach this result.
A more precise stability result of Bellout-Friedman type[8] is also obtained : We prove that the mapping that associates the geometry to the data is locally lipshitzian. This result is optimal. Notice that the stability can be interpreted as the variation of the geometry with respect to the variation of the measurements, which suggests the main tool used : The domain derivative theory [16].
The last section of this presentation is devoted to numerical treatments. The identification process is based on the minimisation of an error functional initially introduced by Kohn and Vogelius in the case of parameter identification.
In this work, this functional is interpreted in the case of line segment cracks identification. Notice that this method can be applied to more general inverse geometrical problems[4].

I Uniqueness results

As pointed out earlier, this work is concerned by inverse geometrical problems defined by Laplacian equation and corresponding to overdetermined data. In this case these data correspond to the trace f of the solution on M , and to its normal derivative $\Phi = \frac{\partial u}{\partial n}$. To establish the uniqueness result one has to answer to the following question :

Does the pair(Φ ,f) uniquely determine the unknown geometry?

To prove the uniqueness, one has to compare two harmonic functions, defined in two different domains and having the same Cauchy data on M : The main tool towards the result is the Holmgren uniqueness theorem. This tool has been widely used in this kind of problems : In the case of an a priori known convex polyhedron by Friedman and Isakov[13], in the case of $C^{2,\alpha}$ inclusions by Bellout and Friedman[8] , in the case of C^2 cracks in Friedman and Vogelius[14] and Bryan and Vogelius[9].

This section is devoted to uniqueness results concerning inverse geometrical problems. Two kinds of inverse geometrical problems are studied : the problem of the identification of inaccessible smooth boundary which is supposed to be islated (this kind of problems can be incoutered in thermal testing of composite materials). The second problem is the identification of a C^2 crack with an a priori known emerging point on the boundary.

II.1 Case of a smooth boundary

Let Ω be an open simply connected set of IR^n with a $C^{1,1}$ boundary $\partial\Omega$.
Γ_Φ , M and Γ a partition of $\partial\Omega$. M is supposed to be C^2. Γ designates the inaccessible part of $\partial\Omega$.
Denote by Ω_Γ the open set Ω . On $\partial\Omega_\Gamma$ a flux Φ ;$\Phi \not\equiv 0$ is imposed ($\int_{\partial\Omega}\Phi = 0$). Furthermore one supposes that supp $\Phi \subset \Gamma_\Phi$.

Consider the direct problem corresponding to an unknown isolated part of the boundary :

$$(II.1)\quad \begin{cases} -\Delta u = 0 & in\ \Omega_\Gamma & (1) \\ \dfrac{\partial u}{\partial n} = \Phi & on\ \partial\Omega_\Gamma \setminus \Gamma & (2) \\ \dfrac{\partial u}{\partial n} = 0 & on\ \Gamma & (3) \\ \int_{\partial\Omega} u = 0 & & (4) \end{cases}$$

THEOREM

let Γ_1 and Γ_2 be two possible $C^{1,1}$ boundaries to identify , Γ_1 and Γ_2 having the same endpoints .

u_i designates the solution of (II.1) for $\Gamma = \Gamma_i$ $i = 1, 2$.

if $u_{1|M} = u_{2|M}$ then $\Gamma_1 = \Gamma_2$

Proof :

The proof is achieved into two steps :

Step1

let $w = u_1 - u_2$; then w satisfies the following Cauchy problem :

$$\begin{cases} -\Delta w = 0 & in\ \Omega_{\Gamma_1} \cap \Omega_{\Gamma_2} \\ w = 0 \ on\ M & \\ \dfrac{\partial w}{\partial n} = 0 & on\ M \end{cases}$$

By the unique continuation theorem :

$$w \equiv 0 \text{ dans } \Omega_{\Gamma_1} \cap \Omega_{\Gamma_2}$$

And therefore :

$$u_1 = u_2 \text{ et } \frac{\partial u_1}{\partial n} = \frac{\partial u_2}{\partial n} \text{ sur } \partial(\Omega_{\Gamma_1} \cap \Omega_{\Gamma_2})$$

Step2

Consider now the open set : $0 = \Omega_{\Gamma_1} \cup \Omega_{\Gamma_2} \setminus \overline{\Omega_{\Gamma_1} \cap \Omega_{\Gamma_2}}$ Suppose that 0 is non empty and let 0_1 be one connected componant of 0.

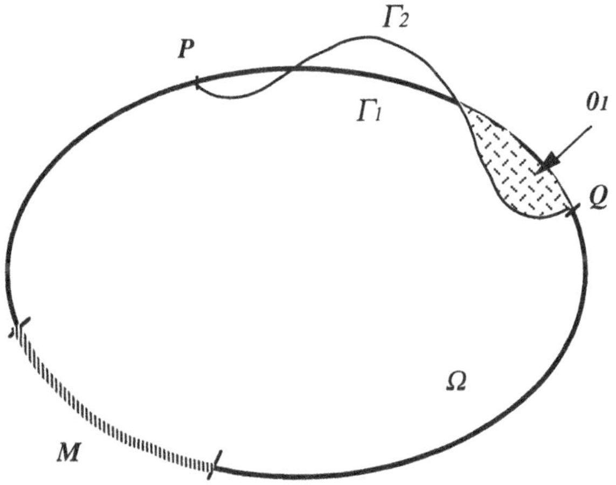

Figure 1

$\partial 0_1$ is constituted from parts of Γ_1 and Γ_2. Suppose for instance that $0_1 \subseteq \Omega_{\Gamma_1} \setminus \Omega_{\Gamma_2}$. One has $\partial 0_1 \cap \Gamma_2 \subseteq \partial(\Omega_{\Gamma_1} \cap \Omega_{\Gamma_2})$ and therefore :

$u_1 = u_2$ et $\frac{\partial u_1}{\partial n} = \frac{\partial u_2}{\partial n}$ sur $\partial 0_1 \cap \Gamma_2$ (where n designates the outer normal to Ω_{Γ_2})

Therefore u_1 extends u_2 across $\partial 0_1 \cap \Gamma_2$ and then

$$\frac{\partial u_1}{\partial n} = \frac{\partial u_2}{\partial n}$$

Since $\partial 0_1 \cap \Gamma_2$ is a smooth part of Ω_{Γ_1}, $\frac{\partial u_1}{\partial n}$ is continuous across $\partial 0_1 \cap \Gamma_2$, it follows that:

$$\frac{\partial u_1}{\partial n} = 0 \text{ on } \partial 0_1 \cap \Gamma_2$$

That is, u_1 is on 0_1 a solution of the following problem:

$$\begin{cases} -\Delta u_1 = 0 & in\ 0_1 \\ \frac{\partial u_1}{\partial n} = 0 & on\ \partial 0_1 \cap \Gamma_1 \\ \frac{\partial u_1}{\partial n} = 0 & on\ \partial 0_1 \cap \Gamma_2 \end{cases}$$

and therefore u_1 = cte on Ω_{Γ_1} by the unique continuation theorem. This is in contradiction with $\Phi \not\equiv 0$. That is $0_1 = \emptyset$ and it follows that $0 = \emptyset$.

Remark:

The same result is proven when the inaccessible boundary is supposed to satisfy a boundary condition of Signorini type, the proof is based on the same ideas. [10]

II.2 Case of a crack initiated at the boundary

The body occupies a simply connected domain, one supposes that Ω contains exactly one crack σ which has a known emerging point S on the boundary $\partial\Omega$ (a crack is a C^2 non selfintersecting curve) $\partial\Omega$. $\partial\Omega$ is parametrized by the arclength s with S as origin.
Consider P,Q,R 3 points of $\partial\Omega$, such that :

$$0<s(R)<s(Q)<s(P)$$

and the flux Φ given by :

$$\Phi = \begin{cases} 1 & on RQ \\ -\frac{|RQ|}{|PR|} & on\ QP \\ 0 & elsewhere \end{cases}$$

The corresponding direct problem is therefore given by :

$$\text{(II.2)} \quad \begin{cases} -\Delta u_\sigma = 0 & in\ \Omega \setminus \sigma \\ \frac{\partial u_\sigma}{\partial n} = \Phi & on\ \partial\Omega \\ \frac{\partial u_\sigma}{\partial n} = 0 & on\ \sigma \end{cases}$$

and one supposes the temperature u_σ being measured on a curve M (mes(M) > 0)

THEOREM

Let σ and σ' be two C^2 curves modelising two cracks having S as an endpoint .One supposes that these two cracks lead to the same measurement on M , for the flux Φ defined previously , then $\sigma = \sigma'$.

Proof :

Let u_σ be the solution corresponding to a crack σ having S.as an endpoint.$u_\sigma \in H^1(\Omega \backslash \sigma)$, Denote by τ_σ the vector ∇u_σ; τ_σ is divergence free in $L^2(\Omega \backslash \sigma)$, by the trace theorem τ_σ has a normal trace on the two sides of σ; $\tau_\sigma . n^+$ and $\tau_\sigma . n^-$.

but

$$\nabla u_\sigma . n^+ = \nabla u_\sigma . n^- = 0 \text{ on } \sigma$$

and therefore

$$\tau_\sigma . n^+ = \tau_\sigma . n^- = 0 \quad \text{on } \sigma$$

now $\nabla . \tau_\sigma = 0$ in Ω, in the sense of distributions, therefore there exists a function $\omega_\sigma \in H^1(\Omega)$ such that :

$$\tau_\sigma = -(\nabla \omega_\sigma)^\perp = \left(\frac{\partial \omega_\sigma}{\partial x_2}, -\frac{\partial \omega_\sigma}{\partial x_1}\right)$$

x_1 and x_2 designates the cartesian coordinates.

ω_σ is uniquely determined up to a constant.

Furthermore, one has :

$$\frac{\partial \omega_\sigma}{\partial \tau} = -(\nabla \omega_\sigma)^\perp . n = \Phi \text{ on } \partial \Omega$$

$$\omega_\sigma = K_\sigma \text{ on } \sigma .$$

That is ω_σ satisfies :

$$\begin{cases} -\Delta \omega_\sigma = 0 & \text{in } \Omega \backslash \sigma \\ \omega_\sigma = K_\sigma & \text{on } \sigma \\ \omega_\sigma = \varphi & \text{on } \partial \Omega \end{cases}$$

and $\omega_\sigma \in H^1(\Omega)$ (because ω_σ is continuous across σ)

For σ' one has also :

$$\begin{cases} -\Delta \omega_{\sigma'} = 0 & \text{in } \Omega \backslash \sigma' \\ \omega_{\sigma'} = K_{\sigma'} & \text{on } \sigma' \\ \omega_{\sigma'} = \varphi & \text{on } \partial \Omega \end{cases}$$

Denote by ω the field $\quad \omega = \omega_\sigma - \omega_{\sigma'}$

ω is harmonic in $\Omega \backslash (\sigma \cup \sigma')$, and satisfies :

$$\omega \equiv 0 \qquad \text{on } M$$

and $\dfrac{\partial \omega}{\partial n} = -(\nabla \omega_\sigma)^\perp . \tau + (\nabla \omega_{\sigma'})^\perp . \tau = \tau_{\sigma'} . \tau - \tau_\sigma . \tau = 0$ on M

Since u_σ and $u_{\sigma'}$ have the same trace on M.

It comes that $\omega \equiv 0$, in the exterior connected component $\Omega \backslash (\sigma \cup \sigma')$, denoted by Ω_e .

By the specific choice made of φ (that is of the flux Φ), $\varphi(S)$ is the minimum of φ ,and φ is constant on the arc PR of $\partial\Omega$ and is equal to $\varphi(S)$.

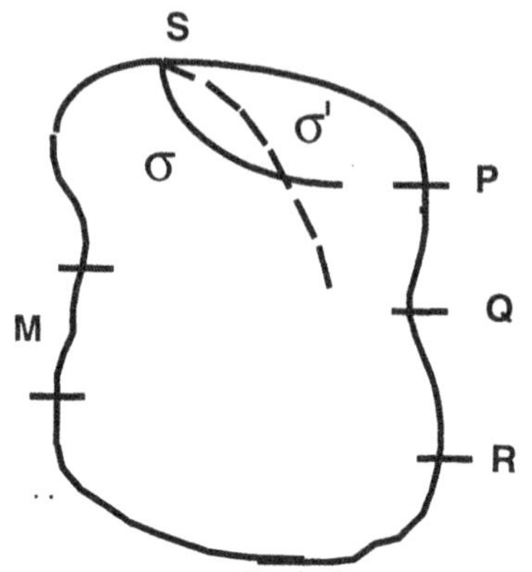

Figure 2

Suppose now that σ and σ' do not coincide, then there exists $z \in \sigma'\backslash\sigma$ (ou $\sigma\backslash\sigma'$). z an interior point of $\Omega\backslash\ \sigma$ where the minimum of ω_σ is achieved(the minimum of ω_σ is exactly the minimum of φ) and therefore $\omega_\sigma \equiv$ cte, this is in contradiction whith the hypothesis $\Phi \not\equiv 0$.

♦

III-STABILITY

III.1 Statement

In this section the problems (II.1) and (II.2)) are reconsidered. In these two cases the overspecified data are supposed to be accessible on an open set **M** of the boundary $\partial\Omega$. Since the measurements are given by experiments, they usually are subjected to errors. The goal of this section is to study the stability of the inverse geometrical problems under consideration, that is, roughly speaking to study if small perturbation in measurements lead to a geometry in the vicinity of the actual geometry.

To formalize this idea, consider:

Γ_{ad} a set of admissible geometries (lcracks or smooth boundaries) the operator η defined for a fixed identifying flux Φ .

$$\eta : \Gamma_{ad} \rightarrow L^2(M)$$
$$\Gamma \rightarrow f = u_{\Gamma}\ _{|M}$$

By the previous section, the operator η is injective, consider the mapping (for simplicity also called η).

$$\eta :\ \Gamma_{ad} \rightarrow \eta(\ \Gamma_{ad})$$
$$\Gamma \rightarrow f = u_{\Gamma}\ _{|M}$$

η is therefore invertible. The stability will be established if one proves the continuity of η^{-1}.

So Γ_{ad}, as well as ($L^2(M)$) has to be equipped with an appropriate topology.

The main resultts of this section are a global "weak " stability result, that is the continuity of the operator η^{-1} the methods followed here to perform these results walk for the smooth unknown boundary problem as well as the line segment crack one. For the reader convenience, this section focuses on the line segment crack problem. The results are completely shown in this case.

III.2. "Weak" stability results

One can see that the compactness of the set of unknown geometries and the uniqueness result lead to the stability. This seems to be general : It was proved for buried cracks in [14], for monotone inclusions in [1]. The next theorem is devoted to this kind of result in the case of straight cracks having an apriori known endpoint on the boundary.

The set Σ is chosen to be compact for the Hausdorff metric :

$$d(\sigma,\sigma') = \overset{s}{(}\ \underset{x \in \sigma}{\text{Max}}\ \ \underset{y \in \sigma'}{\text{Min}}\ \ |\ x-y|\ )$$

($L^2(M)$) is equipped with the L^2- norm. And consider :

$$\eta : \Sigma \to \eta(\Sigma)$$
$$\sigma \to f = u_{\sigma \mid M}$$

Theorem :

The operator η^{-1} is continuous

Proof :

Let σ_n, $\sigma \in \Sigma$ such that the corresponding data $f_{\sigma_n} \to f_\sigma$ in $L^2(M)$

By compactness, σ_n has a subsequence $\sigma_{p(n)}$converging to $\tilde{\sigma} \in \Sigma$

Then $f_{\sigma_{p(n)}} \to f_{\tilde{\sigma}}$ (stability of the direct problem) and therefore $f_{\tilde{\sigma}} = f_\sigma$

By the uniqueness result $\tilde{\sigma} = \sigma$

Then $\sigma = \tilde{\sigma}$ is the unique adherence value of σ_n and d $(\sigma_n, \sigma) \to 0$. ♦

Remark : the same result occurs in the case of an unknown smooth boundary.

Notice that since the set of admissible cracks has been chosen compact, the previous theorem establishes actually that η^{-1} has a modulos of continuity [11]. That is there exists an increasing mapping φ :

$$\varphi : \overline{IR}^+ \to \overline{IR}^+$$

φ continuous in 0 and $\varphi(0) = 0$

and $$| f_\sigma - f_{\sigma'} |_{L^2(M)} \le \varphi(d(\sigma, \sigma'))$$

The goal of the next section is to have more information on φ, that is to "quantify" the continuity of η^{-1}. Actually, one proves that φ is locally lipschitzian. Since the stability is estimating the deviation of the geometry in terms of the deviation of the measurements. This, suggests the tool to use to perform the local stability result : the domain derivative.

III.2. Local lipschitzian stability

III.2. 1 Domain derivative

The method followed to establish the results concerning the domain derivative is based on the results of Murat and Simon [16].

Consider a family of diffeomorphims F_h mapping $\Omega \setminus \sigma_h$ onto $\Omega \setminus \sigma$. The open sets$\Omega \setminus \sigma_h$ are coisen in such a way that $\sigma_h$belong Σ. As in [16] , are chosen as perturbations of the identity :

$$F_h = Id + h\,\theta$$

For h "small" enough, F_h is a set of diffeomorphisms. $\theta \in W^{1,\infty}(\Omega \setminus \sigma)$ and $\theta \equiv 0$ on $\partial\Omega$. The next proposition gives the lagrangian first derivation of the solution of (I) with respect to a variation of the domain.

Proposition

The scalar fild $u^h_{\sigma_h}$defined on $\Omega \setminus \sigma$, has in $H^1(\Omega \setminus \sigma)$, the asymptotic expansion:

$$u^h_{\sigma_h} = u^0_\sigma + hu^1 \text{ .in } H^1(\Omega \setminus \sigma)$$

where u_σ^0 *is the solution of (I),* u^1 *is the solution of the problem:*

$$\int_{\Omega\setminus\sigma} \nabla u^1, \nabla v = \int_{\Omega\setminus\sigma}\left({}^t\frac{\partial\theta}{\partial M}+\frac{\partial\theta}{\partial M}\right)\nabla u_\sigma\, \nabla v - \int_\Omega (\nabla u_\sigma, \nabla v)\ div\ \sigma$$

$$\forall\, v \in H^1(\Omega)$$

Proof :
The proof of this result is similar of the one given in [12] in the case of elasticity. ♦

By this particular choice of F_h , one has :

$$|f_h - f|_M = |u_{\sigma h} - u|_M = |u_{\sigma h} \circ F_h - u|_M$$

Then in ordre to prove a local stability result, it suffices to prove that u^1cannot vanish all over M .

III.2.2 Stability with respect to a length variation
Let σ be a line segment crack with S as an endpoint denote by F the end point of σ belonging to Ω and σ_h a line set cracks $\sigma_h \subset$ (S,F). Such that $|\sigma_n| = (1+h)\,|\sigma|$.
((S,F) line crossing S and F) $|\sigma|$ denote the length of σ.

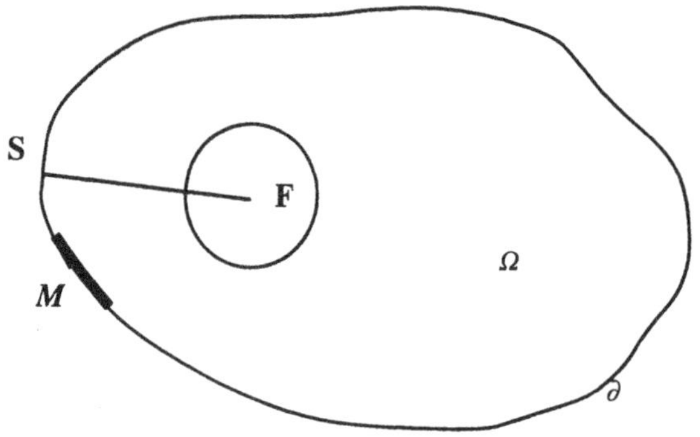

Theorem
Let f_h *(respectively) be the trace of the solution of (I) in* $\Omega\setminus\sigma_h$ *(respectively in* $\Omega\setminus\sigma$*)*
Under the assumption :
(H) *the singularity coefficient of* u_σ *at* ***F*** *(σ= [S,F]) is different of 0*
one has :

$$\lim_{h\to 0} \frac{|f_h - f|}{h} > 0$$

To prove this theorem, one needs some preliminary results.

PRELIMINARY RESULTS
The next result relates the derivative with respect to the crack length of the potential energy to the domain derivative of the heat field u_σ.
Recall that in the case of an insulated crack the solution u_σ is know to be composed by a somooth part u_σ^S and a singular part [15]

$u_\sigma = u^R_\sigma + u^s_\sigma$ where $u^R_\sigma \in H^2(\Omega \backslash \sigma)$

and $u^s_\sigma = c\sqrt{r}\sin\frac{\varphi}{2}$

(r,φ) designate the polar coodinates around F.

Proposition

$$\frac{\partial w}{\partial(\Omega \backslash \sigma)}(\Omega \backslash \sigma) \,.\, \theta = \frac{dw_h}{dh}\Big|_{h=0} =$$

$$\lim_{h\to 0}\left[\frac{1}{2}\int_{\Omega\backslash\sigma} |\nabla u_{\sigma_h}|^2 - \frac{1}{2}\int_{\Omega\backslash\sigma_h} |\nabla u_\sigma|^2\right]$$

$$= \int_{\partial\Omega} \Phi \,.\, u^1$$

(w_h *energy of P1* $\Omega \backslash \sigma_h$)

Proof :

The proof is similar to the one given in [12] in the frame work of elasticity.

The next lemma relates the singularity coefficient C to the domain derivative of the potential energy

Lemma

$$\frac{\partial w}{\partial \Omega}(\Omega) \,.\, \theta = \frac{dw_h}{dh}\Big|_{h=0} = -2\pi C^2$$

Proof of the theorem

By the asymptotic expansion, one has : $\lim_{h\to 0} \frac{|f_h - f|}{h} = |u^1|_{L^2(M)}$

Stability is achieved if one proves that $|u^1|_{L^2(M)} > 0$.

If not $\qquad |u^1|_{L^2(M)} = 0$

that is

$$u^1 \equiv 0 \text{ sur } M$$

and

$$\frac{\partial u^1}{\partial n} \equiv 0 \text{ sur } M \text{ (by the asymptotic expansion)}$$

u^1 is harmonic in a neigbourhood of $\partial\Omega$ (by the particular choice of θ)

By HOLMGREN's theorem

$$u^1 \equiv 0 \qquad \text{on } \partial\Omega$$

and by the lemma ; $\qquad \frac{dw}{dh}\Big|_{h=0} = \int_{\partial\Omega}\Phi \,.\, u^1 = -2\pi C^2$

If $C \neq 0$; u^1 *can not vanish* ; contradiction ♦

III.2.3 Stability with respect to a variation of the angle :

Let V_1 and V_2 by two open neighbourhoods of σ in Ω such that

$\bar{V}_1 \subseteq V_2 \subseteq \bar{V}_2 \subseteq \Omega$ $\quad$ $\bar{V}_1 \cap \partial\Omega = \{s\}$.

and consider F_h, $\quad$ $F_h = I + h\theta$

where θ is given by :

$$\theta = \begin{cases} -y \\ x \end{cases} \quad \text{in } \bar{V}_1$$

$$\theta = 0 \quad \text{in } (\Omega \backslash \sigma) \backslash \bar{V}_2$$

where θ is extended smoothly to all $\Omega\backslash\sigma$. ($\theta \in w^{1,\infty}(\Omega\backslash\sigma)$).

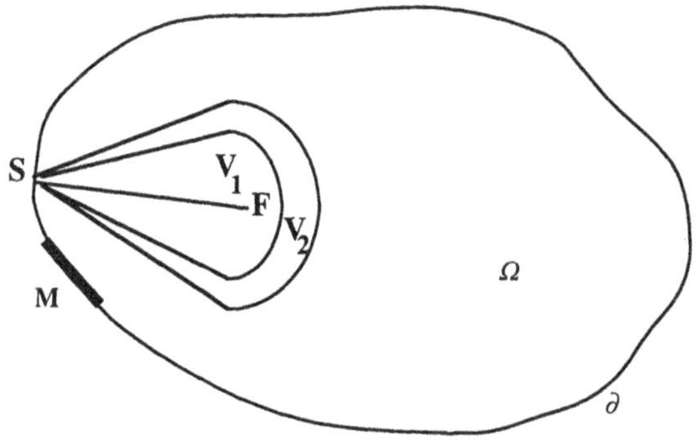

Theorem

Let fh (respectively f) be the trace of the solution of (I) posed on $\Omega \backslash \sigma_h$ (respectively $\Omega \backslash \sigma$) , then

$$\lim_{h\to 0} \frac{|fh - f|}{h} > 0$$

Proof :

Step1 : $u^1 \equiv 0$ on $\partial\Omega$

To prove the stability one has to establish that :

$|u^1|_{L^2(M)} > 0.$

Since here again one has :

$$\lim_{h\to 0} \frac{|fh - f|_{L^2(M)}}{h} = |u^1|_{L^2(M)}$$

If $|u^1|_{L^2(M)} = 0$ one would have :

$$u^1 \equiv 0 \quad \text{on } M$$

$$\frac{\partial u^1}{\partial n} \equiv 0 \quad \text{on } M$$

Then $u^1 \equiv 0$ on $\partial\Omega$ by the uniqueness theorem.

Step 2 : ISOLATE' THE SINGULARITY

Here again, as for the previous result, the singularity of the solution has to be "isolated" is $\Omega \backslash \sigma$.

Consider a bole around F (the endpoint of σ belonging to Ω) and with radius δ.

u^1 is the solution of the problem :

$$(*) \int_{\Omega\backslash\sigma} \nabla u^1 . \nabla v = \int_{\Omega\backslash\sigma}\left(\left({}^t\frac{\partial\theta}{\partial M} + \frac{\partial\theta}{\partial M}\right)\nabla u_\sigma, \nabla v\right) - \int_{\Omega\backslash\sigma} (\nabla u_\sigma, \nabla v)\ div\ \theta$$

Denote by $(\Omega\backslash\sigma)_\delta$ the complementary of B_δ in $\Omega\backslash\sigma$.

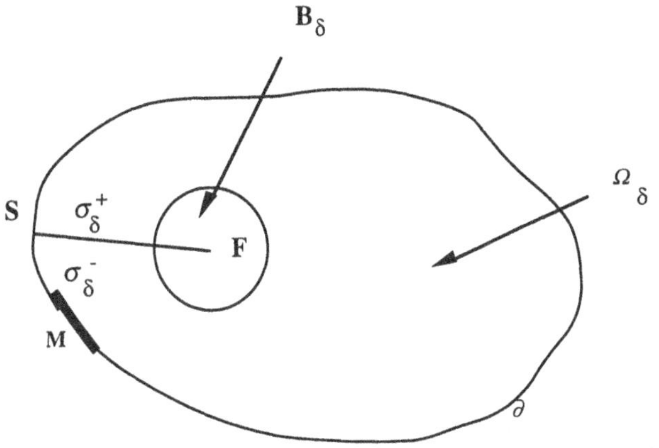

(*) is equivalent to :

$$0 = \int_{(\Omega\backslash\sigma)_\delta} \nabla u^1 . \nabla v - \int_{(\Omega\backslash\sigma)_\delta}\left(\left({}^t\frac{\partial\theta}{\partial M} + \frac{\partial\theta}{\partial M}\right)\nabla u_\sigma, \nabla v\right) + \int_{(\Omega\backslash\sigma)_\delta} (\nabla u_\sigma, \nabla v)\ div\ \theta$$

$$+ \int_{B_\delta} \nabla u^1 . \nabla v - \left(\left({}^t\frac{\partial\theta}{\partial M} + \frac{\partial\theta}{\partial M}\right)\nabla u_\sigma, \nabla v\right) + (\nabla u_\sigma, \nabla v)\ div\ \theta$$

the function integrated on B_δ is in $L^1(\Omega\backslash\sigma)$ and then in $L^1(B_\delta)$, and taking into account that the measure of B_δ goes to 0 when δ goes to 0 , one can derive that :

$$\lim_{\delta\to 0} \int_{B_\delta} \nabla u^1 . \nabla v - \int_{B_\delta}\left(\left({}^t\frac{\partial\theta}{\partial M} + \frac{\partial\theta}{\partial M}\right)\nabla u_\sigma, \nabla v\right) + \int_{B_\delta} (\nabla u_\sigma, \nabla v)\ div\ \theta = 0$$

This proves that (*) is equivalent to :

$$(**)\ 0 = \lim_{\delta\to 0} \int_{(\Omega\backslash\sigma)_\delta} \nabla u^1 . \nabla v - \int_{(\Omega\backslash\sigma)_\delta}\left(\left({}^t\frac{\partial\theta}{\partial M} + \frac{\partial\theta}{\partial M}\right)\nabla u_\sigma, \nabla v\right) - \int_{(\Omega\backslash\sigma)_\delta} (\nabla u_\sigma, \nabla v)\ div\ \theta$$

Using Green's formulae and harmonic test functions one derives that (**) is equivalent to :

$$(***)\ 0 = \lim_{\delta\to 0} \quad + \int_{\partial\Omega} u^1 \frac{\partial v}{\partial n}) - \int_{\partial\Omega} (\nabla u_\sigma, \nabla v)\ \theta.n$$

$$+ \int_{\sigma_\delta^+ \cup \sigma_\delta^-} u^1 \frac{\partial v}{\partial n} - \int_{\sigma_\delta^+ \cup \sigma_\delta^-} (\nabla u_\sigma, \nabla v)\ \theta.n$$

$$+ \int_{\partial B_\delta} u^1 \frac{\partial v}{\partial n} - \int_{\partial B_\delta} (\nabla u_\sigma, \nabla v)\ \theta.n$$

Since $u^1 \equiv 0$ on $\partial\Omega$, the first integral vanishes.
In the same way the second integral is equal to 0 since $\theta \equiv 0$ on $\partial\Omega$.

Step 3 : SELECTION OF THE TEST FUNCTIONS

A specific sequence of test functions $(v_m)_{m \in IN}$ is selected in order to lead to a contradiction.

$v_m = r^m \cos m\varphi$ where (r, φ) are the polar coordinnates around S.
Replace in (***) v by v_m, and examine the terms integrated on ∂B_δ.

since $u^1 \in L^2(\partial B_\delta)$, as well as $\frac{\partial v_m}{\partial n}$, one has :

$$\left| \int_{\partial B_\delta} u^1 \frac{\partial v_m}{\partial n} \right| < c_1(m) \text{ meas } (\partial B_\delta)$$

and therefore $\lim_{\delta \to 0} \left| \int_{\partial B_\delta} u^1 \frac{\partial v_m}{\partial n} \right| = 0$

Consider the decomposition of u_σ into a smooth part and a singular part (as shown in the previous section) :

$u_\sigma = u^R + u^S$

In the vicinity of the crack tip F $u^S = c\delta^{1/2} \sin \Pi\varphi$ and $u^R \in H^2(\Omega)$
then

$$\left| \int_{\partial B_\delta} (\nabla u_\sigma, \nabla v_n)\, \theta.n \right| < c_2(m)\, \delta^{1/2} \text{meas } (\partial B_\delta)$$

and therefore $\lim_{\delta \to 0} \left| \int_{\partial B_\delta} (\nabla u_\sigma, \nabla v_n)\, \theta.n \right| \to 0$

The integrals on ∂B_δ have no contribution in the expression (***).

Furthermore $\frac{\partial v_m}{\partial n} /_\sigma = \frac{\partial v_m}{\partial \varphi} /_{\varphi=0} = \frac{\partial v_m}{\partial \varphi} /_{\varphi=2\Pi} = 0$

Then (***) is equivalent to :

$$0 = \lim_{\delta \to 0} \int_{\sigma_\delta^+ \cup \sigma_\delta^-} (\nabla u_\sigma, \nabla v)\, \theta.n = \int_\sigma \left([\nabla u_\sigma], \nabla v_m \right) \theta.n \quad \forall v_m$$

where [ ] denotes the jump of the function across σ.
this gives:

$$\int_0^l \left[\frac{\partial u_\sigma}{\partial r} \right] r^m = 0 \qquad \forall\, m \geq 1$$

Using now the geometrical compatibility conditions [17] which express that the operators jump [ ] and gradient commute, it turns out that :

$$\int_0^l \frac{\partial}{\partial r} [u_\sigma]\, r^m = 0 \qquad \forall\, m \geq 1$$

using now the continuity of $[u_\sigma]$ (actually u_σ is"smooth"up to the boundary, that is in particular on σ^+ and on σ^-), integrating by parts, the previous equality is equivalent to :

$$\int_0^l m[u_\sigma]\, r^{m-1} = \qquad [r^m [u_\sigma]]_0^l$$

but the crack is "closed" at r = l. (crack tip)
Therefore:

$$\int_0^l [u_\sigma]\, r^m = 0 \quad \forall\, m \geq 0$$

and by density of the set of polynomials in $C^0(0,l)$, one obtains that $[u_\sigma]=0$
This implies that the crack is "closed" and therefore that the flux does not insure the uniqueness, which is in contradiction with the choice of Φ.

♦

Remark :
Recently, the same approach has been successfully followed to prove the stability result for a smooth unknown boundry satisfying a boundary condition of Signorini type [10] , and to prove the stability of the line segment crack problem when the thermal testing is replaced by the elastostatic testing. In this case the result is performed both for a length variation and an angle variation under the condition that the stress intensity factors do not vanish [7].

IV Numerical treatments

IV1 Statement

The main assumption towards the identification result is that the overspecified data (Φ, f) is available on all the known part of the boundary. When this assumption is fullfilled the problem can be splitted into two direct problems :

$$(IV.1N)\quad \begin{cases} -\Delta u_\sigma = 0 & in\ \Omega\setminus\sigma \\ \dfrac{\partial u_\sigma}{\partial n} = \Phi & on\ \partial\Omega \\ \dfrac{\partial u_\sigma}{\partial n} = 0 & on\ \sigma \\ \int_{\partial\Omega} u_\sigma = 0 & \end{cases}$$

$$(IV.1M)\quad \begin{cases} -\Delta u_\sigma = 0 & in\ \Omega\setminus\sigma \\ u_\sigma = f & on\ \partial\Omega \\ \dfrac{\partial u_\sigma}{\partial n} = 0 & on\ \sigma \end{cases}$$

The crack to be identified is exactly the one for whom the solutions of the problems (IV.1.N) and (IV.1.N) coincide.
In terms of vector flux field the problem (IV.1.N) can be rephrased as follows :

$$(IV.2.N)\quad \begin{cases} div\ q = 0 & \text{in } \Omega\setminus\sigma \\ q.n = \Phi_m & \text{on } \partial\Omega \\ q.n = 0 & \text{on } \sigma \end{cases}$$

$$\text{with } \int_{\partial\Omega} \Phi_m = 0$$

and the constitutive law:

$$q= -k\nabla u_\sigma \quad \text{in } \Omega\setminus\sigma$$

where k is the themal conductivity of the body (k will be taken equal to 1).

Denote by :

$$H^1(\Omega\setminus\sigma) = \{u \in L^2(\Omega\setminus\sigma)\ /\nabla u\in (L^2(\Omega\setminus\sigma))^2\}$$
$$V(\Omega\setminus\sigma) = \{u\in H^1(\Omega\setminus\sigma)\ /u_{|\partial\Omega} = f_m\}$$
$$Q(\Omega\setminus\sigma) = \{q\in (L^2(\Omega\setminus\sigma))^2/\ div q\in L^2(\Omega\setminus\sigma)\ q.n_{|\partial\Omega} = \Phi_m \text{ et } q.n_{|\sigma} = 0\}$$

Lastly , denote by Σ the set of admissible cracks.
The Kohn and Vogelius error functional has in this case the following expression :

$$F(\sigma, u_\sigma ,q) = \frac{1}{2}\int_{\Omega\setminus\sigma} (\nabla u_\sigma + q)^2$$
$$(u_\sigma ,q) \in V(\Omega\setminus\sigma)\times Q(\Omega\setminus\sigma)$$

The trick here (as in the case of parameter identification) is to develop this expression :

$$F(\sigma, u_\sigma, q) = \frac{1}{2}\int_{\Omega\setminus\sigma} |\nabla u_\sigma|^2 + \frac{1}{2}\int_{\Omega\setminus\sigma} q^2 + \int_{\Omega\setminus\sigma} q.\nabla u_\sigma$$

Integrating by parts one obtains :

$$\int_{\Omega\setminus\sigma} q.\nabla u_\sigma = \int_{\partial(\Omega\setminus\sigma)} q.n\, u_\sigma - \int_{\Omega\setminus\sigma} \mathrm{div}q \,.\, u_\sigma$$

Using that $u_\sigma \in V(\Omega\setminus\sigma)$ and $q \in Q(\Omega\setminus\sigma)$, one can point out that :

$$\int_{\Omega_\Gamma} q.\nabla u = \int_{\partial\Omega_\Gamma\setminus\Gamma} \Phi_m f_m = \text{constant}.$$

That is minimizing F is equivalent to minimizing $\bar{F}$

$\bar{F}(\sigma, u_\sigma, q) = \frac{1}{2}\int_{\Omega\setminus\sigma} |\nabla u_\sigma|^2 + \frac{1}{2}\int_{\Omega\setminus\sigma} q^2$ In this new expression the Neuman and mixed problems are decoupled :

Denote

$$\tilde{F}(\sigma) = \underset{u_\sigma \in V(\Omega\setminus\sigma)}{\mathrm{Min}} \frac{1}{2}\int_{\Omega\setminus\sigma} |\nabla u_\sigma|^2 + \underset{q \in Q(\Omega\setminus\sigma)}{\mathrm{Min}} \frac{1}{2}\int_{\Omega\setminus\sigma} q^2$$

Lemma

$$\underset{q \in Q(\Omega\setminus\sigma)}{\mathrm{Min}} \frac{1}{2}\int_{\Omega\setminus\sigma} q^2 = -\underset{u_\sigma \in H^1(\Omega\setminus\sigma)}{\mathrm{Min}} \frac{1}{2}\int_{\Omega\setminus\sigma} \nabla u_\sigma^2 + \int_{\partial\Omega\setminus\sigma} \Phi_m u_\sigma$$

Proof :

The proof is given in [3] ♦

Then according to this last lemma $\tilde{F}$ has the following expression :

$$\tilde{F}(\sigma) = \underset{u_\sigma \in V(\Omega\setminus\sigma)}{\mathrm{Min}} \frac{1}{2}\int_{\Omega\setminus\sigma} |\nabla u_\sigma|^2 - \underset{H^1(\Omega\setminus\sigma)}{\mathrm{Min}} \left(\frac{1}{2}\int_{\Omega\setminus\sigma} |\nabla u_\sigma|^2 + \int_{\partial\Omega\setminus\sigma} \Phi_m u_\sigma\right)$$

The numerical trials are performed in the case of a line segment crack. The line containing the crack is supposed to be known. Therefore the identification problem is reduced to a two parameters problem : the determination of the crack endpoints.

Denote by $\sigma=[X_D, X_G]$ the crack to be determined. The Kohn and Vogelius functional is then given by the following expression :

$$F(X_D, X_G) = \underset{H^1(\Omega(X_D, X_G))}{\mathrm{Min}} J_D(T) - \underset{H^1(\Omega(X_D, X_G))}{\mathrm{Min}} J_N(T) + cte$$

J_D corresponds to the energy of the problem (IV.2.M).

J_N that of the problem(IV.2.N).

The functional *F* is to be minimized by a gradient method.

Formally ; the first optimality condition is given by :

$$\begin{cases} \dfrac{\partial F}{\partial X_D} = \dfrac{\partial J_D}{\partial X_D} - \dfrac{\partial J_N}{\partial X_D} = 0 \\ \dfrac{\partial F}{\partial X_G} = \dfrac{\partial J_D}{\partial X_G} - \dfrac{\partial J_N}{\partial X_G} = 0 \end{cases}$$

The terms $\frac{\partial}{\partial X}$ appear therefore as the "Thermical intensity factor **G** ". They are related to the singularity coefficient on the crack tip.

$$\frac{\partial F}{\partial X_D} = G_D^D - G_N^D = -K\,[C_D^2 - C_N^2]$$

where K is a nonnegative coefficient depending on the geometry and (C_D, C_N) are scalars related to the thermal "loading " f_m and Φ_m.

IV.2Numerical trials

The geometry of the structure concerned by the numerical trials is given by the figure 6.

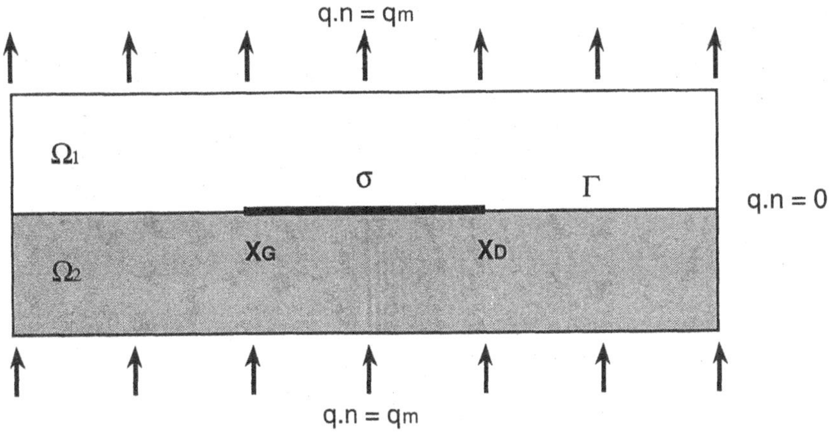

Figure 7

The detail of the mesh is given by the figure 8

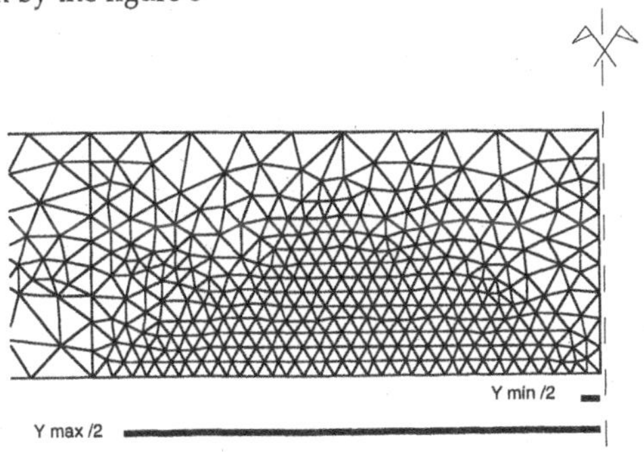

Figure 8

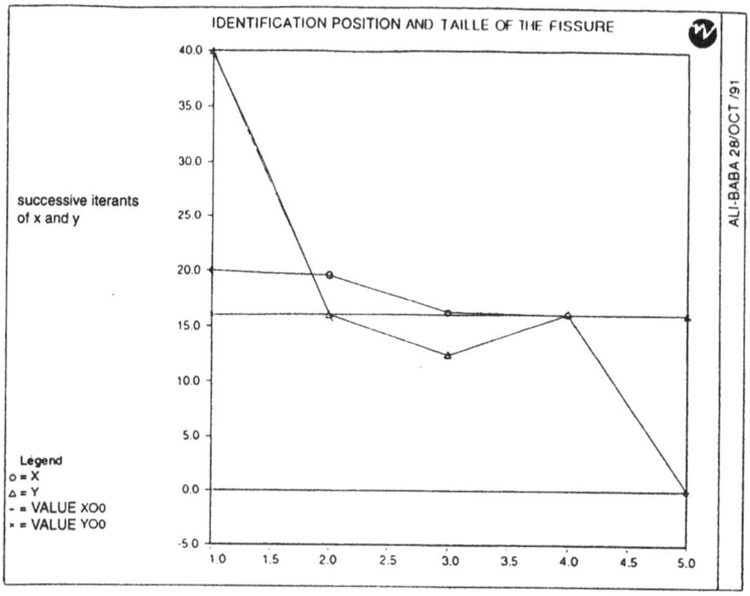
IDENTIFICATION POSITION AND TAILLE OF THE FISSURE
successive iterants
of x and y
Legend
o = X
Δ = Y
- = VALUE XOO
* = VALUE YOO
40.0
35.0
30.0
25.0
20.0
15.0
10.0
5.0
0.0
-5.0
1.0
1.5
2.0
2.5
3.0
3.5
4.0
4.5
5.0
ALI-BABA 28/OCT /91

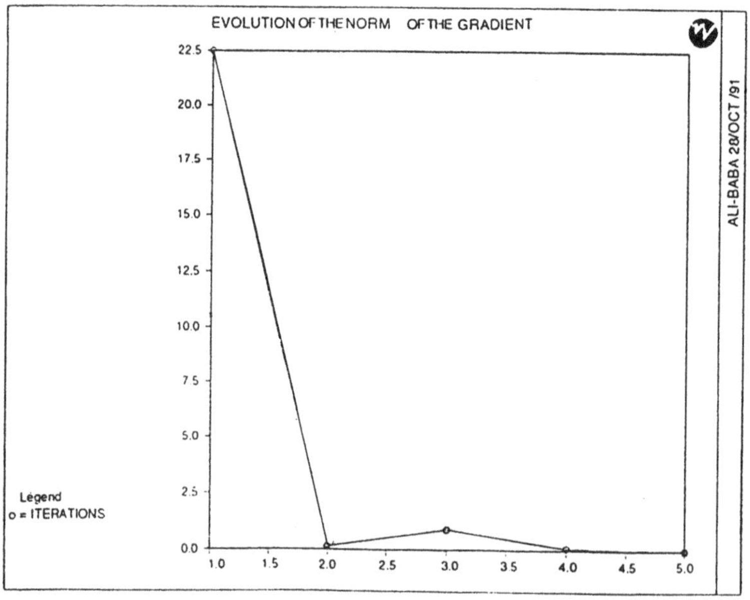
EVOLUTION OF THE NORM OF THE GRADIENT
Legend
o = ITERATIONS
22.5
20.0
17.5
15.0
12.5
10.0
7.5
5.0
2.5
0.0
1.0
1.5
2.0
2.5
3.0
3.5
4.0
4.5
5.0
ALI-BABA 28/OCT /91

Identification of the center X and the length Y of the crack :
The figure 9 gives the evolution of the parameters X and Y and that of the gradient of F with respect to the number of iterations.

Comment :
In the case of planar or line segment crack, an explicit inversion formula determining the plane (or the line) containing the crack was established [3].
That is the problem of a line segment crack determination can be solved in two steps.
Step 1 : The determination of te line containing the crack.
Step 2 : the location of the crack iteratively by minimizing the Kohn and Vogelius functional.

References

1 **G. ALESSANDRINI**. Remark on a paper by Bellout and Friedman. *Bolletino U.M.I (7)3A(1989)*1

2 **G. ALESSANDRINI., A.DIAZ VALENZUELA** Unique determination of multiple cracks by two measurements. *Quaderni Matematici Universita Di Trieste (1994)*

3 **S. ANDRIEUX, A. BEN ABDA**. Identification de fissures planes par une donnée au bord unique : un procédé direct de localisation d'identification.*C.R. Acad. Sci Paris, t.315, Série I (1992)*

4 **S. ANDRIEUX, A. BEN ABDA**. Fonctionnelles de Kohn-Vogelius et identification de domaines.*Collection de notes internes de la Direction des Etudes et Recherches ,92 NI J 0006 (1992)*

5 **S. ANDRIEUX, A. BEN ABDA, M. JAOUA**. Identification de frontière inaccessiible par une unique mesure de surface, *in les Annales Maghrébines de l'ingénieur(1993)*

6 **A. BEN ABDA** Sur quelques problèmes inverses géométriques.... *Thèse E.N.I.Tunis (1993).*

7 **H. BEN AMEUR** *D.E.A Thesis E.N.I.Tunis* (In preparation)

8 **H. BELLOUT, A. FRIEDMAN.** Identification problems in potential theory*Arch. Rat. Mech. Analy. 101 (1988)*

9 **K. BRYAN, M. VOGELIUS** A uniqueness result concerning the identification of a collection of cracks from finitely many electrostatic boundary measurements*SIAM J. Math Anal. Vol23 No4 (1992)*

10 **S. CHABEN,** *D.E.A Thesis E.N.I.Tunis* (In preparation)

11 **G. CHOQUET** Cours de Topologie *Masson (1984)*

12 **Ph. DESTUYNDER, M. JAOUA** Sur une interprétation de l'intégrale de Rice en théorie de la rupture fragile *Math. Meth. Appli. Sci 3 (1981)*

13 **A. FRIEDMAN, V.ISAKOV** On the uniqueness in the inverse conductivity problem with one measurement. *Indiana Yniversity Mathematics Journal Vol 38, No.5(1989)*

14 **A. FRIEDMAN, M. VOGELIUS** Determining cracks by boundary measurements*Indiana Uni. Math. Jou Vol 38 , nbr3 (1989)*

15 **P. GRISVARD Elliptic Problems in nonsmooth domains. *Pitman*(1985)**

16 **F. MURAT, J. SIMON** Quelques résultats sur le contrôle par un domaine géométrique *Université Pierre et Marie Curie(1974)*

17 **C.TRUESDELL, R.TOUPIN** The classical field theories . *In Handbuch der physik, Springer-Verlag, Berlin (1960) VOL III/1l*

S.A : EDF-DER-MMN, Avenue du Général de Gaulle
92141 Clamart-Cedex France

A.B.A , M.J : ENIT, LAMAP, BP 37 1002 TUNIS
& Institut Préparatoire aux études scientifiques et techniques, BP 51, Route de Sidi Bou Saïd
2070 La Marsa, TUNISIA

A Viscosity Solutions Approach to Some Asymptotic Problems in Optimal Control

F. Bagagiolo[(1)], M. Bardi[(2)] and I. Capuzzo Dolcetta[(3)]

Abstract. We present some new asymptotic results on singular perturbations and ergodic optimal control problems for systems with spatial state constraints.

1. Introduction

The aim of this paper is to show how the theory of viscosity solutions and, more precisely, comparison and stability properties, can be succesfully employed in the analysis of some asymptotic problems in optimal control. We focus our attention here on two specific problems, namely a singular perturbation and an ergodic control problem for state constrained system. The same framework, however, can be suitably adapted in order to deal with a variety of limit problems such as, for example,

- regular perturbation problems,
- small random perturbations
- discrete time approximations,
- penalization of constraints and/or costs.

---

(1) Dipartimento di Matematica, Università di Trento, Via Sommarive 14, 38050 Povo (Trento), Italy;
e-mail: bagagiol@alpha.science.unitn.it;
(2) Dipartimento di Matematica Pura e Applicata, Università di Padova, Via Belzoni 7, 35131 Padova, Italy;
e-mail: bardi@pdmat1.math.unipd.it;
partially supported by M.U.R.S.T. project "Problemi non lineari nell'analisi e nelle applicazioni fisiche, chimiche e biologiche";
(3) Dipartimento di Matematica, Università "La Sapienza", P.le A. Moro 2, 00185 Roma, Italy;
e-mail: capuzzo@mat.uniroma1.it;
partially supported by M.U.R.S.T. project "Analisi Non Lineare e Calcolo delle Variazioni".

We refer to [3] and the references therein for the above mentioned problems as well as for more details on the methods presented here. We recall that the theory of viscosity solutions for Hamilton-Jacobi equations goes back to [14], [11] and [10]. It was used for studying state-constrained problems in [15] and [8]. A comprehensive account of the theory and its applications to optimal control, with special attention to discontinuous solutions and the weak limit techniques employed here can be found in [4] and [3].

The plan of the paper is as follows: in section 2 we recall briefly for the convenience of the reader some basic facts about viscosity solutions. Section 3 and 4 are devoted to the presentations of our results on the singular perturbation problem and the ergodic control problem, respectively. Finally, we wish to mention that general references for singular perturbation problems are [12] and [7]; for some earlier results based on viscosity solutions we refer to [6], [13], [9] and [8].

2. Basic facts about viscosity solutions

It is well-known that dynamic programming techniques allow to show that the value function v of deterministic optimal control problems satisfies a nonlinear first order PDE of the form

(HJ) $$F(x, v, Dv) = 0 \quad \text{in} \quad \Omega \subseteq \boldsymbol{R}^N$$

in the global viscosity sense (the first results in this direction can be found in [14]). By this we mean that the following conditions (i), (ii) holds for any $\varphi \in \mathrm{C}^1(\overline{\Omega})$:

(i) v is upper semicontinuos and
$F(x, v(x), D\varphi(x)) \leq 0$ at any local maximum x of $v - \varphi$

(ii) v is lower semicontinuos and
$F(x, v(x), D\varphi(x)) \geq 0$ at any local minimum x of $v - \varphi$;

of course, the validity of (i) and (ii) implies the continuity of v.

The asymptotic problems that we are going to study in the next sections can be casted in the following framework.

Let $\varepsilon > 0$ be a parameter describing some perturbation of a given optimal control problem and $v^\varepsilon \in \mathrm{C}(\Omega)$ be the corresponding value function satisfying the Hamilton-Jacobi equation

(HJ)$^\varepsilon$ $$F^\varepsilon(x, v^\varepsilon, Dv^\varepsilon) = 0 \quad \text{in} \quad \Omega$$

in the viscosity sense.

The typical questions concerning the behaviour of $(\mathrm{HJ})^\varepsilon$ as $\varepsilon \to 0$ are:
- does there exist $u = \lim v^\varepsilon$?
- does u solve an equation of the form

$$F(x, u, Du) = 0 \quad \text{in} \quad \Omega$$

for some F?
- is u the value of some limit control problem determined by F?

Next we describe rather vaguely a "strategy" to answer the above questions. Assume that the uniform bound

$$\|v^\varepsilon\|_\infty \le C, \quad C \text{ independent of } \varepsilon > 0 \tag{2.1}$$

is known and set for $x \in \Omega$

$$\begin{aligned} \overline{v}(x) &:= \limsup_{(y,\varepsilon)\to(x,0+)} v^\varepsilon(y) \\ \underline{v}(x) &:= \liminf_{(y,\varepsilon)\to(x,0+)} v^\varepsilon(y). \end{aligned} \tag{2.2}$$

Observe that by their very definition $\overline{v}$ and $\underline{v}$ are, respectively, upper and lower semicontinuous and

$$\underline{v}(x) \le \overline{v}(x), \quad x \in \Omega.$$

Assume also that one is able to show that

$$F(x, \overline{v}, D\overline{v}) \le 0 \le F(x, \underline{v}, D\underline{v}) \quad \text{in} \quad \Omega, \tag{2.3}$$

in the viscosity sense, for some $F : \Omega \times \mathbb{R} \times \mathbb{R}^N \to \mathbb{R}$ satisfying a comparison principle. By this we mean that from (2.3) (sometimes supplemented with a boundary condition) one can infer that

$$\overline{v}(x) \le \underline{v}(x) \quad x \in \Omega.$$

At this point it is clear that $\overline{v}$ and $\underline{v}$ coincide with some $u \in C(\Omega)$ which is (by comparison, again) the unique viscosity solution of

$$F(x, u, Du) = 0 \quad \text{in} \quad \Omega.$$

If F satisfies some structural conditions (including notably the convexity of the function $p \to F(x, u, p)$), then u can be shown to be the value function v of some optimal control problem.

As for the convergence of v^ε to v, this is a consequence of the following simple fact pointed out in [5] (see also [4] and [3]).

Lemma. *Let $\{v^\varepsilon\}_{\varepsilon>0}$ be a family of functions such that*

$$\sup_{x\in K} |v^\varepsilon(x)| \le C_K$$

on a compact set K. If $\underline{v} = \overline{v}$ on K, then

$$v^\varepsilon \to v := \overline{v} = \underline{v}$$

uniformly on K.

In order to implement the above outlined strategy in specific control problems four steps are to be performed:

(a) prove the uniform estimate (2.1); this is often an easy one to achieve by direct control theoretic techniques (see, however, the ergodic problem in §4);
(b) guess the correct limit Hamiltonian F; in the simplest situations F is the uniform limit of F^ε, but this is not the case for the singular perturbation problem in §3;
(c) prove that the weak limits defined by (2.2) satisfy the differential inequalities (2.3) in the viscosity sense;
(d) prove that F satisfies the comparison property.

The last two steps are where the viscosity techniques play a fundamental role.

3. A singular perturbation problem

We consider a system described by $N+M$ state variables whose evolution is governed by

$$\begin{cases} y'(t) = f(y(t), \zeta(t), \alpha(t)) \\ \varepsilon\zeta'(t) = g(y(t), \zeta(t), \beta(t)) \end{cases} \tag{3.1}$$

with initial conditions $y(0) = x \in \mathbb{R}^N$, $\zeta(0) = z \in \mathbb{R}^M$.

For small values of the parameter $\varepsilon > 0$, the evolution of the y component of the state $(y(t), \zeta(t))$ is comparatively much slower than that of ζ.

Formally, setting $\varepsilon = 0$ reduces (3.1) to

$$\begin{cases} y'(t) = f(y(t), \zeta(t), \alpha(t)) \\ 0 = g(y(t), \zeta(t), \beta(t)), \end{cases}$$

a mixed system of differential and algebraic equations.

The optimal control problem we study here (see [2] for variants and extensions as well) is to minimize

$$J(x, z, \alpha, \beta) := \int_0^{+\infty} l(y^\varepsilon(t), \zeta^\varepsilon(t), \alpha(t))e^{-t}dt$$

where $y^\varepsilon(\cdot), \zeta^\varepsilon(\cdot)$ are the trajectories of (3.1) corresponding to definite choices of the controls α and β and initial positions x, z. We consider also a state constraint on the fast variable, modelised by the requirement that

$$\zeta^\varepsilon(t) \in \overline{\Omega}, \quad \forall t > 0, \tag{3.2}$$

where Ω is an open connected set in $\mathbb{R}^M$ with smooth boundary $\partial\Omega$. The value function is then

$$v^\varepsilon(x,z) := \inf\{J(x,z,\alpha,\beta) : (\alpha,\beta) \in (\mathcal{A}\times\mathcal{B})^\varepsilon_{x,z}\},$$

where

$$(\mathcal{A}\times\mathcal{B})^\varepsilon_{x,z} := \{(\alpha,\beta) : [0,+\infty[\to A\times B, \quad \alpha \quad \text{and} \quad \beta \quad \text{measurable and (3.2) holds}\}$$

Here A, B are given compact metric spaces.

The question is the limit behaviour of v^ε as $\varepsilon \to 0^+$. We shall assume that f, g are continuous and that

$$\text{(3.3)} \qquad \begin{aligned} &|f(x,z,a) - f(x',z',a)| + |g(x,z,b) - g(x',z',b)| \le L(|x-x'| + |z-z'|) \\ &|f(x,z,a)| \le K(1+|x|) \end{aligned}$$

for some $L > 0$ independent of $(a,b) \in A \times B$ and some $K > 0$ independent of $(z,a) \in \overline{\Omega} \times A$. Moreover, we admit the controllability condition

$$\text{(3.4)} \qquad \overline{S} \subseteq \overline{\text{co}}g(x,z,B), \quad \forall x \in \mathbb{R}^N, z \in \overline{\Omega}$$

(here $\overline{S}$ is the closed unit ball in $\mathbb{R}^M$ and co denotes the convex hull).

As for the running cost we assume for simplicity

$$\text{(3.5)} \qquad \begin{cases} l \quad \text{continuous}, \quad 0 \le l(x,z,a) \le M \\ |l(x,z,a) - l(x',z',a)| \le L(|x-x'| + |z-z'|) \end{cases}$$

Under this set of assumptions $v^\varepsilon \in C(\mathbb{R}^N \times \overline{\Omega})$ for $\varepsilon > 0$ (see [2] for a detailed proof) and the viscosity theory for state constrained problems applies (see [15], [8] and [2]). Therefore, for $\varepsilon > 0$, v^ε is the constrained viscosity solution of

$$\text{(HJ)}^\varepsilon \qquad \begin{cases} v^\varepsilon + \sup\limits_{a\in A}\{-f\cdot D_x v^\varepsilon - l\} - \dfrac{1}{\varepsilon}\inf\limits_{b\in B} g\cdot D_z v^\varepsilon \le 0 \quad \text{in} \quad \mathbb{R}^N \times \Omega \\ v^\varepsilon + \sup\limits_{a\in A}\{-f\cdot D_x v^\varepsilon - l\} - \dfrac{1}{\varepsilon}\inf\limits_{b\in B} g\cdot D_z v^\varepsilon \ge 0 \quad \text{in} \quad \mathbb{R}^N \times \overline{\Omega}. \end{cases}$$

By this we mean that condition (i) in the definition of §2 is satisfied at all points of $\mathbb{R}^N \times \Omega$, while (ii) holds in $\mathbb{R}^N \times \overline{\Omega}$.

Since as $\varepsilon \to 0^+$ the state variable ζ evolves faster and faster, the value function v^ε is expected to be less and less sensitive to the variable z. Hence the guess is that if v^ε has limit v, as $\varepsilon \to 0^+$, this should be independent of z. This heuristics is confirmed by the next

Theorem. *Assume* (3.3), (3.4), (3.5). *Then*

$$v^\varepsilon(x,z) \to v(x) \quad \textit{locally uniformly in} \quad \mathbb{R}^N \times \overline{\Omega}$$

with v given by

(3.6) $$v(x) = \inf\left\{ \int_0^{+\infty} l(y(t),\zeta(t),\alpha(t))e^{-t}dt, \quad (\zeta,\alpha) \in \mathcal{Z} \times \mathcal{A} \right\}$$

where $y(\cdot)$ is the solution of

$$\begin{cases} y'(t) = f(y(t),\zeta(t),\alpha(t)) \\ y(0) = x \in \mathbb{R}^N \end{cases}$$

and

$$\mathcal{Z} \times \mathcal{A} := \{(\zeta,\alpha) : [0,+\infty[\to \overline{\Omega} \times A, \quad (\zeta,\alpha) \quad \text{measurable}\}$$

Sketch of the proof. We shall follow the strategy described in §2. Observe first that standard dynamic programming arguments show that $v \in \mathrm{BUC}(\mathbb{R}^N)$ is a viscosity solution of

(HJ) $$v + \sup_{(z,a)\in\overline{\Omega}\times A} \{-f \cdot Dv - l\} = 0 \text{ in } \mathbb{R}^N.$$

It is also well-known that the comparison principle holds for (HJ) (see [10], [4], [3]). Hence v is the unique viscosity solution of (HJ).

The uniform bound

$$0 \le v^\varepsilon(x,z) \le M, \quad \forall\varepsilon > 0, \quad \forall(x,z) \in \mathbb{R}^N \times \overline{\Omega}$$

is easily obtained. We define then the semicontinuous functions $\underline{v}, \overline{v}$ by

$$\underline{v}(x,z) = \liminf_{(x',z',\varepsilon)\to(x,z,0+)} v^\varepsilon(x',z')$$
$$\overline{v}(x,z) = \limsup_{(x',z',\varepsilon)\to(x,z,0+)} v^\varepsilon(x',z').$$

The next steps are to show that:

(3.7) $v(x) \le \underline{v}(x,z)$, $\forall(x,z) \in \mathbb{R}^N \times \overline{\Omega}$;

(3.8) $\overline{v}(x,z)$ is independent of $z \in \Omega$;

(3.9) $\tilde{v}(x) := \overline{v}(x,z)$ is a subsolution of (HJ).

The proof of (3.7) is simple. Observe first that (HJ) implies that v satisfies

$$v + \sup_{a\in A}\{-f(x,z,a) \cdot Dv - l(x,z,a)\} \le 0 \text{ in } \mathbb{R}^N$$

for each fixed $z \in \overline{\Omega}$. On the other hand, since v does not depend on z, then $D_z v \equiv 0$ so that one easily checks that v is a subsolution of $(\mathrm{HJ})^\varepsilon$ for any $\varepsilon > 0$. At this point one concludes that

$$v(x) \le v^\varepsilon(x,z), \ \forall(x,z) \in \mathbb{R}^N \times \overline{\Omega}, \ \forall\varepsilon > 0$$

by comparison results for constrained viscosity solutions (see [15] and [8]). The above yields (3.7).

The proof of (3.8) requires more technicalities. First, one shows that $\overline{v}$ satisfies

$$|D_z\overline{v}(x,z)| \leq 0 \text{ in } \mathbb{R}^N \times \Omega \tag{3.10}$$

in the viscosity sense. The key tools in proving this claim are the fact that v^ε are subsolutions of $(\mathrm{HJ})^\varepsilon$ and the inequality

$$\inf_{\eta \in \overline{co}g(x,z,B)} \eta \cdot p \leq \inf_{\eta \in \overline{S}} \eta \cdot p = -|p|, \ \forall p \in \mathbb{R}^M$$

which is a consequence of assumption (3.4). Next, from (3.10), one can deduce (3.8): this is obvious if $\overline{v}$ is smooth, it follows from the the standard theory of viscosity solutions if $\overline{v}$ is continuous (see [11] and [3]) and it requires a more technical proof in the general case (see [16] and [2]).

To show that (3.9) holds, let $\overline{x}$ be a strict local maximum for $\tilde{v}-\varphi$, $\varphi \in C^1$. Hence $(\overline{x},\overline{z})$ is a strict maximum for $\tilde{v}(x)-\varphi(x)-|z-\overline{z}|^2$ for any fixed $\overline{z} \in \Omega$ and so, by a Lemma of Barles & Perthame (see [5], [4] and [3]), there exist $(x_n,z_n) \to (\overline{x},\overline{z})$, as $\varepsilon_n \to 0^+$, such that $v^{\varepsilon_n}(x_n,z_n) \to \tilde{v}(\overline{x})$ as $\varepsilon_n \to 0^+$, and (x_n,z_n) is a local maximum for $v^{\varepsilon_n}-\varphi$. From $(\mathrm{HJ})^\varepsilon$ we have

$$v^\varepsilon + \sup_{a\in A}\{-f\cdot D\varphi - l\} \leq \frac{1}{\varepsilon}\inf_{b\in B} g\cdot 2(z^\varepsilon-\overline{z}) \ \text{ at } (x^\varepsilon,z^\varepsilon) \tag{3.11}$$

Since $z^\varepsilon - \overline{z} \in \overline{S}$ for small enough ε we obtain, using again (3.4),

$$\inf_{b\in B} g(x^\varepsilon,z^\varepsilon,b)\cdot(z^\varepsilon-\overline{z}) \leq \inf_{\eta\in\overline{S}} \eta\cdot(z^\varepsilon-\overline{z}) \leq -|z^\varepsilon-\overline{z}| \leq 0.$$

We can then let $\varepsilon \to 0^+$ in (3.11) and (3.9) follows.

As a consequence of (3.7), (3.8), (3.9) we conclude that

$$\overline{v}(x,z) = \tilde{v}(x) \leq v(x) \leq \underline{v}(x,z) \leq \overline{v}(x,z) = \tilde{v}(x), \ \forall (x,z) \in \mathbb{R}^N \times \Omega;$$

hence $\overline{v}(x,z) = \underline{v}(x,z) = v(x)$ in $\mathbb{R}^N \times \Omega$. So, by the Lemma in §1,

$$v^\varepsilon(x,z) \to v(x) \text{ locally uniformly in } \mathbb{R}^N \times \Omega.$$

The convergence for $z \in \partial\Omega$ does not follow directly from the arguments above. Indeed, (3.8) holds only for $z \in \Omega$ since v^ε is a subsolution of $(\mathrm{HJ})^\varepsilon$ in $\mathbb{R}^N \times \Omega$. However, the full statement can be proved by a rather technical argument that we do not reproduce here (see [2] for details).

□

4. An ergodic control problem with state constraints

We consider the control system

$$\left\{\begin{array}{l} y'(t) = f(y(t), \alpha(t)),\ t > 0 \\ y(0) = x \end{array}\right. \tag{4.1}$$

and the infinite horizon discounted cost functional

$$J(x,\alpha) = \int_0^{+\infty} l(y_x(t), \alpha(t)) e^{-\varepsilon t} dt,\ \varepsilon > 0, \tag{4.2}$$

where $y_x(\cdot)$ is the solution of (4.1) corresponding to α. The value function is

$$v^\varepsilon(x) := \inf\{J(x,\alpha) :\ \alpha \in \mathcal{A}_x\}; \tag{4.3}$$

the admissible controls $\alpha \in \mathcal{A}_x$ are those which keep the trajectories of (4.1) in $\overline{\Omega}$, Ω being a given bounded open set in $\mathbb{R}^N$ with smooth boundary. Namely

$$\mathcal{A}_x := \{\alpha : [0, +\infty[\to A,\ \text{measurable} : y_x(t) \in \overline{\Omega}\ \forall t > 0\}$$

where A is a given compact set in $\mathbb{R}^N$. We are interested in the limit behaviour of v^ε as $\varepsilon \to 0^+$.

Assume f, l continuous and that, for some constants L, M, the following holds

$$\begin{array}{l} |f(x,a) - f(x',a)| + |l(x,a) - l(x',a)| \le L|x - x'| \\ 0 \le l(x,a) \le M \\ \text{for all } x, x' \in \mathbb{R}^N,\ a \in A. \end{array} \tag{4.4}$$

Assume also that

$$\exists \nu > 0 : f(x,a) \cdot n(x) \le -\nu,\ \forall x \in \partial\Omega,\ \forall a \in A, \tag{4.5}$$

where n denotes the exterior normal to Ω.

It is known (see [15] and [8]) that under these conditions the value function $v^\varepsilon \in C(\overline{\Omega})$ is the unique solution of

$$\left\{\begin{array}{l} v^\varepsilon + \dfrac{1}{\varepsilon} H(x, Dv^\varepsilon) \le 0 \text{ in } \Omega \\ v^\varepsilon + \dfrac{1}{\varepsilon} H(x, Dv^\varepsilon) \ge 0 \text{ in } \overline{\Omega} \end{array}\right. \tag{HJ$^\varepsilon$}$$

in the viscosity sense. Here the hamiltonian H is

$$H(x,p) = \max_{a \in A}\{-f(x,a) \cdot p - l(x,a)\}.$$

Observe that uniform a priori bounds for v^ε are not available in general (in the simplest case $l(x,a) \equiv l_0 > 0$, a direct computation shows that $v^\varepsilon(x) \equiv l_0/\varepsilon$)). We consider then new functions

$$u^\varepsilon(x) := \varepsilon v^\varepsilon(x);\ w^\varepsilon(x) := v^\varepsilon(x) - v^\varepsilon(x_0)$$

where x_0 is a fixed reference point in $\overline{\Omega}$. It is a simple exercise in viscosity solutions theory to show that u^ε, w^ε are, respectively, the unique solutions of

$$(4.6)\qquad \begin{cases} u^\varepsilon + H(x, \frac{1}{\varepsilon} Du^\varepsilon) \leq 0 \text{ in } \Omega \\ u^\varepsilon + H(x, \frac{1}{\varepsilon} Du^\varepsilon) \geq 0 \text{ in } \overline{\Omega} \end{cases}$$

and

$$(4.7)\qquad \begin{cases} \varepsilon w^\varepsilon + H(x, Dw^\varepsilon) + \varepsilon v^\varepsilon(x_0) \leq 0 \text{ in } \Omega \\ \varepsilon w^\varepsilon + H(x, Dw^\varepsilon) + \varepsilon v^\varepsilon(x_0) \geq 0 \text{ in } \overline{\Omega}. \end{cases}$$

It is natural to expect that controllability conditions for system (4.1) such as

$$(4.8)\qquad r\overline{S} \subseteq \overline{\text{co}} f(x, A),\ \forall x \in \Omega \text{ for some } r > 0,$$

(here $\overline{S}$ is the closed unit ball in $\mathbb{R}^N$) should be related to a nice limit behaviour in ergodic problems (see [9] and [1]). This is confirmed by the next result:

Theorem. *Under the assumptions* (4.4), (4.5), (4.8) *there exist a constant* χ^0 *and, for each fixed* $x_0 \in \overline{\Omega}$, *a function* $w^0 \in C(\overline{\Omega})$ *such that, for some sequence* $\varepsilon_n \to 0^+$,

$$u^{\varepsilon_n} \to \chi^0,\ w^{\varepsilon_n} \to w^0 \textit{ uniformly in } \overline{\Omega}$$

and w^0 *is a viscosity solution of*

$$(4.9)\qquad \begin{cases} H(x, Dw^0) + \chi^0 \leq 0 \textit{ in } \Omega \\ H(x, Dw^0) + \chi^0 \geq 0 \textit{ in } \overline{\Omega}. \end{cases}$$

Sketch of the proof. It is straightforward to check that $\|u^\varepsilon\|_\infty \leq C$. From the assumptions made it is not very hard to prove that

$$(4.10)\qquad |v^\varepsilon(x) - v^\varepsilon(z)| \leq C'|x - z|,\ \forall x, z \in \overline{\Omega}$$

for some constant C' depending on M, L, ν and r. Hence

$$(4.11)\qquad |u^\varepsilon(x) - u^\varepsilon(z)| \leq C'\varepsilon|x - z|,\ \forall x, z \in \overline{\Omega}.$$

By Ascoli-Arzelà, there exist a sequence $\varepsilon_n \to 0^+$ and a function χ^0 such that

$$(4.12)\qquad \|u^{\varepsilon_n} - \chi^0\|_\infty \to 0, \text{ as } n \to +\infty.$$

From (4.11) it follows then that χ^0 is a constant.

As a consequence of (4.10) we have also

$$|w^\varepsilon(x)| \leq C'\text{diam}\Omega; \; |w^\varepsilon(x) - w^\varepsilon(z)| \leq C'|x - z|$$

for all $x, z \in \overline{\Omega}$. By Ascoli-Arzelà again, there exist $\varepsilon_n^{'} \to 0^+$ and $w^0 \in C(\overline{\Omega})$ such that

$$\|w^{\varepsilon_n^{'}} - w^0\|_\infty \to 0 \text{ as } n \to +\infty.$$

Observe that equation (4.7) satisfied by w^{ε_n} is of the form

$$F^\varepsilon(x, w^\varepsilon, Dw^\varepsilon) = 0$$

with

$$F^\varepsilon(x, r, p) = \varepsilon r + H(x, p) + u^\varepsilon(x_0).$$

By (4.12), F^{ε_n} converges uniformly to

$$H(x, p) + \chi^0$$

for r in a bounded set. Therefore (4.9) follows (modulo further extraction of subsequences and a diagonal procedure) from standard stability results for viscosity solutions (see e.g. [10], [4] and [3]).

□

References

[1] M. Arisawa, *Ergodic problem for the Hamilton-Jacobi- Bellman equation I. Existence of the ergodic attractor*, to appear in Annales IHP, Analyse Nonlineaire.

[2] F. Bagagiolo, *Soluzioni di viscosità vincolate di equazioni di Bellman e perturbazione singolare in problemi di controllo ottimo*, Thesys, Università di Padova, July 1993.

[3] M. Bardi and I. Capuzzo Dolcetta *Optimal control and viscosity solutions of Hamilton-Jacobi-Bellman equations*, Birkhäuser, to appear.

[4] G. Barles, *Solutions de viscosité des équations de Hamilton-Jacobi*, Mathématiques et Applications 17, Springer-Verlag (1994).

[5] G. Barles and B. Perthame, *Discontinuous solutions of deterministic optimal stopping time problems*, RAIRO Math. Methods and Num. Anal. 21 (1987), 557-579.

[6] E.N. Barron, L.C. Evans and R. Jensen, *Viscosity solutions of Isaacs' equations and differential games with Lipschitz controls*, J. Differential Equations 53 (1984), 213-233.

[7] A. Bensoussan, *Perturbation methods in optimal control*, Wiley/Gauthier, Chichester (1988).

[8] I. Capuzzo Dolcetta and P.L. Lions, *Viscosity solutions of Hamilton-Jacobi equations and state constraints problems*, Trans. Amer. Math. Soc. 318 (1990), 643-683.

[9] I. Capuzzo Dolcetta and J.L. Menaldi *Asymptotic behaviour of the first order obstacle problem* J. Differential Equations 75 (1988), 303-328.

[10] M.G. Crandall, L.C. Evans and P.L. Lions, *Some properties of viscosity solutions of Hamilton-Jacobi equations*, Trans. Amer. Math. Soc. 282 (1984), 487-502.

[11] M.G. Crandall and P.L. Lions *Viscosity solutions of Hamilton-Jacobi equations*, Trans. Amer. Math. Soc. 277 (1983), 1-42.

[12] P.V. Kokotović, *Applications of singular perturbation techniques to control problems*, SIAM Review 26 (1984), 501-550.

[13] P.L. Lions *Equations de Hamilton-Jacobi et solutions de viscosité*, Colloque De Giorgi Paris 1983, Pitman, London (1985).

[14] P.L. Lions, *Generalized Solutions of Hamilton-Jacobi Equations*, Pitman, Boston (1982).

[15] M.H. Soner, *Optimal control with state-space constraints*, SIAM J. Control Optim. 24 (1986), 551-561.

[16] P. Soravia, personal communication.

Homogenization and Continuous Dependence for Dirichlet Problems in L^1

Lucio BOCCARDO

Dipartimento di Matematica, Università di Roma I

Piazza A. Moro 2, 00185, Roma, Italia

tel: (39.6)49913202; e-mail: boccardo@gpxrme.sci.uniroma1.it

ABSTRACT

In this note, the behaviour of the (unique for every n, in the sense of the definition of entropy solution) solutions u_n of the boundary value problem

$$A_n(u_n) = f \in L^1(\Omega) \quad \text{in } \Omega, \qquad u_n = 0 \quad \text{on } \partial\Omega,$$

is studied, if A_n is a sequence of monotone nonlinear operators acting on $W_0^{1,p}(\Omega)$, which G-converges.

1. INTRODUCTION AND ASSUMPTIONS

This note deals with continuous dependence of solutions of the following boundary value problem

$$\begin{cases} A(u) = f \in L^1(\Omega) & \text{in } \Omega \\ u = 0 & \text{on } \partial\Omega \end{cases} \tag{P}$$

with respect to (strong or weak) L^1 perturbations of the right hand side or G-convergence of the differential operator A, defined from $W_0^{1,p}(\Omega)$ into its dual. Here, and in the following, Ω is a bounded set of $\mathbf{R}^N$ and $2 \le p < N$. The partial differential operator A is defined by

$$A(v) = -\text{div}\,(a(x, \nabla v))$$

where $a : \Omega \times R^N \to R^N$ is a Caratheodory function. We assume that there exist two real positive constants α, β such that for every $\xi\ , \eta \in \mathbf{R}^N$, $(\xi \neq \eta)$, and for almost every $x \in \Omega$,

$$a(x, \xi)\xi \ge \alpha|\xi|^p, \tag{1.1}$$

$$|a(x, \xi)| \le \beta(|\xi|^{p-1} + 1), \tag{1.2}$$

$$[a(x,\xi) - a(x,\eta)][\xi - \eta] \geq \alpha|\xi - \eta|^p. \tag{1.3}$$

Concerning the right hand side, we assume that

$$f \in L^1(\Omega). \tag{1.4}$$

Remark that, if $p > N$, then $L^1(\Omega) \subset W^{-1,p'}(\Omega)$ and so the L^1 framework is a variational setting.

For properties of solutions if $1 < p < 2$ see [BBGGPV].

The existence of a weak solution in the Sobolev space $W_0^{1,q}(\Omega)$, for every $q < \tilde{p} = \frac{(p-1)N}{N-1}$ has been studied (also if the right hand side is a Radon measure) by G. Stampacchia (see [St]), if A is linear and $p = 2$, and by the author in some papers in collaboration with T. Gallouët, if A is nonlinear ([BG1], [BG2],[BBGGPV]; see also [Dv]). Of course, u is a weak solution if

$$\begin{cases} u \in W_0^{1,q}(\Omega), \ \forall q < \tilde{p} = \frac{(p-1)N}{N-1} : \\ \displaystyle\int_\Omega a(x,\nabla u)\nabla\varphi = \int_\Omega f\varphi \qquad \forall\varphi \in \mathcal{D}(\Omega). \end{cases} \tag{1.5}$$

Even in the linear case, the solution is not necessarily unique: a counterexample can be found in [Se] (for a complete description see [Pr]). However, the uniqueness of the solution is proved (see [BBGGPV]) by adding to the weak formulation the so called "entropy condition":

Definition 1.1. A function u is an entropy solution of problem(1.5) if

$$\begin{cases} u \in W_0^{1,q}(\Omega), \ \forall q < \tilde{p} = \frac{(p-1)N}{N-1}, \\ T_k(u) \in W_0^{1,p}(\Omega), \ \forall k > 0 : \\ \displaystyle\int_\Omega a(x,\nabla u)\nabla T_k[u - \varphi] \leq \int_\Omega f T_k[u - \varphi] \\ \qquad \forall\varphi \in W_0^{1,p}(\Omega) \cap L^\infty(\Omega), \end{cases} \tag{1.6}$$

where

$$T_k(t) = \begin{cases} t & \text{if } |t| \leq k, \\ k\frac{t}{|t|} & \text{if } |t| > k. \end{cases}$$

Remark that the soputions u does not have finite energy. Thus, while it is not possible to use the solution u as test function in (1.5) (observe that $\frac{(p-1)N}{N-1} < p$) an important role is played by the test functions

$$T_k[u - \varphi], \qquad \varphi \in W_0^{1,p}(\Omega) \cap L^\infty(\Omega).$$

Some L^1-perturbation of Dirichlet problems with solution having finite energy can be found in [B1], [BD], where is studied homogenization of boundary value problems of the type

$$\begin{cases} A(u)+|u|^{s-2}u=f\in L^2(\Omega) & \text{in } \Omega \\ u\in H_0^1(\Omega), u\in L^s(\Omega). \end{cases}$$

In the L^1 framework of the Dirichlet problem, the first results concerning uniqueness and G-convergence have been obtained by A. Dall'Aglio ([Da]). Another approach can be found in a recent paper by F. Murat ([M2]) by means of the "renormalized solutions" (see [BDGM] for the definition). We shall show an homogenization theorem for nonlinear Dirichlet problems in L^1, using the point of view of entropy solution. In this method, important tools are the *a priori* estimate in the *large* Sobolev space $W_0^{1,q}(\Omega)$ (see [BGa1], [BGa 2]) and the following L^1-version of the Minty Lemma.

Lemma 1.2.

The function u is a solution of (1.6) if and only if is a solution of

$$\begin{cases} u\in W_0^{1,q}(\Omega),\ \forall q<\tilde{p}=\frac{(p-1)N}{N-1}, \\ T_k(u)\in W_0^{1,p}(\Omega),\ \forall k>0: \\ \displaystyle\int_\Omega a(x,\nabla\varphi)\nabla T_k[u-\varphi]\le\int_\Omega fT_k[u-\varphi] \\ \qquad \forall\varphi\in W_0^{1,p}(\Omega)\cap L^\infty(\Omega). \end{cases} \tag{1.7}$$

Proof. One half of the result is just the monotonicity of the differential operator A.

Conversely, let u be a solution of (1.7). The choice of $\varphi=T_i(u)+t\,T_k(v-u)$ (i, k in $\mathbf{R}^+$, $t\in(0,1)$) where v is an arbitrary function in $W_0^{1,p}(\Omega)\cap L^\infty(\Omega)$, yields

$$\begin{aligned} I_\Omega &= \int_\Omega a(x,\nabla(T_i(u)+t\,T_k[v-u]))\nabla T_k(t\,T_k[v-u]-G_i(u)) \\ &\ge \int_\Omega f\,T_k(t\,T_k[v-u]-G_i(u))=J_\Omega, \end{aligned} \tag{1.8}$$

where $G_i(u)=u-T_i(u)$. Now

$$\begin{aligned} I_\Omega &= \int_{\{x\in\Omega:|t\,T_k[v-u]-G_i(u)|<k\}} a(x,\nabla T_i(u)+t\nabla T_k[v-u])\nabla(t\,T_k[v-u]-G_i(u)) \\ &= t\int_{\{x\in\Omega:|t\,T_k[v-u]-G_i(u)|<k\}} a(x,\nabla T_i(u)+t\,\nabla T_k[v-u])\nabla T_k[v-u] \\ &\quad -\int_{\{x\in\Omega:|t\,T_k[v-u]-G_i(u)|<k\}} a(x,t\,\nabla T_k[v-u])\nabla G_i(u). \end{aligned}$$

If $i > k + \|v\|_{L^\infty(\Omega)}$, then

$$I_\Omega = t\int_\Omega a(x,\nabla u + t\,\nabla T_k[v-u]))\nabla T_k[v-u]\chi_{\{x\in\Omega:|t\,T_k[v-u]-G_i(u)|<k\}}.$$

Passing to the limit as i tends to infinity and observing that

$$J_\Omega \to t\int_\Omega fT_k[v-u],$$

and that

$$\chi_{\{x\in\Omega:|t\;T_k[v-u]-G_i(u)|<k\}} \to \chi_{\{x\in\Omega:|t\;T_k[v-u]|<k\}}, \qquad \text{a. e. in } \Omega,$$

we have

$$I_\Omega \to t\int_\Omega a(x,\nabla u + t\nabla T_k[v-u]))\nabla T_k[v-u].$$

Thus,

$$t\int_\Omega a(x,\nabla u + t\,\nabla T_k[v-u]))\nabla T_k[v-u] \ge \int_\Omega f\,t\,T_k[v-u], \qquad \forall v\in W_0^{1,p}(\Omega)\cap L^\infty(\Omega).$$

Dividing by t, and passing to the limit with respect to $t\to 0^+$, using again the Lebesgue theorem we obtain (1.6).

2. CONTINUOUS DEPENDENCE OF THE SOLUTIONS WITH RESPECT TO L^1 PERTURBATIONS OF THE RIGHT HAND SIDE

Theorem 2.1. A priori estimate.

Assume that (1.1)-(1.3) hold. Let u_n be the solutions of

$$\begin{cases} u_n \in W_0^{1,q}(\Omega),\ \forall q < \tilde p = \frac{(p-1)N}{N-1}, \\ T_k(u_n)\in W_0^{1,p}(\Omega),\ \forall k>0: \\ \displaystyle\int_\Omega a(x,\nabla u_n)\nabla T_k[u_n-\varphi] \le \int_\Omega f_n T_k[u_n-\varphi] \\ \qquad \forall\varphi\in W_0^{1,p}(\Omega)\cap L^\infty(\Omega), \end{cases} \tag{2.1}$$

If the sequence $\{f_n\}$ is bounded in $L^1(\Omega)$, the sequence $\{u_n\}$ is bounded in $W_0^{1,q}(\Omega)$ for any $q<\tilde p$.

Proof. Taking $k = 1$ and $\varphi = T_h(u_n)$ in the definition of entropy solution we have, recalling (1.1),

$$\alpha\int_{B_h}|\nabla u_n|^p \le \|f_n\|_{L^1(\Omega)}. \tag{2.2}$$

where, for every $h \in \mathbf{N}$,

$$B_h = B_{n,n} = \{x \in \Omega : h \leq |u_n(x)| < h + 1\}.$$

Thus, for every $\lambda > 1$, we have

$$\int_\Omega \frac{|\nabla u_n|^p}{(1+|u_n|)^\lambda} = \sum_{h=0}^{\infty} \int_{B_h} \frac{|\nabla u_n|^p}{(1+|u_n|)^\lambda} \leq \sum_{h=0}^{\infty} \frac{\|f_n\|_{L^1(\Omega)}}{\alpha(1+h)^\lambda} = c_1 \tag{2.3}$$

Since $q < p$, by the Sobolev embedding and the Hölder inequality,

$$c_0 \left(\int_\Omega |u_n|^{q*} \right)^{\frac{q}{q*}} \leq \int_\Omega |\nabla u_n|^q = \int_\Omega \frac{|\nabla u_n|^q}{(1+|u_n|)^{\frac{\lambda q}{p}}} (1+|u_n|)^{\frac{\lambda q}{p}}$$

$$\left(\int_\Omega \frac{|\nabla u_n|^p}{(1+|u_n|)^\lambda} \right)^{\frac{q}{p}} \left(\int_\Omega (1+|u_n|)^{\frac{\lambda q}{p-q}} \right)^{1-\frac{q}{p}} \leq (c_1)^{\frac{q}{p}} c_2 \left(1 + \int_\Omega |u_n|^{\frac{\lambda q}{p-q}} \right)^{1-\frac{q}{p}}.$$

The previous inequalities give an a priori estimate of the sequence $\{u_n\}$ in the Sobolev space $W_0^{1,q}(\Omega)$, for every $q < \tilde{p}$, because

(i) $p < N$ implies $\frac{q}{q^*} > 1 - \frac{q}{p}$;

(ii) $q^* = \frac{\lambda q}{p-q}$ and $q < \tilde{p}$ implies $\lambda > 1$.

□

Theorem 2.2. *Let A, u_n, q be as before, $\{f_n\}$ be a sequence of functions converging to f in $L^1(\Omega)$-weak, and u be the entropy solution of (1.6). Then for every $1 < q < \tilde{p}$, u_n converges to u in $W_0^{1,q}(\Omega)$ strongly.*

Proof. Definition 1.1 and monotonicity of the differential operator A imply that

$$\int_\Omega a(x, \nabla\varphi)\nabla T_k(u_n - \varphi) \leq \int_\Omega f_n T_k(u_n - \varphi), \quad \forall \varphi \in W_0^{1,p}(\Omega) \cap L^\infty(\Omega). \tag{2.4}$$

The a priori estimate of Theorem 2.1 implies that there exists a function $u^* \in W_0^{1,q}(\Omega)$ (for every q such that $1 \leq q < \tilde{p}$) such that

$$u_n \rightarrow u^* \qquad \text{weakly in } W_0^{1,q}(\Omega), \ \forall q < \tilde{p} \tag{2.5}$$

$$u_n \rightarrow u^* \quad \text{strongly in } L^r(\Omega), \ \forall r < \frac{(p-1)N}{N-p}. \tag{2.6}$$

Moreover the use of $\varphi = 0$ as test function in (1.6) implies that the sequence $T_k(u_n)$ is bounded in the Sobolev space $W_0^{1,p}(\Omega)$ and so

$$T_k(u_n) \to T_k(u^*) \qquad \text{weakly in } W_0^{1,p}(\Omega),\ \forall k \geq 0\,. \tag{2.7}$$

We use (2.7) and the Egoroff theorem in order to pass to the limit (for $n \to \infty$) in the previous inequality and we prove that

$$\int_\Omega a(x, \nabla\varphi)\nabla T_k(u^* - \varphi) \leq \int_\Omega f T_k(u^* - \varphi), \quad \forall \varphi \in W_0^{1,p}(\Omega) \cap L^\infty(\Omega).$$

Thus Lemma 1.1 implies that $u = u^*$. In the proof of the uniqueness of the entropy solution in [BBGGPV] it is shown that for every $k > 0$

$$\int_\Omega [a(x, \nabla u_n) - a(x, \nabla u)]\nabla T_k(u_n - u) \leq \int_\Omega (f_n - f)T_k(u_n - u). \tag{2.8}$$

So the inequalities

$$\int_\Omega |\nabla u_n - u|^q = \int_{\{x\in\Omega:|u_n - u|\leq k\}} |\nabla u_n - u|^q + \int_{\{x\in\Omega:|u_n - u|>k\}} |\nabla u_n - u|^q$$

$$\leq \left(\frac{1}{\alpha}\int_\Omega [a(x, \nabla u_n) - a(x, \nabla u)]\nabla T_k(u_n - u)\right)^{\frac{q}{p}} + c(\tilde{q})meas\{x \in \Omega : |u_n - u| > k\}^{1-\frac{q}{\tilde{q}}}$$

(for any $q < \tilde{q} < \tilde{p}$) get the strong convergence. □

If the sequence $\{f_n\}$ converges strongly in L^1, we shall prove the rate of the convergence of the sequence $\{u_n\}$.

Theorem 2.3. *Let A, u_n, q be as before, $\{f_n\}$ be a sequence of functions converging to f in $L^1(\Omega)$, and u be the solution of (1.5). If we assume also (1.3), then for every $1 \leq r < q < \tilde{p}$, we have*

$$\|u_n - u\|_{W_0^{1,r}(\Omega)} \leq c(r, q, p)\, \|f_n - f\|^{\lambda}_{L^1(\Omega)}, \tag{2.9}$$

where

$$\lambda = \frac{q^*(1 - \frac{r}{q})}{pq^*(1 - \frac{r}{q}) + r}.$$

Proof. We use again the inequality (for every $k > 0$)

$$\int_\Omega [a(x, \nabla u_n) - a(x, \nabla u)]\nabla T_k(u_n - u) \leq \int_\Omega (f_n - f)T_k(u_n - u)$$

obtaining

$$\alpha\|T_k(u_n - u)\|^p_{W_0^{1,p}(\Omega)} \le k\ \|f_n - f\|_{L^1(\Omega)}.$$

Then,

$$\begin{aligned}\int_\Omega |\nabla(u_n - u)|^r &= \int_\Omega |\nabla T_k(u_n - u)|^r + \int_{\{x\in\Omega : |u_n - u| > k\}} |\nabla(u_n - u)|^r \\ &\le c_1(k\|f_n - f\|_{L^1(\Omega)})^{\frac{r}{p}} + c_2(r, q, p)\mathrm{meas}\{x \in \Omega : |u_n - u| > k\}^{1-\frac{r}{q}} \\ &\le c_1(k\|f_n - f\|_{L^1(\Omega)})^{\frac{r}{p}} + \frac{c_3(r, q, p)}{k^{q^*(1-\frac{r}{q})}},\end{aligned}$$

and by minimization on k we deduce (2.8). □

3. HOMOGENIZATION OF DIRICHLET PROBLEMS IN L^1

For the sake of simplicity in this Section we restrict ourselves to the case $p = 2$. Let $\{A_n\}$ be a sequence of monotone operators defined by

$$A_n(v) = -\mathrm{div}(a_n(x, \nabla v))$$

where the Caratheodory functions $a_n : \Omega \times \mathbf{R}^N \to \mathbf{R}^N$ satisfy, for almost every $x \in \Omega$ and for every ξ and $\hat{\xi}$ in $\mathbf{R}^N$, $\xi \neq \hat{\xi}$,

$$\begin{cases} (a(x,\xi) - a(x,\hat{\xi}))(\xi - \hat{\xi}) \ge \alpha|\xi - \hat{\xi}|^2, \\ |a(x,\xi) - a(x,\hat{\xi})| \le \beta|\xi - \hat{\xi}|, \\ a(x, 0) = 0, \end{cases} \tag{3.1}$$

where $\alpha > 0$ and $\beta > 0$ are given positive constants.

Definition 3.1. Let $\{a_n\}$ be a sequence of Caratheodory functions satisfying (3.1) and let a be a function satisfying the same hypothesis (possibly with different constants $\tilde{\alpha}, \tilde{\beta}$). The sequence $\{a_n\}$ is said to H-converge to a if for every g in $H^{-1}(\Omega)$ the sequence $\{z_n\}$ of the unique solutions

$$\begin{cases} -\mathrm{div}(a_n(x, Dz_n)) = g, \\ z_n \in H_0^1(\Omega), \end{cases}$$

satisfies

$$\begin{cases} z_n \rightharpoonup z \text{ weakly in } H_0^1(\Omega). \\ a_n(x, Dz_n) \rightharpoonup a(x, Dz) \text{ weakly in } (L^2(\Omega))^N, \end{cases}$$

where z is the unique solution of

$$\begin{cases} -\text{div}(a(x, Dz)) = g, \\ z \in H_0^1(\Omega). \end{cases}$$

This notion was first introduced under the name of G-convergence by S. Spagnolo [Sp] in the linear symmetric case where $a_n(x,\xi) = B_n(x)\xi$. He proved the following fundamental compactness theorem: any sequence of symmetric, uniformly coercive and uniformly bounded matrices $B_n(x)$ admits a subsequence which G-converges to a matrix $B(x)$ of the same type.

These definition and compactness result were then generalized to the linear non symmetric case by L. Tartar and F. Murat (see [T], [M1], [MT]) and by L. Tartar [T] to the present case of nonlinear monotone operators defined by functions a_n which satisfy (3.1). A more general result of compactness was proved by V. Chiado Piat, G. Dal Maso and A. Defranceschi [CDD] for nonlinear, non strictly monotone and possibly multivalued operators acting from $W_0^{1,p}(\Omega)$ into $W^{1,p'}(\Omega)$.

In this Section we consider the problems

$$\begin{cases} -\text{div}(a_n(x, Du_n)) = f \quad \text{in } \Omega, \\ u_n \in W_0^{1,q}(\Omega) \quad \forall q < \frac{N}{N-1}, \end{cases} \tag{3.2}$$

where f is given in $L^1(\Omega)$. For each $n \in \mathbf{N}$ there exists a unique entropy solution u_n of (3.2). We will prove the following:

Theorem 3.2. *Assume that a_n satisfies (3.1) and H-converges to a. The sequence $\{u_n\}$ of solutions of (3.2) satisfies*

$$u_n \rightharpoonup u \ \textit{in } W_0^{1,q}(\Omega) \textit{weak},$$

where u is the unique entropy solution of the problem

$$\begin{cases} -\text{div}(a(x, Du)) = f \quad \textit{in } \Omega, \\ u \in W_0^{1,q}(\Omega) \quad \forall q < \frac{N}{N-1}. \end{cases} \tag{3.3}$$

Proof. We proved in Theorem 2.1 that the sequence $\{u_n\}$ is bounded in $W_0^{1,q}(\Omega)$ for every $q < \frac{N}{N-1}$. Thus there exists a subsequence (still denoted by $\{u_n\}$) and a function $u^* \in W_0^{1,q}(\Omega)$ such that

$$u_n \rightharpoonup u^* \quad \text{in } W_0^{1,q}(\Omega)\text{weak}, \tag{3.4}$$

and (since $T_k(u_n)$ is bounded in $H_0^1(\Omega)$)

$$T_k(u_n) \rightharpoonup T_k(u^*) \quad \text{in } H_0^1(\Omega)\text{weak}. \tag{3.5}$$

Let v be a smooth function and v_n the solutions of the Dirichlet problems

$$v_n \in H_0^1(\Omega) : A_n(v_n) = A(v). \tag{3.6}$$

Since A_n H-converges to A, then v_n converges weakly in $H_0^1(\Omega)$ to v. Thanks to the Stampacchia L^∞-regularity theorem, the sequence $\{v_n\}$ is bounded in $L^\infty(\Omega)$ (the bound depends on v). We are going to use v_n as test function φ in the entropy formulation of problem (3.2). In order to better explain the difficulties of the proof, we study two different cases.

First leg - upwind: linear symmetric case

If

$$a_n(x,\xi) = B_n(x)\xi$$

where $\{B_n(x)\}$ is a sequence of symmetric, uniformly coercive and uniformly bounded matrices, then

$$\int_\Omega B_n(x)\nabla[u_n - v_n]\nabla T_k[u_n - v_n] \le \int_\Omega f\, T_k[u_n - v_n] - \int_\Omega B_n(x)\nabla v_n \nabla T_k[u_n - v_n],$$

that is

$$\int_\Omega B_n(x)\nabla T_k[u_n - v_n]\nabla T_k[u_n - v_n] \le \int_\Omega f\, T_k[u_n - v_n] - \int_\Omega B(x)\nabla v \nabla T_k[u_n - v_n]$$

Now we recall that the G-convergence of $B_n(x)$ implies (see e.g. [BM]) that

$$\forall z_n : z_n \rightharpoonup z \quad \text{in} \quad H_0^1(\Omega) \quad \Rightarrow \quad \liminf_{n\to+\infty} \int_\Omega B_n(x)\nabla z_n \nabla z_n \ge \int_\Omega B(x)\nabla z \nabla z.$$

Then, if we pass to the limit in the previous inequality, we get

$$\int_\Omega f\, T_k[u^* - v] - \int_\Omega B(x)\nabla v \nabla T_k[u^* - v] \ge$$

$$\geq \int_\Omega B(x)\nabla T_k[u^* - v]\nabla T_k[u^* - v] = \int_\Omega B(x)\nabla[u^* - v]\nabla T_k[u^* - v].$$

Hence,

$$\int_\Omega B(x)\nabla[u^*]\nabla T_k[u^* - v] \leq \int_\Omega f\, T_k[u^* - v],$$

that is

(First mark): u^* is an entropy solution of (3.3).

The uniqueness of the entropy solution of (3.3) implies that $u^* = u$ and that the whole sequence $\{u_n\}$ converges to u.

Second leg - downwind: nonlinear monotone case

The use of v_n as test function φ in (2.1), the monotonicity property of a_n, and the definition of v_n imply that

$$0 \leq \int_\Omega [a_n(x, \nabla u_n) - a_n(x, \nabla v_n)]\nabla T_k[u_n - v_n] \leq \int_\Omega f\, T_k[u_n - v_n] - \int_\Omega a(x, \nabla v)\nabla T_k[u_n - v_n].$$

If we pass to the limit, for n tending to infinity, we get

$$\int_\Omega f\, T_k[u^* - v] \geq \int_\Omega a(x, \nabla v)\nabla T_k[u^* - v]$$

for every smooth function v. Then, by density, we have also

$$\int_\Omega a(x, \nabla v)\nabla T_k[v - u^*] \geq \int_\Omega f\, T_k[v - u^*], \qquad \forall v \in H_0^1(\Omega) \cap L^\infty(\Omega). \tag{3.7}$$

The L^1 version of the Minty lemma (see Lemma 1.1) implies that

(Second mark): u^* is an entropy solution of (3.3).

The uniqueness of the solution of (3.3) implies that $u^* = u$ and that the whole sequence u_n converges to u. □

Remark 3.3. All the previous theorems still hold if the data in $L^1(\Omega)$ are replaced by measures which are absolutely continuous with respect to the p-capacity, using the results of [BGO]

Acknowledgements.

I wish to thank Andrea Dall'Aglio (Firenze), Luigi Orsina (Roma), Café del Real (Madrid), Café Vergara (Madrid) for scientific help. This work (partially supported by EURHomogenization) contains the unpublished part of my lecture at the Conference.

REFERENCES

[BBGGPV] Ph. Benilan, L. Boccardo, T. Gallouët, R. Gariepy, M. Pierre, J.L. Vazquez, *An L^1 theory of existence and uniqueness of solutions for nonlinear elliptic equations*, Ann. Scuola Norm. Sup. Pisa, **22** n. 2 (1995), 240–273.

[B1] L. Boccardo *Homogeneisation pour une classe d'equations fortement non lineaires*, C. R. Acad. Sci. Paris, **306** (1988), 253–256.

[B2] L. Boccardo, *L^∞ and L^1 variations on a theme of Γ- convergence* in PDE and Calculus of Variations, Birkhäuser (1989), 135–147.

[B3] L. Boccardo, *The role of truncates in nonlinear Dirichlet problems in L^1*, Proceedings of the Conference at Fes (may 1994), Longman, to appear.

[BD] L. Boccardo, T. Del Vecchio, *Homogenization of strongly nonlinear equations with gradient dependent lower order nonlinearity*, Asymptotic Analysis **5** (1991), 75–90.

[BDGM] L. Boccardo, J. Diaz, D. Giachetti, F. Murat *Existence and regularity of renormalized solutions for some elliptic problems involving derivatives of nonlinear terms*, J. Diff. Eq., **106** (1993), 215–237.

[BG1] L. Boccardo, T. Gallouët, *Nonlinear elliptic and parabolic equations involving measure data*, J. Funct. Anal., **87** (1989), 149–169.

[BG2] L. Boccardo, T. Gallouët, *Nonlinear elliptic equations with right hand side measures*, Comm. Partial Differential Equations, **17** (1992), 641–655.

[BGO] L. Boccardo, T. Gallouët, L. Orsina, *Existence and uniqueness of entropy solutions for nonlinear elliptic equations with right hand side measures*, Ann. Inst. H. Poincaré, Anal. non linéaire, to appear.

[BM] L. Boccardo, P. Marcellini, *Sulle convergenza delle soluzioni delle disequazioni variazionali*, Ann. Mat. Pura Appl., **110** (1976), 137–159.

[CDD] V. Chiado-Piat, G. Dal Maso & A. Defranceschi, *G-convergence of monotone operators*, Ann. Inst. H. Poincaré Anal. non linéaire, **7** (1990), 123–160.

[Da] A. Dall'Aglio, *Approximated solutions of equations with L^1 data. Application to the H-convergence of quasi-linear parabolic equations*, Ann. Mat. Pura Appl., **170** (1996), 207–240.

[Dv] T. Del Vecchio, *Nonlinear elliptic equations with measure data*, Potential Analysis, **4** (1995), 185–203.

[M1] F. Murat, *H*-convergence, *Séminaire d'analyse fonctionnelle et numérique de l'Université d'Alger, 1977-78*, multigraphed; English translation in [MT].

[M2] F. Murat, *Homogenization of renormalized solutions of elliptic equations*, Ann. Inst. H. Poincaré Anal. non linéaire, **8** (1991), 309–332.

[MT] F. Murat & L. Tartar, *H*-convergence, in *Topics in mathematical modelling of composite materials*, ed. by R.V. Kohn, Progress in nonlinear differential equations and their applications, Birkhäuser, Boston, (1994), to appear.

[**P**] A. Prignet, *Remarks on existence and uniqueness of solutions of elliptic problems with right hand side measures*, Rend. Mat., **15** (1995), 321–337.

[**Se**] J. Serrin, *Pathological solutions of elliptic differential equations*, Ann. Scuola Norm. Sup. Pisa, **18** (1964), 385-387.

[**Sp**] S. Spagnolo, *Sulla convergenza di soluzioni di equazioni paraboliche ed ellittiche*, Ann. Scuola Norm. Sup. Pisa, **22** (1968), 517–597.

[**St**] G. Stampacchia, *Équations elliptiques du second ordre à coefficients discontinus* Les Presses de l'Université de Montréal, Montréal (1966).

[**T**] L. Tartar, Cours Peccot au Collège de France, March 1977; partially written in [M1]; English translation in [MT].

A Remark on Regularization of the Wave Equation with Boundary Input

Francesca Bucci

Dipartimento di Matematica Pura e Applicata "E. Vitali", Università di Modena, Via Campi 213/B, 41100 Modena - Italy.

Introduction

In linear quadratic optimal control for partial differential equations, Riccati equations play a fundamental role in estabilishing solvability of the corresponding optimization problems. The abstract theory of operator Riccati equations developed so far cover the concrete situations of main interest, both when control is exercised in the interior of the domain and also when control is exercised through the boundary (as reference, see among others, [1]). Correspondingly, an approximation theory for Riccati equations with unboundend coefficients has been developed, providing the desired convergence results under suitable approximation assumptions (see [9] for the case when the control operator - $\mathcal{B}$ below - is bounded, [11] for the case when $\mathcal{B}$ is genuinely unbounded).

In some recent papers [2, 3, 4] we have studied approximation of controlled wave equations by means of regularized problems of parabolic nature, in order to show convergence of the corresponding optimal pairs. There, the results presented in [9] and [11] play an essential role.

In particular, in [4] we investigated the wave equation with Dirichlet boundary control and quadratic cost over a finite-time horizon, whose associated Riccati equation formally reads as

$$\begin{cases} P' = \mathcal{A}^* P + P\mathcal{A} - P\mathcal{B}\mathcal{B}^* P + \mathcal{C}^*\mathcal{C} \\ P(0) = P_0, \end{cases} \tag{1}$$

where $\mathcal{A}$, $\mathcal{B}$, $\mathcal{C}$, P_0 are suitable operators; $\mathcal{B}$ is unbounded, as it is usual in boundary control. In that paper the wave equation with proportional damping is considered as regularized system, and it is showed that, under a certain regularity assumption on the observation $\mathcal{C}$, a convergence result can be obtained

This work was partially supported by M.U.R.S.T. (National Research Project "Equazioni Differenziali Ordinarie e Applicazioni")

after reducing the original problem to an auxiliary problem with *bounded* control operator. The extension to non-smoothing observations is carried out by means of a double approximation.

In the present note we wish to consider once more the controlled boundary value problem

$$\begin{cases} y_{tt}(t,x) = \Delta y(t,x) & (t,x) \in]0,T[\times\Omega \\ y(0,x) = y_0(x),\ y_t(0,x) = y_1(x) & x \in \Omega \\ y(t,x) = u(t,x) & (t,x) \in]0,T[\times\Gamma, \end{cases} \tag{2}$$

where Ω is a bounded domain in $\mathbb{R}^n$ with smooth boundary $\Gamma = \partial\Omega$, $T > 0$ is a fixed final time, $[y_0, y_1] \in L^2(\Omega) \times H^{-1}(\Omega)$ represent initial data and controls u are taken in the class $L^2(0,T; L^2(\Gamma))$. Here we specialize to minimizing the quadratic cost

$$J(u) = \int_0^T \left(|Cy(t,\cdot)|^2_{L^2(\Omega)} + |u(t,\cdot)|^2_{L^2(\Gamma)} \right) dt, \tag{3}$$

overall $u \in L^2(0,T; L^2(\Gamma))$, with y subject to (2) and observation defined by

$$C\,y(t,\cdot) = \int_\Omega \rho_{x_0}(x) y(t,x) dx, \quad t \in]0,T[, \tag{4}$$

ρ_{x_0} being a weight function centered at a fixed point $x_0 \in \Omega$. It is worth stressing that no observation on the time derivative of the state is present, which is a desirable property from a practical point of view.

Stimulated by a remark made by J.P. Zolesio, our purpose is also to introduce a more natural - and realistic - regularized problem, as an abstract approximation of (2), rather than the one considered in [4] (see (2.1) below). In fact, using the wave equation with proportional damping can lead to some numerical difficulties in the computation of the square root of the elastic operator.

Therefore, the proof of analiticity of the semigroup generated by the new regularized dynamic operator in the energy space $E = L^2(\Omega) \times H^{-1}(\Omega)$ is required first, along with the characterization of the corresponding domains of fractional powers (Lemma 2.1 and Proposition 2.1).
Thus, since it is readily verified that the observation operator satisfies the requirements assumed in [4, §1] (see Remark 1.2), then the expected convergence result follows using the same arguments as in [4], without any need of a double approximation.

1 The abstract setting of the problem

Let us introduce some notations. We denote by $< \cdot >$, $|\cdot|$, $||\cdot||$ inner products in Hilbert spaces, norms and operator norms, respectively. If A is a linear closed operator, we denote by $\varrho(A)$, $\sigma(A)$ and $R(\lambda, A) = (\lambda - A)^{-1}$ the resolvent set, the spectrum and the resolvent operator of A, respectively.

Let $-A$ be the Dirichlet realization of the Laplace operator Δ in $H = L^2(\Omega)$, namely

$$Af = -\Delta f, \quad D(A) = \{f \in H^2(\Omega) : f_{|\Gamma} = 0\}. \tag{1.1}$$

Moreover, define the Dirichlet mapping D as usual:

$$D\phi = w \iff \begin{cases} \Delta w = 0 & \text{in } \Omega \\ w_{|\Gamma} = \phi. \end{cases} \tag{1.2}$$

We recall that that A is self-adjoint, non-negative and boundedly invertible in H and that $D \in \mathcal{L}(L^2(\Gamma), H^{1/2}(\Omega))$ [13].

It is well known that the abstract model of (2) in the energy space $E = L^2(\Omega) \times H^{-1}(\Omega) = H \times D(A^{-1/2})$, endowed with the scalar product

$$< z, w >_E = < z_1, w_1 >_H + < A^{-1/2} z_2, A^{-1/2} w_2 >_H, \tag{1.3}$$

is the linear evolution equation

$$\begin{cases} z'(t) = \mathcal{A}z(t) + \mathcal{B}u(t), & t \in]0, T[, \\ z(0) = z_0 \in E, \end{cases} \tag{1.4}$$

where we have set, as usual, $z(t, x) = [y(t, x), y_t(t, x)]$, $z(t) = z(t, \cdot)$, $z_0 = [y_0, y_1]$. The linear operators $\mathcal{A}$ and $\mathcal{B}$ introduced above are defined, respectively, as follows:

$$\mathcal{A} = \begin{pmatrix} 0 & I \\ -A & 0 \end{pmatrix}, \quad D(\mathcal{A}) = D(A^{1/2}) \times H = H_0^1(\Omega) \times L^2(\Omega), \tag{1.5}$$

$$\mathcal{B} = -\mathcal{A}G = \begin{bmatrix} 0 \\ AD \end{bmatrix}, \quad Gu = \begin{bmatrix} Du \\ 0 \end{bmatrix}. \tag{1.6}$$

Remark 1.1 In (1.6) A denotes the usual extension of A defined in (1.1), from H to $D(A)'$ (see [11]).

Hence the optimization problem consists in minimizing the quadratic functional

$$J(u) = \int_0^T (|\mathcal{C}z(t)|_E^2 + |u(t)|_U^2)dt \tag{1.7}$$

overall controls $u \in L^2(0, T; U)$, $U = L^2(\Gamma)$, with z subject to (1.4). In (1.7) we denoted by $\mathcal{C}$ the linear bounded operator in E given by

$$\mathcal{C} = \begin{pmatrix} C & 0 \\ 0 & 0 \end{pmatrix}. \tag{1.8}$$

Since problem (1.4)-(1.7) satisfies all the conditions requested in [10], then it admits a unique optimal pair (u^*, z^*), with the feedback representation of u^* in terms of z^* given by

$$u^*(t) = -\mathcal{B}^* P(t) z^*(t) \quad \text{a.e. in } [0,T], \tag{1.9}$$

$P(\cdot)$ being the corresponding Riccati operator (see, for instance, [11, §7.1]).

Remark 1.2 We remark that $\mathcal{CA}$, once closed, admits a continuous extension on E, which we still denote by $\mathcal{CA}$. In order to prove that, it is equivalent to show that $CA^{1/2} \in \mathcal{L}(H)$, as it readily follows from

$$\mathcal{CA} = \begin{pmatrix} C & 0 \\ 0 & 0 \end{pmatrix} \begin{pmatrix} 0 & I \\ -A & 0 \end{pmatrix} = \begin{pmatrix} 0 & C \\ 0 & 0 \end{pmatrix}.$$

In fact, for any $y \in D(A^{1/2})$, we have

$$CA^{1/2}y(t,\cdot) = \int_\Omega \rho_{x_0}(x) A^{1/2} y(t,x)dx =< \rho_{x_0}, A^{1/2}y >_H,$$

and since $A^{1/2}$ is a simmetric operator and $\rho_{x_0} \in H_0^1(\Omega) = D(A^{1/2})$, then

$$|CA^{1/2}y(t,\cdot)| \leq k|y(t,\cdot)|_{L^2(\Omega)}$$

for some constant $k > 0$, hence $CA^{1/2} \in \mathcal{L}(H)$. As a result,

$$\exists\, c_T > 0 : \quad \int_0^T |\mathcal{C}^* \mathcal{C} e^{t\mathcal{A}} \mathcal{B} u|_E \, dt \leq c_T |u|_U, \qquad \forall u \in U,$$

hence the Riccati operator $P(t)$ in (1.9) is in fact the unique solution to the Riccati equation

$$\begin{cases} P' = \mathcal{A}^* P + P\mathcal{A} - P\mathcal{B}\mathcal{B}^* P + \mathcal{C}^*\mathcal{C} \\ P(0) = 0 \end{cases} \tag{1.10}$$

associated to problem (1.4)-(1.7) (see [7]).

Therefore, following [4], we can introduce the Riccati equation formally verified by $\mathcal{A}^* P(\cdot) \mathcal{A}$, namely

$$\begin{cases} R' = \mathcal{A}^* R + R\mathcal{A} - RGG^* R + (\mathcal{CA})^* \mathcal{CA} \\ R(0) = 0. \end{cases} \tag{1.11}$$

As we already noticed in [4], such equation may be viewed as the one corresponding to an optimal control problem of the following form:

$$\inf_{v \in L^2(0,T;U)} \overline{J}(v), \qquad \overline{J}(v) = \int_0^T (|\mathcal{CA}w(t)|_E^2 + |v(t)|_U^2)dt, \tag{1.12}$$

with w subject to

$$\begin{cases} w'(t) = \mathcal{A}w(t) + Gv(t), & 0 < t < T, \\ w(0) = \mathcal{A}^{-1}z_0 \in E. \end{cases} \tag{1.13}$$

Moreover the operators G and $\mathcal{C}\mathcal{A}$ which appear in (1.11) are *bounded.* Hence it is readily verified that $(\mathcal{A}, G, \mathcal{C}\mathcal{A})$ satisfy all the conditions which ensure that the equation (1.11) admits a unique mild solution $R(\cdot)$ and that for any $z_0 \in E$ there exists a unique optimal pair (v^*, w^*) of (1.12)-(1.13). See, for instance, [1].

The connection between problem (1.12)-(1.13) and the original one is specified in the following lemmas (see [4] for the proofs):

Lemma 1.1 *Let $P(t)$ and $R(t)$ be the solutions to the Riccati equations (1.10) and (1.11), respectively. Then*

$$P(t) = \mathcal{A}^{*-1}R(t)\mathcal{A}^{-1}, \quad t \in [0, T]. \tag{1.14}$$

Lemma 1.2 *Let (u^*, z^*) be the optimal pair of problem (1.4)-(1.7). Consider the control problem (1.12)-(1.13) and denote by (v^*, w^*) the corresponding optimal pair. Then*

$$v^* = u^* \qquad w^* = \mathcal{A}^{-1}z^*. \tag{1.15}$$

2 The approximating problem

We want to introduce now an abstract approximating problem, endowed with a parabolic-like behaviour. Following usual techniques, we shall proceed by perturbing the second order controlled equation which models (2), namely

$$y''(t) + Ay(t) = ADu(t), \qquad t \in]0, T[,$$

by means of a "strong" damping. Thus, given $\epsilon \in (0, 1)$, it seems natural to consider the equation

$$y''(t) + \epsilon Ay'(t) + Ay(t) = ADu(t), \qquad t \in]0, T[, \tag{2.1}$$

or, in terms of a first order system,

$$\begin{cases} z_\epsilon'(t) = \mathcal{A}_\epsilon z_\epsilon(t) + \mathcal{B}u(t), & t \in]0, T[, \\ z_\epsilon(0) = z_0 \in E, \end{cases} \tag{2.2}$$

where we have set

$$\mathcal{A}_\epsilon = \begin{pmatrix} 0 & I \\ -A & -\epsilon A \end{pmatrix}, \tag{2.3}$$

with domain

$$D(\mathcal{A}_\epsilon) = \left\{[y_1, y_2] \in H \times H : \; y_1 + \epsilon y_2 \in D(A^{1/2})\right\}, \tag{2.4}$$

and $\mathcal{B}$ is the same as in (1.6),

$$\mathcal{B} = -\mathcal{A}_\epsilon G = \begin{bmatrix} 0 \\ AD \end{bmatrix}, \qquad Gu = \begin{bmatrix} Du \\ 0 \end{bmatrix}. \tag{2.5}$$

The cost functional to be minimized is now

$$J_\epsilon(u) = \int_0^T (|Cz_\epsilon(t)|_E^2 + |u(t)|_U^2)dt, \tag{2.6}$$

with z_ϵ subject to (2.2).

Remark 2.1 Second order abstract equation of type

$$y'' + \rho A^\alpha y' + Ay = 0, \quad \rho > 0,\ \alpha \in [0,1],$$

have been widely investigated by G. Chen & R. Triggiani, in order to show analiticity of the underlying semigroup on the energy space $E = D(A^{1/2}) \times L^2(\Omega)$, when $\alpha \geq 1/2$. In particular, it is well known that the operator (2.3) is the infinitesimal generator of an analytic semigroup in both $H_0^1(\Omega) \times L^2(\Omega)$ and $L^2(\Omega) \times L^2(\Omega)$ (see [5], [6] respectively). Here we shall need similar results, even though the dynamic operator $\mathcal{A}_\epsilon$ is viewed now on the space $E = L^2(\Omega) \times D(A^{-1/2})$.

First, we state some elementary properties of $\mathcal{A}_\epsilon$.

Lemma 2.1 *Let $\mathcal{A}_\epsilon$ be the operator defined by (2.3) and (2.4), $\epsilon > 0$. Then we have the following:*

(a) $\mathcal{A}_\epsilon$ is the infinitesimal generator of a strongly continuous semigroup of contractions $e^{t\mathcal{A}_\epsilon}$ on E;

(b) For any $\lambda \in \mathbb{C}$, with Re $\lambda > 0$, and any $z \in E$, we have

$$\begin{array}{ll} (i) & R(\lambda;\mathcal{A}_\epsilon)z \to R(\lambda;\mathcal{A})z, \\ (ii) & R(\lambda;\mathcal{A}_\epsilon^*)z \to R(\lambda;\mathcal{A}^*)z, \end{array} \tag{2.7}$$

as ϵ tends to zero.

Proof. First we show that $\mathcal{A}_\epsilon$ is dissipative on $E = H \times D(A^{-1/2})$ for any $\epsilon > 0$. Given $y = \begin{bmatrix} y_1 \\ y_2 \end{bmatrix} \in D(\mathcal{A}_\epsilon)$, we have

$$< \mathcal{A}_\epsilon y, y > = < y_2, y_1 > - < y_1, y_2 > -\epsilon |y_2|^2,$$

hence

$$Re < \mathcal{A}_\epsilon y, y > \leq 0.$$

Moreover, since range$(\lambda - \mathcal{A}_\epsilon) = E$ for any λ with Re $\lambda > 0$, then, by the Lumer-Phillips theorem, $\mathcal{A}_\epsilon$ is the generator of a C_0-semigroup of contractions $e^{t\mathcal{A}_\epsilon}$ on E (see [14]).

As for point (b), it is sufficient to recall the explicit expression of the resolvent operators $R(\lambda;\mathcal{A})$, $R(\lambda;\mathcal{A}_\epsilon)$, $R(\lambda;\mathcal{A}^*)$, $R(\lambda;\mathcal{A}_\epsilon^*)$ in terms of $R(\mu;-A)$. We have

$$R(\lambda,\mathcal{A}) = \begin{pmatrix} \lambda R(\lambda^2;-A) & R(\lambda^2;-A) \\ -AR(\lambda^2;-A) & \lambda R(\lambda^2;-A) \end{pmatrix}$$

for any $\lambda \in \varrho(\mathcal{A}) = \{\lambda \in \mathbb{C} : \quad \lambda \neq \pm i\sqrt{\lambda_k},\ \lambda_k \in \sigma(A)\}$, and, respectively,

$$R(\lambda,\mathcal{A}_\epsilon) = \frac{1}{\lambda\epsilon+1}\begin{pmatrix} (\lambda+\epsilon A)R(\frac{\lambda^2}{\lambda\epsilon+1};-A) & R(\frac{\lambda^2}{\lambda\epsilon+1};-A) \\ -AR(\frac{\lambda^2}{\lambda\epsilon+1};-A) & \lambda R(\frac{\lambda^2}{\lambda\epsilon+1};-A) \end{pmatrix}$$

for $\lambda \in \varrho(\mathcal{A}_\epsilon) = \{\lambda \in \mathbb{C} : \lambda \neq -1/\epsilon,\ \lambda \neq \frac{-\epsilon\lambda_k \pm i\sqrt{\epsilon^2\lambda_k^2-4\lambda_k}}{2},\ \lambda_k \in \sigma(A)\}$. Hence the condition (2.7)(i) is a trivial consequence of continuity of the function $\mu \to R(\mu,-A)$.
The relation (2.7)(ii) can be proved exactly in the same way. □

What is more inportant,

Proposition 2.1 *The semigroup $e^{t\mathcal{A}_\epsilon}$ generated by $\mathcal{A}_\epsilon$ is analytic on E and there exists $\theta \in (0,1)$ such that*

$$G \in \mathcal{L}(U, D((-\mathcal{A}_\epsilon)^\theta)). \tag{2.8}$$

Proof. In order to prove analiticity of the semigroup generated by $\mathcal{A}_\epsilon$ we shall show that for any λ, with Re $\lambda > 0$, the resolvent operator satisfies

$$||\lambda R(\lambda,\mathcal{A}_\epsilon)||_{\mathcal{L}(E)} \le c_\epsilon, \tag{2.9}$$

for some positive constant c_ϵ. Thus, conclusion immediately follows from the sufficiency part of Hille's characterization.
Inequality (2.9) is equivalent to the following set of three inequalities:

$$\left\| \frac{\lambda^2}{\lambda\epsilon+1} R\left(\frac{\lambda^2}{\lambda\epsilon+1};-A\right)\right\|_{\mathcal{L}(H)} \le M_\epsilon,$$

$$\left\| \frac{\lambda}{\lambda\epsilon+1} A^{1/2} R\left(\frac{\lambda^2}{\lambda\epsilon+1};-A\right)\right\|_{\mathcal{L}(H)} \le M_\epsilon,$$

$$\left\| \frac{1}{\lambda\epsilon+1} AR\left(\frac{\lambda^2}{\lambda\epsilon+1};-A\right)\right\|_{\mathcal{L}(H)} \le M_\epsilon,$$

for some positive constant M_ϵ, which in turn immediately follow from similar estimates proved in [5] (see (3.1) – (3.3) in [5, page 24]), by simply rewriting

$$V^{-1}(\lambda) = \frac{1}{\lambda\epsilon+1} R(\frac{\lambda^2}{\lambda\epsilon+1};-A),$$

for any λ with Re $\lambda > 0$.

It remains to show that (2.8) is fulfilled. In order to do that, we study the real interpolation spaces $(D(\mathcal{A}_\epsilon), E)_{\theta,2}$, which are isomorphic to the domains of fractional powers $D((-\mathcal{A}_\epsilon)^{1-\theta})$ (see [12]). Recall that

$$D(\mathcal{A}_\epsilon) = \left\{ \begin{bmatrix} y_1 \\ y_2 \end{bmatrix} \in L^2(\Omega) \times L^2(\Omega) : y_1 + \epsilon y_2 \in H_0^1(\Omega) \right\},$$

$$E = L^2(\Omega) \times H^{-1}(\Omega).$$

We shall use a result pertaining to complex interpolation[1] of subspaces, which is due to Baiocchi (see, for instance, [13, Chapter 1, §14.3]). Hence, with the same notation as in [13], we introduce the following spaces:

$$\begin{array}{lll} X = L^2(\Omega) \times L^2(\Omega), & Y = L^2(\Omega) \times H^{-1}(\Omega), & \\ \mathcal{X} = H_0^1(\Omega), & \mathcal{Y} = H^{-1}(\Omega), & \\ \widetilde{\mathcal{X}} = L^2(\Omega), & \widetilde{\mathcal{Y}} = H^{-1}(\Omega), & \Psi = H^{-1}(\Omega); \end{array}$$

$$\partial : \begin{bmatrix} y_1 \\ y_2 \end{bmatrix} \longrightarrow y_1 + \epsilon y_2,$$

$$\mathcal{G} : \chi \longrightarrow \begin{bmatrix} 0 \\ \frac{1}{\epsilon}\chi \end{bmatrix}, \quad r \equiv 0.$$

Using the above definitions we can rewrite

$$D(\mathcal{A}_\epsilon) = \left\{ \begin{bmatrix} y_1 \\ y_2 \end{bmatrix} \in X : \partial \begin{bmatrix} y_1 \\ y_2 \end{bmatrix} \in \mathcal{X} \right\}, \quad E = \left\{ \begin{bmatrix} y_1 \\ y_2 \end{bmatrix} \in Y : \partial \begin{bmatrix} y_1 \\ y_2 \end{bmatrix} \in \mathcal{Y} \right\},$$

and it is easy to show that Baiocchi's theorem applies to the present situation yielding the equality

$$(D(\mathcal{A}_\epsilon), E)_{\theta,2} = \left\{ \begin{bmatrix} y_1 \\ y_2 \end{bmatrix} \in L^2(\Omega) \times (L^2(\Omega), H^{-1}(\Omega))_{\theta,2} : y_1 + \epsilon y_2 \in (H_0^1(\Omega), H^{-1}(\Omega))_{\theta,2} \right\}.$$

Thus, taking $\theta = \frac{1}{4} + \rho$, with $\rho > 0$ sufficiently small, by well known interpolation results we finally get

$$(D(\mathcal{A}_\epsilon), E)_{\frac{1}{4}+\rho,2} = \left\{ \begin{bmatrix} y_1 \\ y_2 \end{bmatrix} \in L^2(\Omega) \times H^{-1/4-\rho}(\Omega) : y_1 + \epsilon y_2 \in H^{1/2-2\rho}(\Omega) \right\}. \tag{2.10}$$

It is now immediate to deduce (2.8), since $D \in \mathcal{L}(U, H^{1/2-2\rho}(\Omega))$. This concludes the proof. □

[1] However, in the present case real and complex interpolation spaces coincide, since we are concerned with Hilbert spaces and L^2-norms (see [15, Remark 3 p.143]).

Proposition 2.1 shows that the triplet $(\mathcal{A}_\epsilon, \mathcal{B}, \mathcal{C})$ satisfies all the hypotheses which guarantee existence and uniqueness of the solution $P_\epsilon(t)$ to the Riccati equation

$$\begin{cases} P_\epsilon' = \mathcal{A}_\epsilon^* P_\epsilon + P_\epsilon \mathcal{A}_\epsilon - P_\epsilon \mathcal{B}\mathcal{B}^* P_\epsilon + \mathcal{C}^*\mathcal{C} \\ P_\epsilon(0) = 0 \end{cases} \tag{2.11}$$

associated to (2.2)-(1.7) and of the corresponding optimal pair $(u_\epsilon^*, z_\epsilon^*)$, plus several regularity properties for them (see [8], [11]).

Remark 2.2 It is to be noted that $\mathcal{C}\mathcal{A}_\epsilon$ admits a continuous extension, too. In fact $\mathcal{C}\mathcal{A}_\epsilon = \mathcal{C}\mathcal{A}$, for any $\epsilon > 0$. Hence we can introduce an auxiliary optimal control problem and its associated Riccati equation, whose control operator is bounded.

To show convergence of the optimal pair of problem (2.2)-(2.6) to the optimal pair of problem (1.4)-(1.7), as $\epsilon \to 0$, we can now use similar arguments as in [4], by repeating the steps carried out in Lemma 2.2, Lemma 2.3, Theorem 2.1 and Proposition 2.2 therein. We omit the proofs.

We finally get

Theorem 2.1 *Let P_ϵ, P the solutions to the Riccati equations associated to the problems (2.2)-(1.7), (1.4)-(2.6), respectively, $(u_\epsilon^*, z_\epsilon^*)$, (u^*, z^*) the relative optimal pairs. Then, as $\epsilon \to 0$, it follows*

$$\begin{array}{lll} (i) & |P_\epsilon(\cdot)x - P(\cdot)x|_{C([0,T];E)} \to 0,\ x \in E; & J_\epsilon(u_\epsilon^*) \to J(u^*); \\ (ii) & |u_\epsilon^*(\cdot) - u^*(\cdot)|_{C([0,T];U)} \to 0; & |z_\epsilon^*(\cdot) - z^*(\cdot)|_{C([0,T];E)} \to 0. \end{array} \tag{2.12}$$

Acknowledgements I wish to thank Prof. Luciano Pandolfi for making useful comments.

References

[1] A. Bensoussan, G. Da Prato, M.C. Delfour and S.K. Mitter, Representation and Control of Infinite Dimensional Systems, vol. II, Birkhäuser, 1993.

[2] F. Bucci, *Singular perturbation for controlled wave equations*, J. Mathematical Systems, Estimation and Control, (to appear).

[3] F. Bucci, *A boundary control problem for the wave equation with boundary dissipation: a solution through parabolic regularization*, Proceedings of the workshop "Control of partial differential equations" held in Trento, January 4-9, 1993, G. Da Prato & L. Tubaro Eds., Dekker (to appear).

[4] F. Bucci, *Regularization and approximation of the wave equation with boundary input: an abstract approach*, Dynamic Systems and Appl. (to appear).

[5] S. Chen and R. Triggiani, *Proof of extension of two conjectures on structural damping for elastic system: the case $\frac{1}{2} \leq \alpha \leq 1$*, Pacific J. Mathematics **136** (1989), 15-55.

[6] S. Chen and R. Triggiani, *Characterization of domains of fractional powers of certain operators arising in elastic systems, and applications*, J. Diff. Eq. **88** (1990), 279-293.

[7] G. Da Prato, I. Lasiecka and R. Triggiani, *A direct study of Riccati equations arising in boundary control problems for hyperbolic equations*, J. Diff. Eqns. (1) **64** (1986), 26-47.

[8] F. Flandoli, *On the direct solutions of Riccati equations arising in boundary control theory*, Ann. Mat. Pura e Appl. (IV) **163** (1993), 93-131.

[9] J. S. Gibson, *The Riccati integral equations for optimal control problems on Hilbert spaces*, SIAM J. on Control and Optimiz. **17** (1979), 537-565.

[10] I. Lasiecka, J.L.Lions and R. Triggiani, *Non-homogeneous boundary value problems for second order hyperbolic operators* J. Math. Pure et Appl. **65** (1986), 149-192.

[11] I. Lasiecka and R. Triggiani, Differential and Algebraic Riccati Equations with Application to Boundary/Point Control Problems: Continuous Theory and Approximation Theory, Springer-Verlag Lecture Notes LNCIS n.164, 1991.

[12] J.L. Lions, *Espaces d'interpolation et domaines de puissances fractionnaires d'operateurs*, J. Math. Soc. Japan **14** (1962), 233- 241.

[13] J.L. Lions and E. Magenes, Non-homogeneous boundary value problems and applications, I, Springer Verlag, 1971.

[14] A. Pazy, Semigroups of Linear Operators and Applications to Partial Differential Equations, Springer Verlag, 1983.

[15] H.Triebel, Interpolation Theory, Function Spaces, Differential Operators, North Holland, 1978.

A Pontryagin's Principle for Boundary Control Problems of Quasilinear Elliptic Equations

Eduardo Casas
Departamento de Matemática Aplicada y Ciencias de la Computación
E.T.S.I. Caminos – Universidad de Cantabria
39071–Santander, Spain

1 Introduction

This paper deals with optimal control problems governed by quasilinear elliptic equations. The aim is to derive the first order conditions for optimality. To carry out this goal, some sensitivity analysis of the state with respect to the control is necessary. Casas and Fernández proved that the relation control-state is Gâteaux differentiable for some quasilinear elliptic equations and they provided some nondifferentiability examples for some others. These results allowed them to derive the optimality conditions in an integral form; see [8], [9] and [12]. They also considered state constrained optimal control problems associated to these equations (mainly integral state constraints) in [10], [11] and [13].

Recently, Casas and Yong [14] proved a Pontryagin's principle for pointwise state-constrained control problems governed by elliptic quasilinear equations. There, the key point of the proof was a Lieberman's result [23] about the boundedness of the gradient of the state. It was possible to use this result because the control was distributed in the domain.

An important feature of the proof of Pontryagin's principle is the use of some special variations of the controls different of the classical spike perturbations. This idea was first employed by Li and Yao [21] for evolution problems, and later it was adapted to elliptic problems in [30], [14] and [7]; see Proposition 1 and Corollary 1.

We formulate two principles of Pontryagin's type, called weak and strong respectively. The difference between both is that the second one is formulated in a qualified form, which makes necessary an additional assumption. This assumption is a stability condition of minimum cost functional with respect to small perturbations of the feasible state set; see Definition 1. We also prove that "almost all" problems are stable.

For the case of control problems governed by evolution equations, the reader is referred to Fattorini [18], [19], Fattorini and Frankowska [20], Li and Yong [22].

This research was partially supported by Dirección General de Investigación Científica y Técnica (Madrid)

2 Setting of the control problem

Let Ω be a bounded open subset of $\mathbb{R}^n$, $n > 1$, with a Lipschitz boundary Γ. In this domain we consider the following boundary values problem

$$\begin{cases} -\text{div } a(x, \nabla y(x)) + a_0(x, y(x)) = 0 & \text{in } \Omega \\ a(x, \nabla y(x)) \cdot \nu(x) = b(x, y(x), u(x)) & \text{on } \Gamma, \end{cases} \tag{1}$$

$\nu(x)$ denoting the outward unit normal vector to Γ at the point x.

The assumptions on the data involved in this problem are the following: functions $a : \Omega \times \mathbb{R}^n \longrightarrow \mathbb{R}^n$ and $a_0 : \Omega \times \mathbb{R} \longrightarrow \mathbb{R}$ are of class C^1 with respect to the second variable and measurable with respect to the first one; $b : \Omega \times \mathbb{R} \times \mathbb{K} \longrightarrow \mathbb{R}$ is of class C^1 with respect to the second variable, $(\mathbb{K}, d)$ being a metric space;

$$\begin{cases} a_0(\cdot, 0) \in L^{\hat{r}}(\Omega), \ \hat{r} > \dfrac{n}{\alpha}; \\ \exists \eta_1 \in L^{\hat{p}}(\Gamma), \ \hat{p} > \max\left\{\dfrac{n-1}{\alpha - 1}, \dfrac{2(n-1)}{n}\right\}, \text{ such that} \\ |b(x, 0, u)| \leq \eta_1(x) \text{ a.e.}[\sigma] \ x \in \Gamma, \ u \in \mathbb{K}; \\ \text{exists an increasing function } h : \mathbb{R}_+ \longrightarrow \mathbb{R}_+ \text{ such that} \\ \left|\dfrac{\partial a_0}{\partial y}(x, s)\right| + \left|\dfrac{\partial b}{\partial y}(x', s, u)\right| \leq h(|s|) \text{ a.e. } x \in \Omega, \text{ a.e.}[\sigma] \ x' \in \Gamma, \ s \in \mathbb{R}, \ u \in \mathbb{K}; \end{cases} \tag{2}$$

$$\begin{cases} \exists \mu_1 \in L^\infty(\Omega) \text{ and } \exists \mu_2 \in L^\infty(\Gamma), \text{ with } \|\mu_1\|_{L^\infty(\Omega)} + \|\mu_2\|_{L^\infty(\Gamma)} > 0, \text{ such that} \\ \dfrac{\partial a_0}{\partial y}(x, y) \geq \mu_1(x) \geq 0 \text{ a.e. } x \in \Omega, \ \forall y \in \mathbb{R}; \\ -\dfrac{\partial b}{\partial y}(x, y, u) \geq \mu_2(x) \geq 0 \text{ a.e.}[\sigma] \ x \in \Gamma, \ y \in \mathbb{R}, \ u \in \mathbb{K}; \end{cases} \tag{3}$$

$$\begin{cases} \exists \alpha \in (1, +\infty), \ \Lambda_1, \Lambda_2 > 0, \text{ and } \kappa \in [0, 1] \text{ such that} \\ \displaystyle\sum_{i,j=1}^n \frac{\partial a_j}{\partial \eta_i}(x, \eta)\xi_i\xi_j \geq \Lambda_1(\kappa + |\eta|)^{\alpha-2}|\xi|^2, \ \forall \xi \in \mathbb{R}^n, \text{ a.e. } x \in \Omega; \\ \left|\displaystyle\sum_{i,j=1}^n \frac{\partial a_j}{\partial \eta_i}(x, \eta)\right| \leq \Lambda_2(\kappa + |\eta|)^{\alpha-2}; \\ a_j(x, 0) = 0, \ 1 \leq j \leq n. \end{cases} \tag{4}$$

In (1), y denotes the state and u is the control. Examples of operators satisfying the conditions (2)-(4) can be found in [9] and [12]. This type of equations appears in some physical models (steady laminar flow of non-Newtonian fluids, some reaction-diffusion problems, magnetostatics, glaciology, etc.; see Ames [1], Aris [2], Marrocco and Pironneau [24], Pelissier [27]).

The cost functional is given by

$$J(u) = \int_\Omega L(x, y_u(x))dx + \int_\Gamma l(x, y_u(x), u(x))d\sigma(x),$$

where y_u is the solution of (1) corresponding to the control u and $L : \Omega \times \mathbb{R} \longrightarrow \mathbb{R}$ and $l : \Omega \times \mathbb{R} \times \mathbb{K} \longrightarrow \mathbb{R}$ are functions of class C^1 with respect to the second variable, L is measurable with respect to the first one, and satisfy

$$\begin{cases} \exists \eta_2 \in L^1(\Gamma) \text{ such that } |l(x,0,u)| \le \eta_2(x) \text{ a.e.}[\sigma]\ x \in \Gamma,\ u \in \mathbb{K}; \\ \forall M > 0\ \exists \eta_3^M \in L^{\hat{q}}(\Gamma),\ \hat{q} \ge \dfrac{2(n-1)}{n} \text{ if } n \ge 3,\ \hat{q} > 1 \text{ if } n = 2, \text{ such that} \\ \left|\dfrac{\partial l}{\partial y}(x,y,u)\right| \le \eta_3^M(x) \text{ a.e.}[\sigma]\ x \in \Gamma,\ |y| \le M,\ u \in \mathbb{K}. \end{cases} \tag{5}$$

$$\begin{cases} L(\cdot, 0) \in L^1(\Omega); \\ \forall M > 0\ \exists \eta_4^M \in L^{\hat{s}}(\Omega),\ \hat{s} \ge \dfrac{2n}{n+2} \text{ if } n \ge 3,\ \hat{s} > 1 \text{ if } n = 2, \text{ such that} \\ \left|\dfrac{\partial L}{\partial y}(x,y)\right| \le \eta_4^M(x) \text{ a.e. } x \in \Omega,\ |y| \le M; \end{cases} \tag{6}$$

The set of controls, denoted by $\mathcal{U}$, is formed by the measurable functions $u : \Gamma \longrightarrow \mathbb{K}$ such that the mapping

$$x \in \Gamma \longrightarrow (b(x, y, u(x)), l(x, y, u(x))) \in \mathbb{R}^2$$

is measurable for every $y \in \mathbb{R}$. In this set, we consider the so-called Ekeland's distance [17]

$$d_E(u, v) = \sigma(\{x \in \Gamma : u(x) \neq v(x)\}). \tag{7}$$

Under the previous assumptions, (1) has a unique solution in $W^{1,\alpha}(\Omega) \cap L^\infty(\Omega)$ for every $u \in \mathcal{U}$. Therefore the notation y_u makes sense. More precisely the following theorem can be proved by classical methods.

Theorem 1 *Under the assumptions (2)-(4), (1) has a unique solution in $W^{1,\alpha}(\Omega) \cap L^\infty(\Omega)$ for every $u \in \mathcal{U}$ and there exists a constant $M > 0$ such that*

$$\|y_u\|_{W^{1,\alpha}(\Omega)} + \|y_u\|_{L^\infty(\Omega)} \le M \quad \forall u \in \mathcal{U}. \tag{8}$$

*Moreover, if $\{u_k\}_{k=1}^\infty \subset \mathcal{U}$ is a sequence converging to u in $\mathcal{U}$, i.e. $d_E(u, u_k) \to 0$, then $\{y_{u_k}\}_{k=1}^\infty$ converges to y_u strongly in $W^{1,\alpha}(\Omega)$ and *weakly in $L^\infty(\Omega)$.*

Now the state constraints are defined through the functions $G_j : L^\infty(\Omega) \longrightarrow \mathbb{R}$, $1 \leq j \leq m$, given by

$$G_j(z) = \int_\Omega g_j(x, z(x))dx,$$

with $g_j : \Omega \times \mathbb{R} \longrightarrow \mathbb{R}$ being a Carathéodory function of class C^1 with respect to the second variable satisfying

$$g_j(\cdot, 0) \in L^1(\Omega) \quad \text{and} \quad \left|\frac{\partial g_j}{\partial y}(x,y)\right| \leq \eta_4^M(x) \text{ a.e. } x \in \Omega \text{ and } |y| \leq M; \tag{9}$$

where η_4^M is as in (5).

Finally, given $\delta = \{\delta_j\}_{j=1}^m \subset \mathbb{R}^m$, we can formulate the control problem as follows

$$(\mathrm{P}_\delta)\begin{cases} \text{Minimize } J(u) \\ u \in \mathcal{U} \text{ and } G_j(y_u) \leq \delta_j,\ 1 \leq j \leq m. \end{cases}$$

3 Sensitivity analysis of the state equation

Now we focus on the study of the sensitivity analysis of the state with respect to the control. In [13] an example was provided, with $\kappa = 0$, where the relation control-state was not Gâteaux differentiable. However, if $\kappa > 0$ and $\alpha \geq 2$ the Gâteaux differentiability can be proved. The differentiability in the case $\alpha < 2$ and $\kappa > 0$ is an open problem for us. Consequently, here we only consider the case $\kappa > 0$ and $\alpha \geq 2$.

Since our aim is to prove a Pontryagin's principle for the control problem (P_δ), we are interested in studying the variations of the state with respect to some pointwise perturbations of the controls. The following proposition is crucial to accomplish these perturbations.

Proposition 1 *Let $\rho \in (0,1)$, then there exists a sequence of σ-measurable sets $\{E_k\}_{k=1}^\infty$, with $E_k \subset \Gamma$ and $\sigma(E_k) = \rho\sigma(\Gamma)$, such that $\frac{1}{\rho}\chi_{E_k} \rightarrow 1$ *weakly in $L^\infty(\Gamma)$ when $k \rightarrow \infty$.*

Corollary 1 *Given $f \in L^1(\Gamma)$ and $g \in L^{\hat{p}}(\Gamma)$, there exists a family $\{E_\rho\}_{\rho>0}$ of σ-measurable sets, with $\sigma(E_\rho) = \rho\sigma(\Gamma)$, satisfying*

$$\left|\int_\Gamma \left(1 - \frac{1}{\rho}\chi_{E_\rho}(x)\right) f(x)d\sigma(x)\right| + \left\|\left(1 - \frac{1}{\rho}\chi_{E_\rho}\right) g\right\|_{H^{-1/2}(\Gamma)} < \rho, \quad \forall \rho > 0. \tag{10}$$

A detailed proof of this technical result will be provided in a paper to appear.

Now we study the variations of the state and the cost functional with respect to pointwise perturbations of the controls localized in the sets E_ρ. But we need first to introduce some notation. Let us assume that $\kappa > 0$ in (4), then, given $y \in W^{1,\alpha}(\Omega)$, we denote by $H^y(\Omega)$ the space completed of $C^\infty(\bar{\Omega})$ with respect to the norm

$$\|z\|_{H^y(\Omega)} = \Big(\int_\Omega (\kappa + |\nabla y(x)|)^{\alpha-2}|\nabla z(x)|^2 dx + \int_\Omega \mu_1(x)z^2(x)dx + \int_\Gamma \mu_2(x)z^2(x)d\sigma(x)\Big)^{1/2}. \tag{11}$$

It may be easily checked that $H^y(\Omega)$ is a Hilbert space with the inner product

$$(z_1, z_2)_{H^y(\Omega)} = \int_\Omega (\kappa + |\nabla y(x)|)^{\alpha-2} \nabla z_1(x) \nabla z_2(x) dx \\ + \int_\Omega \mu_1(x) z_1(x) z_2(x) dx + \int_\Gamma \mu_2(x) z_1(x) z_2(x) d\sigma(x). \tag{12}$$

Moreover we have

$$W^{1,\alpha}(\Omega) \subset H^y(\Omega) \subset H^1(\Omega) \ \text{ if } \alpha \geq 2 \tag{13}$$

and

$$H^1(\Omega) \subset H^y(\Omega) \subset W^{1,\alpha}(\Omega) \ \text{ if } \alpha < 2, \tag{14}$$

with continuous imbeddings. More general spaces of this type have been studied by Murthy and Stampacchia [25], Coffman *et al.* [16] and Trudinger [29].

Proposition 2 *Let $\kappa > 0$, $\alpha \geq 2$ and $u, v \in \mathcal{U}$. There exist σ-measurable sets $E_\rho \subset \Gamma$, with $\sigma(E_\rho) = \rho\sigma(\Gamma)$, such that if we define*

$$u_\rho(x) = \begin{cases} u(x) & \text{if } x \in \Gamma \setminus E_\rho \\ v(x) & \text{if } x \in E_\rho, \end{cases}$$

and if we denote by y_ρ and y the states corresponding to u_ρ and u, respectively, then the following equalities hold

$$y_\rho = y + \rho z + r_\rho, \quad \lim_{\rho \to 0} \frac{1}{\rho} \|r_\rho\|_{H^1(\Omega)} = 0, \tag{15}$$

and

$$J(u_\rho) = J(u) + \rho z^0 + r_\rho^0, \quad \lim_{\rho \to 0} \frac{1}{\rho} r_\rho^0 = 0, \tag{16}$$

where $z \in H^y(\Omega)$ satisfies

$$\begin{cases} -\text{div}\left(\dfrac{\partial a}{\partial \eta}(x, \nabla y_u(x)) \nabla z\right) + \dfrac{\partial a_0}{\partial y}(x, y_u(x)) z = 0 \quad \text{in } \Omega \\[2ex] \dfrac{\partial a}{\partial \eta}(x, \nabla y_u(x)) \nabla z \cdot \nu(x) = \dfrac{\partial b}{\partial y}(x, y(x), u(x)) z \\[2ex] +b(x, y(x), v(x)) - b(x, y(x), u(x)) \ \text{on } \Gamma \end{cases} \tag{17}$$

and

$$z^0 = \int_\Omega \frac{\partial L}{\partial y}(x, y(x)) z(x) dx \\ + \int_\Gamma \left\{ \frac{\partial l}{\partial y}(x, y(x), u(x)) z(x) + l(x, y(x), v(x)) - l(x, y(x), u(x)) \right\} d\sigma(x). \tag{18}$$

Proof. Let us define

$$f(x) = l(x, y(x), v(x)) - l(x, y(x), u(x))$$

and

$$g(x) = b(x, y(x), v(x)) - b(x, y(x), u(x)).$$

Then, thanks to hypotheses (2) and (5) we have that $f \in L^1(\Gamma)$ and $g \in L^{\hat{p}}(\Gamma)$. Then we can apply Corollary 1 to obtain the existence of σ-measurable sets $E_\rho \subset \Gamma$, with $\sigma(E_\rho) = \rho\sigma(\Gamma)$ such that (10) is satisfied.

Putting

$$z_\rho = \frac{y_\rho - y}{\rho},$$

if we prove that $z_\rho \to z$ in $H^y(\Omega)$, then we will have obtained (15). Moreover, in such a case, using the convergence of $\{y_\rho\}_{\rho>0}$ to y provided by Theorem 1, hypotheses (2), (5) and (6), we deduce

$$\frac{1}{\rho}|r_\rho^0| = \left|\frac{J(u_\rho) - J(u)}{\rho} - z^0\right|$$

$$\leq \int_\Omega \left|\int_0^1 \frac{\partial L}{\partial y}(x, y(x) + \theta[y_\rho(x) - y(x)])d\theta z_\rho(x) - \frac{\partial L}{\partial y}(x, y(x))z(x)\right| dx$$

$$+ \int_\Gamma \left|\int_0^1 \frac{\partial l}{\partial y}(x, y(x) + \theta[y_\rho(x) - y(x)])d\theta z_\rho(x) - \frac{\partial l}{\partial y}(x, y(x), u(x))z(x)\right| d\sigma(x)$$

$$+ \left|\int_\Gamma \left(1 - \frac{1}{\rho}\chi_{E_\rho}(x)\right) g(x)dx\right| \longrightarrow 0 \quad \text{when } \rho \to 0,$$

which proves (16).

To conclude the proof it remains to establish the convergence $z_\rho \to z$ in $H^y(\Omega)$. This can be made by following the proof of Theorem 3.1 of Casas Fernández [12], with the obvious modifications. □

4 The Weak Pontryagin's Principle

The main result to be proved in this section is the following

Theorem 2 *Let us assume that $\alpha \geq 2$ and $\kappa > 0$ in (4). If $\bar{u}$ is a solution of (P_δ), then there exist real numbers $\{\bar{\lambda}_j\}_{j=0}^m \subset [0, +\infty)$ and functions $\bar{\varphi} \in H^1(\Omega)$ and $\bar{y} \in W^{1,\alpha}(\Omega) \cap L^\infty(\Omega)$ such that*

$$\sum_{j=0}^m \bar{\lambda}_j > 0 \quad \text{and} \quad \bar{\lambda}_j(G_j(\bar{y}) - \delta_j) = 0, \ 1 \leq j \leq m; \tag{19}$$

$$\begin{cases} -\text{div}\, a(x, \nabla\bar{y}(x)) + a_0(x, \bar{y}(x)) = 0 & \text{in } \Omega; \\ a(x, \nabla\bar{y}(x)) \cdot \nu(x) = b(x, \bar{y}(x), \bar{u}(x)) & \text{on } \Gamma; \end{cases} \tag{20}$$

$$\left\{\begin{array}{l} -\mathrm{div}\left(\left[\frac{\partial a}{\partial \eta}(x,\nabla\bar{y}(x))\right]^T \nabla\bar{\varphi}\right) + \frac{\partial a_0}{\partial y}(x,\bar{y}(x))\bar{\varphi} \\ = \bar{\lambda}_0\frac{\partial L}{\partial y}(x,\bar{y}(x)) + \sum_{j=1}^{m}\bar{\lambda}_j\frac{\partial g_j}{\partial y}(x,\bar{y}(x)) \ \ in\ \Omega; \\ \left[\frac{\partial a}{\partial \eta}(x,\nabla\bar{y}(x))\right]^T \nabla\bar{\varphi}\cdot\nu(x) = \frac{\partial b}{\partial y}(x,\bar{y}(x),\bar{u}(x))\bar{\varphi} + \bar{\lambda}_0\frac{\partial l}{\partial y}(x,\bar{y}(x),\bar{u}(x)) \ \ on\ \Gamma; \end{array}\right. \tag{21}$$

$$\int_\Gamma H_{\bar{\lambda}_0}(x,\bar{y}(x),\bar{u}(x),\bar{\varphi}(x))d\sigma(x) = \min_{v\in\mathcal{U}}\int_\Gamma H_{\bar{\lambda}_0}(x,\bar{y}(x),v(x),\bar{\varphi}(x))d\sigma(x); \tag{22}$$

where $H_\lambda : \Omega\times\mathbb{R}\times\mathbb{K}\times\mathbb{R}\longrightarrow\mathbb{R}$ *is defined by*

$$H_\lambda(x,y,u,\varphi) = \lambda l(x,y,u) + \varphi b(x,y,u) \ \ \forall\lambda\in\mathbb{R}. \tag{23}$$

Moreover, if one of the following assumptions is satisfied:
(A1) $\mathbb{K}$ *is a separable metric space and functions* b *and* l *are continuous with respect to the third variable on* $\mathbb{K}$*;*
(A2) *There exists a set* $\Gamma_0\subset\Gamma$, *with* $\sigma(\Gamma_0)=\sigma(\Gamma)$, *such that the function*

$$x\in\Gamma\longrightarrow(b(x,y,u),l(x,y,u))\in\mathbb{R}^2$$

is continuous in Γ_0 *for every* $(y,u)\in\mathbb{R}\times\mathbb{K}$*;*
then the following pointwise relation holds

$$H_{\bar{\lambda}_0}(x,\bar{y}(x),\bar{u}(x),\bar{\varphi}(x)) = \min_{u\in\mathbb{K}} H_{\bar{\lambda}_0}(x,\bar{y}(x),u,\bar{\varphi}(x)) \ \ a.e.[\sigma]\ x\in\Gamma. \tag{24}$$

Proof. We follow a penalty method to deal with the state constraints, where the penalty objective function is a slight differentiable modification of that one used by Clarke in [15]. Let $\bar{u}$ be the solution of (P_δ). For every $\epsilon>0$, let us consider the problem

$$(\mathrm{P}_{\delta,\epsilon})\left\{\begin{array}{l}\text{minimize } J_\epsilon(u)\\ \text{such that } u\in\mathcal{U},\end{array}\right.$$

where

$$J_\epsilon(u) = \left\{[(J(u)-J(\bar{u})+\epsilon)^+]^2 + \sum_{j=1}^{m}[(G_j(y_u)-\delta_j)^+]^2\right\}^{1/2}.$$

It is known that $(\mathcal{U},d_E)$, d_E being the Ekeland's distance, is a complete metric space. Also it is easy to check with the help of Theorem 1 that $J_\epsilon:(\mathcal{U},d_E)\longrightarrow\mathbb{R}$ is a continuous function. Furthermore we have

$$J_\epsilon(\bar{u}) = \epsilon \le \epsilon + \inf_{u\in\mathcal{U}} J_\epsilon(u).$$

Then we can apply Ekeland's variational [17] principle to deduce the existence of an element $u^\epsilon\in\mathcal{U}$ such that

$$d_E(u^\epsilon,\bar{u}) \le \sqrt{\epsilon} \tag{25}$$

and u^ϵ is a solution of problem

$$(Q_{\delta,\epsilon})\left\{\begin{array}{l}\text{minimize } J_\epsilon(u)+\sqrt{\epsilon}d_E(u^\epsilon,u)\\ \text{such that } u\in\mathcal{U}.\end{array}\right.$$

Let $v \in \mathcal{U}$ be given. Then we take the sets $\{E_\rho\}_{\rho>0}$ as in Proposition 2 and define the functions

$$u^\epsilon_\rho(x)=\left\{\begin{array}{ll}u^\epsilon(x) & \text{if } x\in\Gamma\setminus E_\rho\\ v(x) & \text{if } x\in E_\rho.\end{array}\right.$$

Using (15) and (16) we get

$$-\sqrt{\epsilon}\sigma(\Gamma)\leq-\sqrt{\epsilon}\frac{d_E(u^\epsilon,u^\epsilon_\rho)}{\rho}\leq\frac{J_\epsilon(u^\epsilon_\rho)-J_\epsilon(u^\epsilon)}{\rho}$$

$$=\frac{[(J(u^\epsilon_\rho)-J(\bar{u})+\epsilon)^+]^2-[(J(u^\epsilon)-J(\bar{u})+\epsilon)^+]^2}{\rho[J_\epsilon(u^\epsilon_\rho)+J_\epsilon(u^\epsilon)]}$$

$$+\frac{\sum_{j=1}^m\left\{[(G_j(y^\epsilon_\rho)-\delta_j)^+]^2-[(G_j(y^\epsilon)-\delta_j)^+]^2\right\}}{\rho[J_\epsilon(u^\epsilon_\rho)+J_\epsilon(u^\epsilon)]}\xrightarrow{\rho\to 0}$$

$$\xrightarrow{\rho\to 0}\left\{(J(u^\epsilon)-J(\bar{u})+\epsilon)^+z^{0,\epsilon}+\sum_{j=1}^m(G_j(y^\epsilon)-\delta_j)^+\int_\Omega\frac{\partial g_j}{\partial y}(x,y^\epsilon)z^\epsilon dx\right\}/J_\epsilon(u^\epsilon)$$

$$=\lambda^\epsilon_0 z^{0,\epsilon}+\sum_{j=1}^m\lambda^\epsilon_j\int_\Omega\frac{\partial g_j}{\partial y}(x,y^\epsilon)z^\epsilon dx, \tag{26}$$

where $y^\epsilon \in W^{1,\alpha}(\Omega)$ is the sate associated to u^ϵ, $z^\epsilon \in H^{y_\epsilon}(\Omega)$ is the solution of

$$\left\{\begin{array}{l}-\operatorname{div}\left(\dfrac{\partial a}{\partial\eta}(x,\nabla y^\epsilon(x))\nabla z^\epsilon\right)+\dfrac{\partial a_0}{\partial y}(x,y^\epsilon(x))z^\epsilon=0 \quad\text{in }\Omega\\[2ex] \dfrac{\partial a}{\partial\eta}(x,\nabla y^\epsilon(x))\nabla z^\epsilon\cdot\nu(x)=\dfrac{\partial b}{\partial y}(x,y^\epsilon(x),u^\epsilon(x))z^\epsilon\\[2ex] +b(x,y^\epsilon(x),v(x))-b(x,y^\epsilon(x),u^\epsilon(x)) \text{ on }\Gamma,\end{array}\right.$$

$$z^{0,\epsilon}=\int_\Omega\frac{\partial L}{\partial y}(x,y^\epsilon(x))z^\epsilon(x)dx$$

$$+\int_\Gamma\left\{\frac{\partial l}{\partial y}(x,y^\epsilon(x),u^\epsilon(x))z^\epsilon(x)+l(x,y^\epsilon(x),v(x))-l(x,y^\epsilon(x),u^\epsilon(x))\right\}d\sigma(x),$$

and

$$\lambda^\epsilon_0=\frac{(J(u^\epsilon)-J(\bar{u})+\epsilon)^+}{J_\epsilon(u^\epsilon)},\quad\lambda^\epsilon_j=\frac{(G_j(y^\epsilon)-\delta_j)^+}{J_\epsilon(u^\epsilon)},\quad j=1,\dots,m.$$

So we have that

$$\lambda^\epsilon_j\geq 0,\quad 0\leq j\leq m,\quad\text{and}\quad\sum_{j=0}^m\lambda^\epsilon_j=1. \tag{27}$$

Let us denote by φ^ϵ the unique variational solution in $H^{y^\epsilon}(\Omega)$ of the problem

$$\begin{cases} -\text{div}\left(\left[\frac{\partial a}{\partial \eta}(x,\nabla y^\epsilon(x))\right]^T \nabla\varphi^\epsilon\right) + \frac{\partial a_0}{\partial y}(x,y^\epsilon(x))\varphi^\epsilon \\ = \lambda_0^\epsilon \frac{\partial L}{\partial y}(x,y^\epsilon(x)) + \sum_{j=1}^m \lambda_j^\epsilon \frac{\partial g_j}{\partial y}(x,y^\epsilon(x)) \text{ in } \Omega; \\ \left[\frac{\partial a}{\partial \eta}(x,\nabla y^\epsilon(x))\right]^T \nabla\varphi^\epsilon \cdot \nu(x) \\ = \frac{\partial b}{\partial y}(x,y^\epsilon(x),u^\epsilon(x))\varphi^\epsilon + \lambda_0^\epsilon \frac{\partial l}{\partial y}(x,y^\epsilon(x),u^\epsilon(x)) \text{ on } \Gamma; \end{cases} \tag{28}$$

the existence and uniqueness of φ^ϵ is an immediate consequence of Lax-Milgram theorem and assumptions (2)-(4). By using the standard argumentation, we can deduce easily from (26) and (28) that

$$\int_\Gamma H_{\lambda_0^\epsilon}(x,y^\epsilon(x),u^\epsilon(x),\varphi^\epsilon(x))d\sigma(x) \le \int_\Gamma H_{\lambda_0^\epsilon}(x,y^\epsilon(x),v(x),\varphi^\epsilon(x))d\sigma(x).$$

Since $v \in \mathcal{U}$ was arbitrary, it follows

$$\int_\Gamma H_{\lambda_0^\epsilon}(x,y^\epsilon(x),u^\epsilon(x),\varphi^\epsilon(x))d\sigma(x) = \min_{v\in\mathcal{U}} \int_\Gamma H_{\lambda_0^\epsilon}(x,y^\epsilon(x),v(x),\varphi^\epsilon(x))d\sigma(x); \tag{29}$$

Now we pass to the limit in (27)-(29) to obtain (19), (21) and (22). Taking subsequences if necessary, it is immediate to obtain (19) from (27) and the definition of λ_j^ϵ. System (20) is a consequence of Theorem 1 and (25). The boundedness of $\{\varphi^\epsilon\}_{\epsilon>0}$ in $H^1(\Omega)$ is a consequence of (2)-(4) and (28), it is enough to argue as in the proof of Proposition 2, Step 1. Then we can take a subsequence of $\{\varphi^\epsilon\}_{\epsilon>0}$ converging weakly to $\bar\varphi \in H^1(\Omega)$ and pass to the limit in (28)-(29).

Finally (24) follows from the assumptions and (22); see Bonnans and Casas [4] and Casas [7]. □

5 The Strong Pontryagin's Principle

Theorem 2 holds with $\bar\lambda_0 = 1$ for "almost all" control problems. We will state this clearly below. The key to achieve this result is the introduction of a stability assumption of the optimal cost functional with respect to small perturbations of the set of feasible controls. This stability allows to accomplish an exact penalization of the state constraints.

Definition 1 *We say that* (P_δ) *is strongly stable if there exist* $\epsilon > 0$ *and* $C > 0$ *such that*

$$\inf(\mathrm{P}_\delta) - \inf(\mathrm{P}_{\delta'}) \le C\sum_{j=1}^m(\delta'_j - \delta_j) \quad \forall \delta' = \{\delta'_j\}_{j=1}^m, \text{ with } \delta'_j \in [\delta_j, \delta_j + \epsilon]. \tag{30}$$

This concept, tightly related to the calmness notion of Clarke [15], was first introduced in relation with optimal control problems by Bonnans [3]; see also Bonnans and Casas [5]. A weaker stability concept was used by Casas [6] to analyze the convergence of the numerical discretizations of optimal control problems. The following proposition states that almost all problems (P_δ) are strongly stable.

Proposition 3 *Let us assume that* (P_{δ_0}) *has feasible controls for some* $\delta_0 = \{\delta_{0j}\}_{j=1}^m \subset \mathbb{R}^m$. *Then* (P_δ) *is strongly stable for all* $\delta \geq \delta_0$ *except at most a zero Lebesgue measure set.*

Proof. It is enough to consider the function $h : [\delta_{01}, +\infty) \times \cdots \times [\delta_{0m}, +\infty) \longrightarrow \mathbb{R}$ defined by

$$h(\delta) = \inf (P_\delta)$$

and remark that it is a nonincreasing monotone function at each variable δ_j and, consequently, differentiable at every point of its domain except at a zero measure set; see Saks [28]. Now it is obvious to check that (P_{δ_0}) is strongly stable at every point where h is differentiable. □

Finally we are ready to state the strong Pontryagin's principle.

Theorem 3 *Under the hypotheses of Theorem 2 and assuming that* (P_δ) *is strongly stable, then Theorem 2 remains to be true with* $\bar{\lambda}_0 = 1$.

We conclude the paper by proving Theorem 3. Firstly we show how the stability of problem (P_δ) makes possible an exact penalization of state constraints.

Proposition 4 *If* (P_δ) *is strongly stable and* $\bar{u}$ *is a solution of this problem, then there exists* $\gamma_0 > 0$ *such that* $\bar{u}$ *is also a solution of*

$$\inf_{u \in \mathcal{U}} J_\gamma(u) = J(u) + \gamma \sum_{j=1}^m (G_j(y_u) - \delta_j)^+ \tag{31}$$

for every $\gamma \geq \gamma_0$.

Proof. Let us suppose that it is false. Then there exists a sequence $\{\gamma_k\}_{k=1}^\infty$ of real numbers, with $\gamma_k \to +\infty$, and elements $\{u_k\}_{k=1}^\infty \subset \mathcal{U}$ such that

$$J(u_k) + \gamma_k \sum_{j=1}^m (G_j(y_k) - \delta_j)^+ < J(\bar{u}) \quad \forall k \geq 1,$$

where y_k is the state corresponding to u_k. From here we obtain that

$$0 < \sum_{j=1}^m (G_j(y_k) - \delta_j)^+ < \frac{J(\bar{u}) - J(u_k)}{\gamma_k} \longrightarrow 0 \quad \text{when } k \to +\infty$$

because J is uniformly bounded in $\mathcal{U}$ due to hypotheses (5) and (6). Let us take $\delta_{jk} = \max\{\delta_j, G_j(y_k)\}$. Since $\delta_k \to \delta$, we can use (30) to deduce

$$C \sum_{j=1}^m (\delta_{jk} - \delta_j) \geq \inf (P_\delta) - \inf (P_{\delta_k}) \geq J(\bar{u}) - J(u_k)$$

$$> \gamma_k \sum_{j=1}^{m} (G_j(y_k) - \delta_j)^+ = \gamma_k \sum_{j=1}^{m} (\delta_{jk} - \delta_j) \quad \forall k \geq k_\epsilon,$$

which is not possible. □

Since the penalty term of J_γ is not Gâteaux differentiable, we are going to modify slightly this functional to attain the differentiability necessary for the proof.

Proposition 5 *Let us take $\gamma \geq \gamma_0$ and for every $\epsilon > 0$ let us consider the problem*

$$(\mathrm{P}_{\delta,\epsilon}) \ \inf_{u \in \mathcal{U}} J_{\gamma,\epsilon}(u) = J(u) + \gamma \sum_{j=1}^{m} \left\{ [(G_j(y_u) - \delta_j)^+]^2 + \epsilon^2 \right\}^{1/2}.$$

Then $\inf(\mathrm{P}_{\delta,\epsilon}) \to \inf(\mathrm{P}_\delta)$ *when* $\epsilon \to 0$.

Proof. It is an immediate consequence of the inequality

$$J_\gamma(u) \leq J_{\gamma,\epsilon}(u) \leq J_\gamma(u) + m\gamma\epsilon \quad \forall u \in \mathcal{U}.$$

□

Proof of Theorem 3. Propositions 4 and 5 imply that $\bar{u}$ is a θ_ϵ^2–solution of $(\mathrm{P}_{\delta,\epsilon})$, with $\theta_\epsilon \to 0$ when $\epsilon \to 0$, i.e.

$$J_{\gamma,\epsilon}(\bar{u}) \leq \inf(\mathrm{P}_{\delta,\epsilon}) + \theta_\epsilon^2.$$

Then we can apply again Ekeland's principle and deduce the existence of an element $u^\epsilon \in \mathcal{U}$ such that

$$d(u^\epsilon, \bar{u}) \leq \theta_\epsilon, \quad J_{\gamma,\epsilon}(u^\epsilon) \leq J_{\gamma,\epsilon}(\bar{u}),$$

and

$$J_{\gamma,\epsilon}(u^\epsilon) \leq J_{\gamma,\epsilon}(u) + \theta_\epsilon d_E(u^\epsilon, u) \quad \forall u \in \mathcal{U}.$$

Now we argue as in the proof of Theorem 3 and replace (26) by

$$-\theta_\epsilon \sigma(\Gamma) \leq \lim_{\rho \to 0} \frac{J_{\gamma,\epsilon}(u_\rho^\epsilon) - J_{\gamma,\epsilon}(u^\epsilon)}{\rho} = z^{0,\epsilon} + \sum_{j=1}^{m} \lambda_j^\epsilon \int_\Omega \frac{\partial g_j}{\partial y}(x, y^\epsilon) z^\epsilon dx,$$

where

$$\lambda_j^\epsilon = \frac{\gamma(G_j(y^\epsilon) - \delta_j)^+}{\{[(G_j(y^\epsilon) - \delta_j)^+]^2 + \epsilon^2\}^{1/2}}, \quad 1 \leq j \leq m.$$

Therefore we have $0 \leq \lambda_j^\epsilon \leq \gamma$ for every $\epsilon > 0$. Then we can take subsequences that converge to elements $\bar{\lambda}_j \geq 0$, $j = 1, \ldots, m$. The rest is as in the proof of Theorem 3, taking $\lambda_0^\epsilon = 1$. □

References

[1] W.F. Ames. *Nonlinear Partial Differential Equations in Engineering.* Academic Press, New York-London, 1965.

[2] R. Aris. *The Mathematical Theory of Diffusion and Reaction in Permeable Catalysts. Vols. 1 and 2.* Clarendon Press, Oxford, 1975.

[3] J.F. Bonnans. Pontryagin's principle for the optimal control of semilinear elliptic systems with state constraints. In *30th IEEE Conference on Control and Decision*, pages 1976–1979, Brighton, England, 1991.

[4] J.F. Bonnans and E. Casas. Un principe de Pontryagine pour le contrôle des systèmes elliptiques. *J. Differential Equations*, 90(2):288–303, 1991.

[5] J.F. Bonnans and E. Casas. An extension of Pontryagin's principle for state-constrained optimal control of semilinear elliptic equations and variational inequalities. *SIAM J. Control Optim.*, To appear.

[6] E. Casas. Finite element approximations for some state-constrained optimal control problems. In P. Borne and S.G. Tzafestas, editors, *Mathematics of the Analysis and Design of Process Control*,

[7] E. Casas. Pontryagin's principle for optimal control problems governed by semilinear elliptic equations. In F. Kappel and K. Kunisch, editors, *International Conference on Control and Estimation of Distributed Parameter Systems: Nonlinear Phenomena*, Int. Series Num. Analysis. Birkhäuser, To appear.

[8] E. Casas and L.A. Fernández. Optimal control of quasilinear elliptic equations. In A. Bermúdez, editor, *Control of Partial Differential Equations*, pages 92–99, Berlin-Heidelberg-New York, 1989. Springer-Verlag. Lecture Notes in Control and Information Sciences 114.

[9] E. Casas and L.A. Fernández. Optimal control of quasilinear elliptic equations with non differentiable coefficients at the origin. *Rev. Mat. Univ. Complut. Madrid*, 4(2–3):227–250, 1991.

[10] E. Casas and L.A. Fernández. State-constrained control problems of quasilinear elliptic equations. In K.H. Hoffmann and W. Krabs, editors, *Optimal Control of Partial Differential Equations*, pages 11–25, Berlin-Heidelberg-New York, 1991. Springer-Verlag. Lecture Notes in Control and Information Sciences 149.

[11] E. Casas and L.A. Fernández. Optimality conditions for state-constrained control problems of quasilinear elliptic equations. In *30th IEEE Conference on Control and Decision*, pages 1991–1995, Brighton, England, 1991.

[12] E. Casas and L.A. Fernández. Distributed control of systems governed by a general class of quasilinear elliptic equations. *J. Differential Equations*, 104(1):20–47, 1993.

[13] E. Casas and L.A. Fernández. Dealing with integral state constraints in control problems of quasilinear elliptic equations. *SIAM J. Control Optim.*, To appear.

[14] E. Casas and J. Yong. Maximum principle for state–constrained optimal control problems governed by quasilinear elliptic equations. *Differential Integral Equations*, To appear.

[15] F.H. Clarke. A new approach to Lagrange multipliers. *Math. Op. Res.*, 1(2):165–174, 1976.

[16] C.V. Coffman, V. Duffin, and V.J. Mizel. Positivity of weak solutions of non-uniformly elliptic equations. *Ann. Mat. Pura Appl.*, 104:209–238, 1975.

[17] I. Ekeland. Nonconvex minimization problems. *Bull. Amer. Math. Soc.*, 1(3):76–91, 1979.

[18] H.O. Fattorini. Optimal control problems for distributed parameter systems governed by semilinear parabolic equations in L^1 and L^∞ spaces. In K.H. Hoffmann and W. Krabs, editors, *Optimal Control of Partial Differential Equations*, pages 60–80, Berlin-Heidelberg-New York, 1991. Springer-Verlag. Lecture Notes in Control and Information Sciences 149.

[19] H.O. Fattorini. Optimal control problems for distributed parameter systems in banach spaces. *Appl. Math. Optim.*, 4:225–257, 1993.

[20] H.O. Fattorini and H. Frankowska. Necessary conditions for infinite dimensional control problems. *Math. Control Signals Systems*, 4:41–67, 1991.

[21] X. Li and Y. Yao. Maximum principle of distributed parameter systems with time lags. In *Distributed Parameter Systems*, pages 410–427, New York, 1985. Springer-Verlag. Lecture Notes in Control and Information Sciences 75.

[22] X. Li and J. Yong. Necessary conditions of optimal control for distributed parameter systems. *SIAM J. Control Optim.*, 29:1203–1219, 1991.

[23] G.M. Lieberman. Boundary regularity for solutions of degenerate elliptic equations. *Nonlinear Anal.*, 12(11):1203–1219, 1988.

[24] A. Marrocco and O. Pironneau. Optimum design with Lagrangian finite methods: design of an electromagnet. *Comp. Meth. Appl. Mech. Eng.*, 15:277–308, 1978.

[25] M.K.V. Murthy and G. Stampacchia. Boundary value problems for some degenerate elliptic operators. *Ann. Mat. Pura Appl.*, 80:1–122, 1968.

[26] J. Nečas. *Les Méthodes Directes en Théorie des Equations Elliptiques*. Editeurs Academia, Prague, 1967.

[27] M.C. Pelissier. *Sur quelques problémes non linéaires en glaciologie*. PhD thesis, Université d'Orsay, 1975.

[28] S. Saks. *Theory of the Integral*. Dover, New York, 1964.

[29] N.S. Trudinger. Linear elliptic equations with measurable coefficients. *Ann. Scuola Norm. Sup. Pisa*, 27:265–308, 1973.

[30] J. Yong. Pontryagin maximum principle for semilinear second order elliptic partial differential equations and variational inequalities with state constraints. *Differential Integral Equations*, To appear.

Computation of Shape Gradients for Mixed Finite Element Formulation

MICHEL C. DELFOUR
Centre de recherches mathématiques et
Département de mathématiques et de statistique,
Université de Montréal, C.P. 6128 Succ. A,
Montréal, Québec, Canada, H3C 3J7

ZOUBIDA MGHAZLI
Département de Mathématiques
et d'Informatique
Université Ibn Tofail, Faculté des Sciences
Kénitra, Maroc

JEAN-PAUL ZOLÉSIO
Institut Non Linéaire de Nice
CNRS-INLN
136 Route des Lucioles
06904 Sophia Antipolis Cédex, France

ABSTRACT. Many physical processes can be modelled by a set of partial differential equations and boundary conditions over an N-dimensional domain. Very often those equations characterize the minimum of an energy functional or the saddle point of a Lagrangian functional corresponding to a constrained minimum energy problem. For instance the Neumann problem, the non-homogeneous Dirichlet problem, or the Stokes problem. The saddle point formulation is also fundamental to the numerical approximation of the solution of boundary value problems by mixed finite element methods.

The object of this paper is to compute shape gradients of such functionals with respect to the domain. Such approximations are extremely interesting for the computation of optimal shapes since they provide a relaxation of the smoothness conditions on the solution of the finite element problem. Computations of higher derivatives based on classical finite elements might require twice continuously differentiable or even smoother elements while in mixed formulations each derivative is approximated by low order finite elements.

The special energy functionals under consideration will not require an adjoint state. We shall use theorems on the differentiability of saddle points and function space parametrization techniques.

This type of computations finds applications in Free Boundary Problems and in Optimal Griding techniques, where it is required to find the best positions of the interior points defining the triangulation of the domain.

1. INTRODUCTION

Many physical processes are governed by a set of partial differential equations over a domain Ω in $\mathbb{R}^N$ which are the stationary conditions which characterize the minimum $J(\Omega)$ of an energy functional $E(\Omega, \varphi)$ over functions φ in a space of functions $V(\Omega)$ defined on Ω. An example of this type of process is the Neumann problem

$$(1.1) \qquad -\Delta u + u = f \quad \text{in } \Omega, \quad \frac{\partial u}{\partial n} = 0 \quad \text{on } \Gamma \text{ (boundary of } \Omega),$$

where

$$(1.2) \qquad J(\Omega) = \inf_{\varphi \in H^1(\Omega)} E(\Omega, \varphi),$$

The research of the first author has been supported in part by a Killam fellowship from Canada Council, National Sciences and Engineering Research Council of Canada operating grant A-8730 and by a FCAR grant from the Ministère de l'Education du Québec.

$H^1(\Omega)$ is the usual Sobolev space and

$$E(\Omega, \varphi) = \frac{1}{2} \int_\Omega \{|\nabla\varphi|^2 + \varphi^2 - 2 f\varphi\}\, dx. \tag{1.3}$$

Sometimes it is useful to consider one of the boundary conditions or one of the equations as a constraint on the minimization of an appropriate energy functional. For instance the non-homogeneous Dirichlet problem

$$-\nabla u + u = f \quad \text{in } \Omega, \quad u = g \quad \text{on } \Gamma \text{ (boundary of } \Omega) \tag{1.4}$$

can be reformulated as

$$J(\Omega) = \inf_{\varphi \in H^1(\Omega)} \sup_{\mu \in H^{-\frac{1}{2}}(\Gamma)} E(\Omega, \varphi) - \langle \mu, \varphi - g \rangle_{H^{\frac{1}{2}}(\Gamma)}, \tag{1.5}$$

where the "multiplier" μ coincides with $\frac{\partial u}{\partial n}$. Another example is the saddle point formulation of Stokes' problem.

The above problems are special cases of a more general family of Shape Optimization problems, where the solution of the state equation does not coincide with the minimizing element of an energy functional. In general the shape derivative can be expressed in terms of the state and the adjoint state arising from the constraint imposed by the state equation or a non-homogeneous boundary condition as in (1.4)– (1.5). However in the special case (1.1)–(1.2)–(1.3) the state equations are not a constraint since they naturally arise as a necessary condition which characterizes the minimum. The advantageous consequence of this situation is that the shape derivative can be completely expressed in terms of the state. The adjoint state is not necessary.

The saddle point formulations are becoming extremely important in Numerical Analysis since they are the basis of many mixed finite element methods (cf. BREZZI–FORTIN [1]). In this paper we give a comprehensive approach to the computation of shape derivatives for problems which are characterized by the saddle point of an energy functional and approximated by mixed finite element methods. We use and extend techniques such as *Function Space Parametrization* developped by DELFOUR–ZOLÉSIO [1, 2] for the minimization of a functional (cf. DELFOUR–PAYRE–ZOLÉSIO [1]). We provide the details for the mixed formulation of the Dirichlet problem. We also give typical examples for linear elasticity and the $\psi - \omega$ formulation of the biharmonic problem.

2. Shape gradient of the minimum of an energy functional revisited

In this section we briefly review the basic constructions and results for the shape derivative of a quadratic energy functional. Then we consider its finite element approximation and show how they are related. It turns out that the approximation error is directly related to the minimum of the energy functional corresponding to the finite element approximation of the problem. Therefore it is possible to reduce the approximation error by minimizing the minimum of the energy functional with respect to the parameters of the finite element approximation such as the positions of the internal nodes. For the interested readers the complete theory and specific numerical examples can be found in Delfour–Payre–Zolésio [1].

2.1. A quadratic minimization problem.

Let $V = V(\Omega)$ be a Hilbert space over a domain Ω in the N-dimensional space $\mathbb{R}^N$, $N \geq 1$, and $u = u(\Omega)$ be the solution of the minimization problem

$$J(\Omega) = \inf_{\varphi \in V} E(\Omega, \varphi), \tag{2.1}$$

where

$$E(\Omega, \varphi) = \frac{1}{2} a(\Omega\,;\, \varphi, \varphi) - \langle F, \varphi \rangle_V, \tag{2.2}$$

$a(\Omega\,;\, \varphi, \psi)$ is a coercive continuous bilinear form in (φ, ψ) on $V(\Omega) \times V(\Omega)$

$$\exists \alpha > 0 \text{ (independent of } \Omega), \quad \forall v \in V(\Omega), \quad a(\Omega\,;\, v, v) \geq \alpha \|v\|^2_{V(\Omega)}. \tag{2.3}$$

$\| \cdot \|_V$ is the norm in $V(\Omega)$, F is an element of the dual $V'(\Omega)$ of $V(\Omega)$ and $\langle \cdot, \cdot \rangle_V$ is the duality pairing.

Let Ω a polygonal domain in $\mathbb{R}^N$ and $V_h(\Omega)$ (usually a finite element subspace over a triangulation τ_h) be a closed linear subspace of the Sobolev space $V(\Omega)$. For instance $V_h(\Omega)$ can be a finite element approximation subspace over a given triangulation τ_h of Ω. Let u_h be the solution of the minimization problem

$$J_h(\Omega) = \inf_{\varphi \in V_h(\Omega)} E(\Omega, \varphi). \tag{2.4}$$

It turns out that for any two h and h'

$$\|u_{h'} - u\|^2_V = \|u_h - u\|^2_V + 2\,[J(u_{h'}) - J(u_h)]. \tag{2.5}$$

So to decrease the approximation error, it is sufficient to choose a triangulation τ_h which minimizes $J(u_h)$ over the chosen control parameters of the finite element approximation.

2.2. Function Space Parametrization.

To compute the first order derivative of $J(\Omega)$ we perturb the domain Ω by a velocity field V which generates a family of transformations $\{T_t : 0 \le t \le \tau\}$ of $\mathbb{R}^N$ and a family of domains $\{\Omega_t = T_t(\Omega) : 0 \le t \le \tau\}$, $\tau > 0$. At t

$$J(\Omega_t) = \inf_{\varphi \in V(\Omega_t)} E(\Omega_t, \varphi), \tag{2.6}$$

where $V(\Omega_t)$ is the same function space $V(\Omega)$ but defined on Ω_t. In general the space $V(\Omega_t)$ will be related to the original space $V(\Omega)$ through a specific parametrization. This is what we call *Function space parametrization* (cf. DELFOUR– ZOLÉSIO [1, 2]).

For the Sobolev spaces $H^1(\Omega)$ or $H_0^1(\Omega)$ an appropriate parametrization is

$$V(\Omega_t) = \{\varphi \circ T_t^{-1} : \varphi \in V(\Omega)\}. \tag{2.7}$$

Notice that since T_t is a homeomorphism which transports the open domain Ω onto the open domain Ω_t and sends the boundary Γ of Ω onto the boundary Γ_t of Ω_t. In particular when V is sufficiently smooth for all φ in $H_0^1(\Omega)$, $\varphi \circ T_t^{-1} \in H_0^1(\Omega_t)$ and conversely for all ψ in $H_0^1(\Omega_t)$, $\psi \circ T_t \in H_0^1(\Omega)$. This parametrization does not affect the value of the minimum $J(\Omega_t)$ but will change the functional E

$$J(\Omega_t) = \inf_{\varphi \in V(\Omega)} E\left(T_t(\Omega), \varphi \circ T_t^{-1}\right). \tag{2.8}$$

This type of parametrization seems to be unique to *Shape Analysis*. It amounts to introducing the new energy functional

$$\tilde{E}(t, \varphi) = E(T_t(\Omega), \varphi \circ T_t^{-1}), \quad \varphi \in V(\Omega). \tag{2.9}$$

We shall see later that each problem has a natural function space parametrization which might differ from the one introduced for $H^1(\Omega)$ or $H_0^1(\Omega)$.

We want to compute the derivative

$$dj(0) = \lim_{t \searrow 0} \frac{j(t) - j(0)}{t} \tag{2.10}$$

of the function

$$j(t) = J(\Omega_t). \tag{2.11}$$

We need a theorem which gives the derivative of a Min with respect to a parameter $t \ge 0$ at $t = 0$. As an illustration consider the homogeneous Dirichlet boundary problem (1.4) with $g = 0$. The minimizing element y^t in $H_0^1(\Omega)$ of

$$\tilde{E}(t, \varphi) = \int_{\Omega_t} \left[\frac{1}{2}|\nabla(\varphi \circ T_t^{-1})|^2 - f(\varphi \circ T_t^{-1})\right] dx \tag{2.12}$$

is the solution of

$$\text{(2.13)} \qquad \begin{cases} y^t \in H_0^1(\Omega), \text{ and } \quad \forall \varphi \in H_0^1(\Omega) \\ \displaystyle\int_{\Omega_t} \{\nabla(y^t \circ T_t^{-1}) \cdot \nabla(\varphi \circ T_t^{-1}) - f(\varphi \circ T_t^{-1})\} \, dx = 0. \end{cases}$$

This expression is to be compared with the characterization of the minimizing element y_t of $E(\Omega_t, \varphi)$ on $H_0^1(\Omega_t)$:

$$\text{(2.14)} \qquad \begin{cases} y_t \in H_0^1(\Omega_t), \text{ and } \quad \forall \varphi \in H_0^1(\Omega_t) \\ \displaystyle\int_{\Omega_t} \{\nabla y_t \cdot \nabla \varphi - f\varphi\} \, dx = 0. \end{cases}$$

It is easy to verify that

$$\text{(2.15)} \qquad y_t = y^t \circ T_t^{-1} \text{ and } y^t = y_t \circ T_t.$$

In view of the above considerations we can rewrite expression (2.12) on the fixed domain Ω

$$\text{(2.16)} \qquad \tilde{E}(t, \varphi) = \int_{\Omega} \left\{ \frac{1}{2} \, [A(t) \, \nabla \varphi] \cdot \nabla \varphi - (f \circ T_t) \, \varphi \, J_t \right\} \, dx,$$

where for t in $[0, \tau]$ small

$$\text{(2.17)} \qquad DT_t = \text{Jacobian matrix of } T_t,$$

$$\text{(2.18)} \qquad J_t = \det DT_t \ (\det DT_t = |\det DT_t| \ \text{ for } t \geq 0 \, \text{small}),$$

$$\text{(2.19)} \qquad A(t) = J_t (DT_t)^{-1} \, {}^*(DT_t)^{-1},$$

where ${}^*(M)$ denotes the transposed of a matrix M. With this change of variable y^t is now characterized by the variational equation

$$\text{(2.20)} \qquad \begin{cases} y^t \in H_0^1(\Omega), \quad \forall \varphi \in H_0^1(\Omega) \\ \displaystyle\int_{\Omega} [A(t) \, \nabla y^t \cdot \nabla \varphi - J_t (f \circ T_t) \varphi] \, dx = 0. \end{cases}$$

2.3. Differentiability of a Min with respect to a parameter.

Consider a functional

$$G : [0, \tau] \times X \to \mathbb{R} \tag{2.21}$$

for some $\tau > 0$ and some set X. For each t in $[0, \tau]$ define

$$g(t) = \inf \{G(t, x) : x \in X\} \tag{2.22}$$

and the set

$$X(t) = \{x \in X : G(t, x) = g(t)\}. \tag{2.23}$$

We wish to study the existence and characterization of the limit

$$dg(0) = \lim_{t \searrow 0} [g(t) - g(0)]/\, t \tag{2.24}$$

when $X(t)$ is not empty for $0 \le t \le \tau$.

When $X(t) = \{x^t\}, 0 \le t \le \tau$, and the derivative

$$\dot{x} = \lim_{\substack{t>0 \\ t \to 0}} \left[x^t - x^0\right] / t \tag{2.25}$$

of x is known, then it is easy to obtain $dg(0)$ under appropriate differentiability of the functional G with respect to t and x. When $\dot{x}$ is not readily available or when the sets $X(t)$ are not singletons, this direct approach fails or becomes very intricate. We use a theorem which gives an explicit expression for $dg(0)$, the derivative of the Min of the functional G with respect to t at $t = 0$. Its originality is that the differentiability of x^t is replaced by a continuity hypothesis on the set valued function and the existence of the partial derivative of the functional G with respect to the parameter t. In other words this technique does not require a priori knowledge of the derivative $\dot{x}$ of the minimizing elements x^t with respect to t.

Theorem 2.1. (CORREA–SEEGER [1]) *Let $\tau > 0$ be a real number, X an arbitrary set and $G : [0, \tau] \times X \to \mathbb{R}$ a well-defined functional. Assume that the following assumptions are verified:*

(H1) *for all $t \in [0, \tau]$, $X(t) \neq \phi$,*

(H2) *for all x in $X(0)$, $\partial_t G(t, x)$ exists everywhere in $[0, \tau]$,*

(H3) *there exists a topology $\mathcal{T}_X$ on X such that for any sequence $\{t_n\} \subset]0, \tau]$, $t_n \to t_0 = 0$, $\exists x_0 \in X(0)$, $\exists$ a subsequence $\{t_{n_k}\}$ of $\{t_n\}$, and for each $k \ge 1$, $\exists x_{n_k} \in X(t_{n_k})$ such that*

(i) *$x_{n_k} \to x^0$ in the $\mathcal{T}_X$-topology,*

(ii) *and*

$$\liminf_{\substack{t \searrow 0 \\ k \to \infty}} \partial_t G(t, x_{n_k}) \ge \partial_t G(0, x^0),$$

(H4) *for all x in $X(0)$, the map $t \mapsto \partial_t G(t, x)$ is upper semicontinuous at $t = 0$.*

Then there exists $x^0 \in X(0)$ *such that*

$$dg(0) = \lim_{t \searrow 0} \frac{g(t) - g(0)}{t} = \inf_{x \in X(0)} \partial_t G(0, x) = \partial_t G(0, x^0). \quad \square \tag{2.26}$$

2.4. Application of the theorem.

It is not possible to provide here all the details of the application of the theorem. We specifically refer to the initial paper by M.C. DELFOUR and J.P. ZOLÉSIO [1] or the recent lecture notes by M.C. DELFOUR [1] .

For the example the t-derivative of expression (2.16) becomes

$$\partial_t \tilde{E}(t, \varphi) = \int_\Omega \left\{ \frac{1}{2} [A'(t) \nabla \varphi] \cdot \nabla \varphi - [\text{div } V_t(f \circ T_t) + J_t \nabla f \cdot V_t] \, \varphi \right\} dx, \tag{2.27}$$

where

$$V_t(X) = V(t, T_t(X)), \; DV_t(X) = DV(t, T_t(X)), \tag{2.28}$$

and

$$A'(t) = (\text{div } V_t) \, I -^* (DV_t) - (DV_t). \tag{2.29}$$

Finally for V in $C^0([0, \tau]\,; \mathcal{D}^2(\mathbb{R}^N, \mathbb{R}^N))$ and f in $H^1(\mathbb{R}^N)$,

$$dJ(\Omega\,; V) = \int_\Omega \left\{ \frac{1}{2} \; A'(0) \nabla y \cdot \nabla y - [\text{div } \; V(0) f + \nabla f \cdot V(0)] \; y \right\} dx. \tag{2.30}$$

3. MIXED FINITE ELEMENT CONTEXT

In §2 we have considered the Dirichlet problem for the Laplace equation as the minimizing element of an energy functional over $H_0^1(\Omega)$. This is the classical way of setting up this problem in view of its finite element approximation.

We now turn to another formulation of the same physical problem. It is based on a variational principle expressing an equilibrium (saddle point) condition rather than on minimization principle. Approximating thoses equations by finite elements methods lead to the so-called mixed finite elements methods. We refer the reader to BREZZI–FORTIN [1] for a clear expository treatment of this subject.

In many applications described by equations (2.14), ∇u rather than u is the interesting variable. This is the case for thermo-diffusion for instance: ∇u will be the heat flux which is very often more important to know than the temperature. So we introduce the auxiliary variable $p = \nabla u$ to transform the Dirichlet problem (2.14) into the system

$$\begin{cases} p = \nabla u, \quad u \in H_0^1(\Omega), \\ \operatorname{div} p + f = 0 \text{ in } \Omega. \end{cases} \tag{3.1}$$

In order to get a weak formulation of those equations, we introduce the function space

$$H(\mathrm{div}\,;\,\Omega) = \{q \in L^2(\Omega)^2\,;\, \mathrm{div}\; q \in L^2(\Omega)\} \tag{3.2}$$

and its norm

$$||q||^2_{H(\mathrm{div}\,;\,\Omega)} = ||q||^2_{0,\Omega} + ||\,\mathrm{div}\; q||^2_{0,\Omega} \tag{3.2}$$

which makes it a Hilbert space, where

$$\mathrm{div}\; q = \frac{\partial q_1}{\partial x_1} + \frac{\partial q_2}{\partial x_2}. \tag{3.4}$$

This space is specially adapted to the study of mixed methods. If we choose $f \in L^2(\Omega)$, a weak formulation of (3.1) is

$$\inf_{q\in H(\mathrm{div}\,;\,\Omega)} \sup_{v\in L^2(\Omega)} \left\{ \frac{1}{2}\int_\Omega |q|^2\,dx + \int_\Omega fv\,dx + \int_\Omega v\,\mathrm{div}\; q\,dx \right\} \tag{3.5}$$

and (u, p) is the saddle point, satisfying the variational system:

$$\begin{cases} \displaystyle\int_\Omega p\cdot q\,dx + \int_\Omega u\;\mathrm{div}\; q\,dx = 0 \quad \forall q \in H(\mathrm{div}\,;\,\Omega), \\ \displaystyle\int_\Omega (\mathrm{div}\; p + f)v\,dx = 0 \quad \forall v \in L^2(\Omega). \end{cases} \tag{3.6}$$

Let $b(\cdot,\cdot)$ be the bilinear form on $H(\mathrm{div}\,;\,\Omega) \times L^2(\Omega)$ defined by

$$b(q,v) = \int_\Omega v\;\mathrm{div}\; q\,dx.$$

It corresponds to the divergence operator $B : H(\mathrm{div}\,;\,\Omega) \to L^2(\Omega)$ which is a linear surjective operator. The existence and uniqueness of a solution $u \in L^2(\Omega)$ of (3.6) follows by coercivity of the bilinear form

$$a(p,q) = \int_\Omega p\cdot q\,dx$$

on $\mathrm{Ker}\, B$ (even if it is not coercive on $H\,(\mathrm{div}\,;\,\Omega)$). The existence of $p \in H(\mathrm{div}\,;\,\Omega)$ is ensured by the inf sup condition:

$$\sup_{q\in H(\mathrm{div}\,;\,\Omega)} \frac{b(q,v)}{||q||_{H(\mathrm{div}\,;\,\Omega)}} \geq k_0\,||v||_{L^2(\Omega)} \quad \forall v \in L^2(\Omega) \tag{3.7}$$

and p is unique if and only if B is surjective. Therefore we have existence and uniqueness of $(u,p) \in L^2(\Omega) \times H(\mathrm{div}\,;\,\Omega)$ solution of (3.6).

Let V_h and W_h be finite dimensional subspaces of $H(\text{div}\,;\Omega)$ and $L^2(\Omega)$ respectively, and consider the problem

$$(3.8)\qquad \begin{cases} \text{Find} \quad u_h \in W_h \quad \text{and} \quad p_h \in V_h \quad \text{such that} \\ \displaystyle\int_\Omega p_h \cdot q\,dx + \int_\Omega u_h \,\text{div}\, q\,dx = 0 \quad \forall q \in V_h, \\ \displaystyle\int_\Omega v(\text{div}\; p_h + f)v\,dx = 0 \quad \forall v \in W_h. \end{cases}$$

The first basic step in mixed methods is to construct appropriate function spaces to approximate $H(\text{div}\,;\Omega)$. Denote by $\mathcal{T}_h$ a triangulation of Ω. The first approximation of $H(\text{div}\,;\Omega)$ was given in RAVIART–THOMAS [1] and is characterized by the following approximation subspace

$$(3.9)\qquad V_h = \left\{q_h \in H(\text{div}\,;\Omega)\ :\ q_{h|K} \in \left(P_k(K)\right)^2 \oplus xP_k(K) \quad \forall K \in \mathcal{T}_h\right\},$$

where $P_k(K)$ is the space of polynomials of order less or equal to $k \geq 1$ on the triangle K. If

$$W_h = \left\{v_h \in L^2(\Omega)\ :\ v_{h|K} \in P_k(K) \quad \forall K \in \mathcal{T}_h\right\}$$

they showed that there exists a unique $(u_h, p_h) \in W_h \times V_h$ solution of (2.8) and if $p \in (H^s(\Omega))^2$ and $u \in H^s(\Omega)$

$$(3.11)\qquad \begin{aligned} ||p - p_h||_{H(\text{div}\,;\Omega)} &\leq C_1 h^s |p|_{s,\Omega} \\ ||u - u_h||_{0,\Omega} &\leq C_2 h^s (|p|_{s,\Omega} + |u|_{s,\Omega}) \end{aligned}$$

for $s \leq k+1$, where C_1 and C_2 are constants independent of h. The solution $(p_h, u_h) \in V_h \times W_h$ of the system (3.8) coincides with the saddle point of the Lagrangian

$$(3.12)\qquad G(q,v) = \frac{1}{2}\int_\Omega |q|^2, dx + \int_\Omega (\text{div}\; q + f)v\,dx$$

over $V_h \times W_h$.

We want now to determine the derivative of $G(\cdot,\cdot)$ with respect to the nodal coordinate of the finite element mesh $\mathcal{T}_h$. We use the "Velocity (or Speed) Method" to generate domain variations and define derivatives. The displacement of the point X of $\mathbb{R}^n$ is governed by the differential equation

$$(3.13)\qquad \begin{aligned} &\frac{dx(t)}{dt} = V\left(t, x(t)\right) \quad t \geq 0, \\ &x(0) = V. \end{aligned}$$

which generates a transformation $T_t(V)$ (noted also T_t) of $\mathbb{R}^n$ defined as

$$T_t(X) = x(t) \quad t \geq 0 \quad X \in \mathbb{R}^n.$$

We denote the domain transported by V after the time t by

$$\Omega_t = T_t(\Omega).$$

It is well known that if T_t and T_t^{-1} are mapping of class C^0, then for all scalar function v we have

$$v \in L^2(\Omega) \Longleftrightarrow v \circ T_t^{-1} \in L^2(\Omega_t). \tag{3.14}$$

So it is natural to consider the following transformation

$$\begin{aligned} \mathcal{T}_t : L^2(\Omega) &\longrightarrow L^2(\Omega_t) \\ v &\longrightarrow \mathcal{T}_t(v) = v \circ T_t^{-1} \end{aligned} \tag{3.15}$$

for the scalar functions. But if q is a vectorial function of $H(\text{div}\,;\Omega)$, $q \circ T_t$ is not necessary an element of $H(\text{div}\,;\Omega_t)$. Indeed $H(\text{div}\,;\Omega)$ is characterized by the functions which have a "continuous" normal trace at the interfaces of the triangulation of Ω. This is an essential point for mixed finite element approximation. The above transformation does not preserve normal components and does not map $H(\text{div}\,;\Omega)$ into $H(\text{div}\,;\Omega_t)$. So we have to introduce a new transformation which will be adapted to the mixed formulation.

The use of a reference element, and therefore of coordinates changes is an essential ingredient of finite element methods, either for convergence studies or for practical implementation. In order to transport base functions on Ω onto base functions on Ω_t, we use a reference element for the definition of the transformation T_t. Let $\widehat{K}$ be a reference element, K be a triangle of $\mathcal{T}_h$ and F_K be the mapping from $\mathbb{R}^2$ into $\mathbb{R}^2$ such that

$$\begin{aligned} F_K(\widehat{K}) &= K, \\ F_K(\hat{x}) &= B_K\hat{x} + x_0^K, \end{aligned} \tag{3.16}$$

where B_K is a matrix. F_K is then an affine mapping and the Jacobian matrix is B_K which is constant. We write

$$J_K = |\det B_K|. \tag{3.17}$$

We define the transformation T_t by using the reference element, as follows

$$\begin{aligned} T_t^K &= T_{t|K}, \\ T_t^K &= F_{K_t} \circ F_K^{-1}, \end{aligned} \tag{3.18}$$

where K_t is the transported triangle. We have then

$$DT_t^K = B_{K_t} \cdot B_K^{-1}. \tag{3.19}$$

We denote

$$I_t^K = |\det B_{K_t}| \cdot |\det B_K^{-1}|. \tag{3.20}$$

We now define a transformation, denoted $\mathcal{G}_t$, from $H(\text{div}\,;\Omega)$ into $H(\text{div}\,;\Omega_t)$ in the following way:

$$\mathcal{G}_t^K = \mathcal{G}_{t|K}$$

and if $q_t^K := \mathcal{G}_t^K(q^K)$, where $q^K = q_{|K}$, q_t^K is defined by

$$q_t^K \circ (T_t^K)^{-1} = \frac{1}{I_t^K} DT_t^K q^K \tag{3.21}$$

and we have

$$(\text{div}\ q_t) \circ (T_t^K)^{-1} = \frac{1}{I_t^K} \text{div}\ q^K. \tag{3.22}$$

Taking account with (3.15), this transformation satisfies the following property:

$$\int_{K_t} \text{div}\ q_t^K v_t^K dx_t = \int_K \text{div}\ q^K v^K\, dx \quad \forall K \in \mathcal{T}_h. \tag{3.23}$$

The transformation $\mathcal{G}_t$ is built from a special isomorphism, $\mathcal{G}^K$, known as the Piola's transformation which is defined from $H(\text{div}\,;\widehat{K})$ into $H(\text{div}\,;K)$ by

$$q(x) = \frac{1}{J_K} B_K \hat{q}(\hat{x}). \tag{3.24}$$

It satisfies

$$\text{div}\ q = \frac{1}{J_K} \text{div}\ \hat{q} \tag{3.25}$$

preserves the normal trace, and makes it possible to define approximations of $H(\text{div}\,;K)$ through the reference element. We refer to THOMAS [1] and RAVIART–THOMAS [1] for more details.

In fact, $\mathcal{G}_t$ is given by

$$\mathcal{G}_t^K = \mathcal{G}^{K_t} \circ (\mathcal{G}^K)^{-1}, \tag{3.26}$$

where $\mathcal{G}^K$ is defined by (3.24).

Under the action of the field V associated with the transformation T_t, the domain Ω is transformed into a new domain $\Omega_t = T_t(\Omega)$. Let (p_t, u_t) be the saddle point on Ω_t of the Lagrangian

$$G(t, q, v) = \frac{1}{2} \int_{\Omega_t} |q|^2\, dx_t + \int_{\Omega_t} \text{div}\ q\ \ v\, dx_t + \int_{\Omega_t} f v\, dx_t$$

over $V_h(\Omega_t) \times W_h(\Omega_t)$, and introduce the functional

$$g(t) = \inf_{q \in V_h(\Omega_t)} \sup_{v \in W_h(\Omega_t)} G(t, q, v).$$

To get the derivative of $g(t)$ with respect to t, we use the theorem of DELFOUR–ZOLÉSIO [1] related to the differentiality of a Min Max. To do so, we parametrize the elements of $V_h(\Omega_t)$ and $W_h(\Omega_t)$ with the help of the transformation T_t and get spaces independent of t. Indeed with the use of coordinates changes and (3.23) we have

$$\begin{aligned} G(t, q_t, v_t) &= \frac{1}{2} \int_{\Omega_t} q_t \circ T_t^{-1} \cdot q_t \circ T_t^{-1} \, dx_t \\ &\quad + \int_{\Omega_t} (\text{div } q_t) \circ T_t^{-1} v \circ T_t^{-1} \, dx_t \\ &\quad + \int_{\Omega_t} f \circ T_t^{-1} v \circ T_t^{-1} \, dx \\ &= \frac{1}{2} \int_{\Omega} A(t) q \cdot q \, dx + \int_{\Omega} \text{div } q \, v \, dx + \int_{\Omega} f \circ T_t^{-1} v \, I_t \, dx, \end{aligned}$$

where

$$\begin{aligned} A^K(t) &= A(t) \,|_K, \\ A^K(t) &= \frac{1}{I_t^K} {}^\top(DT_t^K) \cdot DT_t^K. \end{aligned}$$

So

$$g'(0) = \inf_{g \in V_h} \sup_{v \in W_h} \left\{ \frac{1}{2} \int_{\Omega} A'(0) q \cdot q \, dx + \int_{\Omega} \text{div} \left(f V(0) \right) v \, dx \right\}$$

and

$$A'(0) = -\, \text{div } V(0) \cdot I + DV(0) + {}^\top DV(0).$$

Remark. If we perturb the position of the nodal point M_j in the direction ℓ_k

$$M_j^t = M_j + t\ell_k,$$

where $\ell_k, k = 1, 2$, is the Euclidean basis of $\mathbb{R}^2$, then we find that the vector fields which transport triangles of $\mathcal{T}_h$ onto triangles of $\mathcal{T}_h^t$ and shape functions for $\mathcal{T}_h$ onto shape function for $\mathcal{T}_h$, is the same as in ZOLÉSIO [1] and DELFOUR– ZOLÉSIO [3], that is

$$V(t, x) = \phi_j^t(x) \ell_k,$$

where ϕ_j^t is the shape function corresponding to C^0 and linear approximation (P^1) at M_j.

Let $\widehat{K}$ be the reference element given by its vertices:

$$\hat{a}^{(1)} = (1,0), \quad \hat{a}^{(2)} = (0,1) \quad \text{and } \hat{a}^{(3)} = (0,0).$$

Let $\partial\widehat{K}^{(i)}$ denote the opposite side to vertex $\hat{a}^{(i)}$ and let

$$\hat{x}_3 = 1 - \hat{x}_1 - \hat{x}_2.$$

Then $\partial K^{(i)}$ belongs to the line whose equation is $\hat{x}_i = 0$, and $(\hat{x}_1, \hat{x}_2, \hat{x}_3)$ are the barycentric coordinates of $\hat{x} = (\hat{x}_1, \hat{x}_2)$.

Let $K \in \mathcal{T}_h$ be specified by its vertices $a^{(i)} = \left(a_1^{(i)}, a_2^{(i)}\right)$ for $i = 1,2,3$. If F_K is the mapping such that

$$F_K\left(\hat{a}^{(i)}\right) = a^{(i)} \text{ for } i = 1,2,3,$$

then F_K is given by

$$B_K = \begin{pmatrix} A_1^{(1)} - A_1^{(3)} & A_1^{(2)} - A_1^{(3)} \\ A_2^{(1)} - A_2^{(3)} & A_2^{(2)} - A_2^{(3)} \end{pmatrix}, \quad x_0^K = \begin{pmatrix} A_1^{(3)} \\ A_2^{(3)} \end{pmatrix}.$$

Let $(\lambda_1, \lambda_2, \lambda_3)$ be the barycentric coordinates of $x \in K$, with respect to $a^{(i)}, i = 1,2,3$:

$$\begin{cases} x = \lambda_1 a^{(1)} + \lambda_2 a^{(2)} + \lambda_3 a^{(3)}, \\ \lambda_1 + \lambda_2 + \lambda_3 = 1. \end{cases}$$

We have $\lambda_i = \hat{x}_i$.

We denote the vertices of the transported triangle K_t by

$$a^{(i)}(t) = \left(a_1^{(i)}(t), a_2^{(i)}(t)\right).$$

If

$$a^{(i)}(t) = a^{(i)} + tw^{(i)},$$

where $w^i \in \mathbb{R}^2$. Then

$$\begin{aligned} T_t^K(x) &= F_{K_t}\left(F_{K(x)}^{-1}\right) \\ &= F_{K_t}(\hat{x}) \\ &= B_{K_t}\hat{x} + x_0^{K_t} \\ &= B_K\hat{x} + x_0^K + t\sum_{i=1}^{3} \hat{x}_i w^{(i)} \\ &= x + t\sum_{i=1}^{3} \lambda_i(x) w^{(i)}. \end{aligned}$$

So

$$V_1^K(t,x) = \sum_{i=1}^{3} \lambda_i(x) w^{(i)}$$

and λ_i corresponds to the basis function associated with $a^{(i)}$. □

4. EXAMPLES

4.1 Linear elasticity.

The small displacement $\underline{u} = (u_1, u_2)$ of an elastic, isotropic and homogeneous material under the action of some external forces is governed by the equation:

$$(4.1) \qquad \begin{cases} \underline{\underline{\sigma}} = 2\mu\underline{\underline{\epsilon}}(\underline{u}) + \lambda \operatorname{tr}(\underline{\underline{\epsilon}}(\underline{u}))\underline{\underline{I}} & \text{in } \Omega \\ \underline{\operatorname{div}}\, \underline{\underline{\sigma}} + \underline{f} = 0 & \text{in } \Omega \\ \underline{u} = 0 & \text{on } \Gamma. \end{cases}$$

$\underline{\underline{\sigma}}$ is the stress tensor and $\underline{\underline{\epsilon}}(\underline{u})$ is the linearized strain tensor. The constants λ and μ are the Lamé coefficients.

We introduce the space

$$(4.2) \qquad \mathbb{H}_s(\operatorname{div}; \Omega) = \left\{ \underline{\underline{\tau}} \in \left(L^2(\Omega)\right)^4 ;\ \underline{\operatorname{div}}\ \underline{\underline{\tau}} \in \left(L^2(\Omega)\right)^2, \quad \tau_{12} = \tau_{21} \right\}.$$

A mixed formulation of (3.1) is given by the saddle point problem in

$$\mathbb{H}_s(\operatorname{div}; \Omega) \times \left(L^2(\Omega)\right)^2 :$$

$$G(\underline{\underline{\sigma}}, \underline{v}) = \inf_{\underline{\underline{\sigma}}} \sup_{\underline{v}} G\left(\underline{\underline{\sigma}}, \underline{v}\right)$$

$$(4.3) \qquad = \frac{1}{2}\left(\frac{1}{4(\lambda+\mu)} \int_\Omega |\operatorname{tr} \underline{\underline{\sigma}}|^2 dx + \frac{1}{2\mu} \int_\Omega |\underline{\underline{\sigma}}^D|^2\, dx \right) + \int_\Omega \left(\underline{\operatorname{div}}\, \underline{\underline{\sigma}} + \underline{f} \right) \underline{v}\, dx.$$

The solution $(\underline{\underline{\sigma}}, \underline{u})$ of this saddle point problem is characterized by the system:

$$(4.4) \qquad \begin{cases} \operatorname{div}\ \underline{\underline{\sigma}} + \underline{f} = 0 \\ \operatorname{tr} \underline{\underline{\sigma}} = (\lambda + \mu) \operatorname{tr} \underline{\underline{\epsilon}}(\underline{u}) \\ \underline{\underline{\sigma}}^D = 2\,\mu\, \underline{\underline{\epsilon}}^D(\underline{u}) \end{cases}$$

which corresponds to the equilibrium condition and the constitutive law. The deviator $\underline{\underline{\epsilon}}^D$ of the tensor $\underline{\underline{\epsilon}}$ is defined as

$$\underline{\underline{\epsilon}}^D = \underline{\underline{\epsilon}} - \frac{1}{2} \left(\operatorname{tr} \underline{\underline{\epsilon}}\right) \underline{\underline{I}}.$$

Consider an approximation of (4.3) by finite element method. Following the description above, to determine the derivative of $G(\underline{\underline{\tau}}, \underline{v})$ with respect to the nodal coordinate of the finite element mesh $\mathcal{T}_h$, we want to find a transformation from $\mathbb{H}_s(\operatorname{div}; \Omega)$ into $\mathbb{H}_s(\operatorname{div}; \Omega_t)$ such that

$$(4.5) \qquad \int_{K_t} \underline{\operatorname{div}}\, \underline{\underline{\tau}}_t^K \cdot \underline{v}_t^K\, dx_t = \int_t \operatorname{div}\ \underline{\underline{\tau}}^K \cdot v\, dx \quad \forall K \in \mathcal{T}_h.$$

This transformation must preserve the normal components and the symmetry property. In view of the notation of §3 let

$$\mathcal{H}_t : \mathbb{H}_s(\text{div}\,;\Omega) \longrightarrow \mathbb{H}_s(\text{div}\,;\Omega_t)$$

defined by

$$\mathcal{H}_t^K(\underline{\underline{\tau}}^K) = (\underline{\underline{DT}}_t^K)\,\underline{\underline{\tau}}^K\,(\underline{\underline{DT}}_t^K)^\top \tag{4.6}$$

and

$$\mathcal{G}_t : \left(L^2(\Omega)\right)^2 \longrightarrow \left(L^2(\Omega_t)\right)^2$$

defined by

$$\mathcal{G}_t^K(\underline{v}^K) = \frac{1}{I_t^K}(\underline{\underline{DT}}_t^K)^{-\top}\underline{v}^K. \tag{4.7}$$

Then we get (4.5) with $\underline{\underline{\tau}}_t^K = \mathcal{H}_t^K(\underline{\underline{\tau}}^K)$ and $v_t^K = \mathcal{G}_t(v^K)$ and

$$\underline{\text{div}\,\underline{\underline{\tau}}_t^K} = (\underline{\underline{DT}}_t^K)\cdot\underline{\text{div}\,\underline{\underline{\tau}}^K}.$$

So

$$\begin{aligned}
&G(t, \underline{\underline{\tau}}_t, \underline{v}_t) \\
&= \frac{1}{2}\left\{\frac{1}{2\mu}\int_\Omega \underline{\underline{\tau}}_t \circ T_t^{-1} : \underline{\underline{\tau}}_t \circ T_t^{-1}\,dx_t + \frac{1}{4(\lambda+\mu)}\int_\Omega [(\text{tr}\,\underline{\underline{\tau}}_t)\circ T^{-1}]^2 dx_t\right. \\
&\quad \left. + \int_\Omega \text{div}\,\underline{\underline{\tau}}_t \circ T_t^{-1}\cdot \underline{v}\circ T_t^{-1} + f\circ T_t^{-1}\cdot \underline{v}\,dx_t\right\}.
\end{aligned}$$

$$\int_{\Omega_t} \underline{\underline{\tau}}_t \circ T_t^{-1} : \underline{\underline{\tau}}_t \circ T_t^{-1}\,dx_t = \sum_{K\in\mathcal{T}_h}\int_K I_t^K (\underline{\underline{DT}}_t^K)^\top (\underline{\underline{DT}}_t^K)\underline{\underline{\tau}}^K (\underline{\underline{DT}}_t^K)^\top (\underline{\underline{DT}}_t^K) : \underline{\underline{\tau}}^K\,dx$$

$$\begin{aligned}
\int_{\Omega_t} (\text{tr}\,\underline{\underline{\tau}}_t)^2\,dx_t &= \int_{\Omega_t} (\underline{\underline{\tau}}_t : \underline{\underline{I}})^2\,dx_t, \text{ where } \underline{\underline{I}} = \begin{pmatrix} 1 & 0 \\ 0 & 1 \end{pmatrix} \\
&= \sum_{K\in\mathcal{T}_h}\int_K I_t^K \left(\underline{\underline{DT}}_t^K \underline{\underline{\tau}}^K (\underline{\underline{DT}}_t^K)^\top : \underline{\underline{I}}\right)^2 dx \\
&= \sum_{K\in\mathcal{T}_h}\int_K \left(\underline{\underline{\tau}}^K : (\underline{\underline{DT}}_t^K)^\top \underline{\underline{I}}(\underline{\underline{DT}}_t^K)\right)^2 I_t^K\,dx \\
&= \sum_{K\in\mathcal{T}_h}\int_K I_t^K \left((\underline{\underline{DT}}_t^K)^\top \underline{\underline{DT}}_t^K \underline{\underline{\tau}}^K : \underline{\underline{I}}\right)^2 dx.
\end{aligned}$$

Let

$$A^K(t) = I_t^K \left(\underline{\underline{DT}}_t^K\right)^\top \underline{\underline{DT}}_t^K.$$
$$B^K(t) = \left(\underline{\underline{DT}}_t^K\right)^\top \underline{\underline{DT}}_t^K.$$
$$C^K(t) = \sqrt{I_t^K} \left(\underline{\underline{DT}}_t^K\right)^\top \underline{\underline{DT}}_t^K.$$

Then

$$\begin{aligned} G(t, \underline{\underline{\tau}}_t, \underline{v}_t) = \sum_{K \in \mathcal{T}_h} \Big\{ \frac{1}{2} \Big\{ \frac{1}{2\mu} \int_K A^K(t) \underline{\underline{\tau}}^K B^K(t) : \underline{\underline{\tau}}\, dx \\ + \frac{1}{4(\lambda+\mu)} \int_K \left(\operatorname{tr}(C^K(t) \underline{\underline{\tau}}^K)\right)^2 dx \Big\} \\ + \int_K \operatorname{div} \underline{\underline{\tau}}^K \cdot v^K \, dx + \int_K f \circ T_t^{-1} (\underline{\underline{DT}}_t^K)^{-\top} \underline{v}^K \, dx \Big\} \Big\}. \end{aligned} \tag{4.9}$$

So if $\mathbb{H}_h$ and V_h are the respective approximations of $\mathbb{H}_s(\operatorname{div}; \Omega)$ and $(L^2(\Omega))^2$,

$$\begin{aligned} g'(0) = \inf_{\underline{\underline{\tau}} \in \mathbb{H}_h} \sup_{\underline{v} \in V_h} \Big\{ \frac{1}{2} \Big(\frac{1}{2\mu} \int_\Omega (A'(0) \underline{\underline{\tau}} + \underline{\underline{\tau}} B'(0)) : \underline{\underline{\tau}}\, dx \\ + \frac{1}{4(\lambda+\mu)} \int_\Omega \left(\operatorname{tr}(C'(0) \underline{\underline{\tau}})\right)^2 dx \\ - \int_\Omega \left(\nabla \underline{f} \cdot V(0) \underline{v}^K - \underline{f} \cdot {}^\top DV(0) \underline{v}^K \right) dx \Big\}, \end{aligned} \tag{4.10}$$

where

$$A'(0) = \operatorname{div} V(0) I_d + {}^\top DV(0) + DV(0).$$
$$B'(0) = {}^\top DV(0) + DV(0).$$
$$C'(0) = \frac{1}{2} \operatorname{div} V(0)\, I_d + {}^\top DV(0) + DV(0).$$

4.2 The $\psi - \omega$ biharmonic problem.

Consider the problem of the deflection of a thin clamped plate under a distributed loading f. The vertical deflection u is solution of the minimization problem

$$\inf_{v \in H_0^2(\Omega)} \frac{1}{2} \int_\Omega |\Delta v|^2 \, dx - \int_\Omega f v \, dx. \tag{4.11}$$

The unique solution is characterized by the boundary value problem

$$\begin{cases} -\Delta^2 \psi &= f \quad \text{in } \Omega, \\ \psi_{|\Gamma} &= 0, \\ \frac{\partial \psi}{\partial n}_{|\Gamma} &= 0. \end{cases} \tag{4.12}$$

The problem (4.11) can be transformed into a saddle point problem (cf. BREZZI–FORTIN [1] as follows:

$$\inf_{\mu\in L^2(\Omega)} \sup_{\phi\in H_0^1(\Omega)} \frac{1}{2}\int_\Omega |\mu|^2\,dx - \int_\Omega \nabla\mu\cdot\nabla\phi\,dx + \int_\Omega f\phi\,dx. \tag{4.13}$$

The solution (ψ,ω) of (4.13) is the solution of

$$\begin{cases} -\Delta\psi &= \omega, \\ \psi_{|\Gamma} &= 0, \\ -\Delta\omega &= f. \end{cases}$$

Let

$$G(\phi,\mu) = \frac{1}{2}\int_\Omega |\mu|^2\,dx - \int_\Omega \nabla\mu\cdot\nabla\phi\,dx + \int_\Omega f\phi\,dx$$

and let T_t be the transformation defined, in the classical way by (3.15).

Then

$$G(t,\phi_t,\mu_t) = \frac{1}{2}\int_\Omega I_t|\mu|^2\,dx - \int_\Omega A(t)\nabla\mu\cdot\nabla\phi\,dx + \int_\Omega f\circ T_t^{-1}\phi\, I_t\,dx,$$

where

$$A(t) = I_t\,{}^\top(DT_t^{-1})DT_t^{-1}.$$

So

$$g'(0) = \frac{1}{2}\int_\Omega \mu^2 \operatorname{div} V(0)\,dx - \int_\Omega A'(0)\nabla\mu\cdot\nabla\phi\,dx + \int_\Omega \operatorname{div}\big(fV(0)\big)\phi\,dx.$$

REFERENCES

F. BREZZI and M. FORTIN [1], *Mixed and Hybrid Finite Element Methods*, Springer-Verlag, New York, Berlin, Heidelberg, 1991.

R. CORREA and A. SEEGER [1], *Directional derivative of a minimax function*, Nonlinear Analysis, Theory, Methods and Applications **9** (1985), 13–22.

M.C. DELFOUR [1], *Shape derivatives and differentiability of Min Max*, Shape optimization and free boundaries (M.C. Delfour, ed.), Kluwer Academic Publishers, Dordrecht, Boston, London, 1992, pp. 35–111.

M.C. DELFOUR, G. PAYRE and J.P. ZOLÉSIO [1], *An optimal triangulation for second-order elliptic problems*, Comput. Methods Appl. Mech. Engrg. **50** (1985), 231–261.

M.C. DELFOUR and J.P. ZOLÉSIO [1], *Shape sensitivity analysis via MinMax differentiability*, SIAM J. Control Optim. **26** (1988), 834–862.

[**2**], *Velocity method and Lagrangian formulation for the computation of the shape Hessian*, SIAM J. Control Optim. **29** (1991), 1414–1442.

[**3**], *An optimal triangulation for second-order elliptic problems*, Comput. Methods Appl. Mech. Engrg. **50** (1985), 231–261.

P.A. RAVIART and J.M. THOMAS [1], *A mixed finite element method for 2nd order elliptic problems*, Mathematical Aspects of Finite Element Methods, Lecture Notes in Mathematics series, Vol. 606, Springer-Verlag, Berlin, Heidelberg, New York, Chichester, Tokyo, 1975.

J.M. THOMAS [1], *Sur l'analyse numérique des méthodes d'éléments finis hybrides et mixtes*, Université Pierre et Marie Curie, Paris, France, 1977.

J.P. ZOLÉSIO [1], *Les dérivées par rapport aux nœuds des triangulations et leur utilisation en identification de domaine*, Ann. Sci. Math. Québec **8** (1984), 97–120.

On a Geometrical Bang-Bang Principle for Some Compliance Problems

MICHEL C. DELFOUR
Centre de recherches mathématiques et
Département de mathématiques et de statistique,
Université de Montréal, C.P. 6128 Succ. Centre-ville,
Montréal, Québec, Canada, H3C 3J7

JEAN-PAUL ZOLÉSIO
Institut Non Linéaire de Nice
CNRS
1361 Route des lucioles
06904 Sophia Antipolis Cédex, France

ABSTRACT. In this paper we consider a variation of the optimal design problem for the compliance as studied by Céa and Malanowski in 1970. It consists in finding the optimal distribution of two materials in a fixed domain. Here the support of the force is applied only in the region occupied by the strong material. When this problem is relaxed to the closed convex hull of characteristic functions, it often leads to the occurence of a microstructure or a composite material. We complete the characterization of the solutions and establish the existence of optimal partitions into two measurable subdomains. Moreover we show that the set of solutions of the relaxed problem always contains measurable partitions. This seems to be a "geometrical bang-bang principle", where the extreme points of the convex set of measurable functions with values in $[0,1]$ are the characteristic functions of measurable sets. The numerical results on a specific example considerably differ with the ones for the original problem. Here the gradient of the solution is zero in the region occupied by the strong material. The techniques readily extend to higher order elliptic operators and to variational inequalities over closed convex sets.

1. INTRODUCTION

In 1970 CÉA-MALANOWSKI [1] considered a minimum compliance problem where the optimization variable was the distribution of two materials with different physical characteristics within a fixed domain D. In this paper we consider a variation of this problem where we assume that the force is only applied to the strong material. The behaviour of the solution is different from the one of the original problem, but the general techniques techniques are essentially the same. This is solved by characterizing the solutions of the problem relaxed to the closed convex hull of characteristic functions. Then we establish that any solution to the relaxed problem can be suitably modified to yield the characteristic function of a measurable domain which is also an optimal solution. This is a "geometrical bang-bang principle", where the extreme points of the convex set of measurable functions with values in $[0,1]$ are the characteristic functions of measurable sets. It parallels results in Optimal Control. A numerical example is presented to illustrate the results. It is to be compared with the computations made for the original problem in DELFOUR and ZOLÉSIO [1]. Here the gradient of the solution is zero in the region occupied by the strong material.

The research of the first author has been supported in part by a Killam fellowship from Canada Council, National Sciences and Engineering Research Council of Canada operating grant A-8730 and infrastructure grant INF-7939 and by a FCAR grant from the Ministère de l'Education du Québec.

2. The minimum compliance problem with variable support for the exterior body forces

2.1. Problem formulation in terms of characteristic functions.

Let $D \subset \mathbb{R}^N$ be a bounded domain with Lipschitzian boundary ∂D and let Ω be a subset of D. Assume that the domain D is partitioned into two subdomains $\Omega_1 = \Omega$ and $\Omega_2 = \complement_D \overline{\Omega}$ separated by a smooth boundary $\partial\Omega_1 \cap \partial\Omega_2$. Domain Ω_1 (resp. Ω_2) is made of a material characterized by a constant $k_1 > 0$ (resp. $k_2 > 0$). In practice k_1 will be the strong (resp. weak) material ($k_1 > k_2$). Let y be the solution of the following *transmission problem* where the support of the exterior body forces is in Ω_1

$$\text{(2.1)} \qquad \begin{cases} -k_1 \, \Delta y = f \ \text{ in } \ \Omega_1 \\ -k_2 \, \Delta y = 0 \ \text{ in } \ \Omega_2 \\ y = 0 \ \text{ on } \ \partial D \\ k_1 \dfrac{\partial y}{\partial n_1} + k_2 \dfrac{\partial y}{\partial n_2} = 0 \ \text{ on } \ \partial\Omega_1 \cap \partial\Omega_2, \end{cases}$$

where n_1 (resp. n_2) is the unit outward normal to Ω_1 (resp. Ω_2) and f is a given function in $L^2(D)$. Our objective is to maximize the equivalent of the *compliance*

$$\text{(2.2)} \qquad C(\Omega_1) = -\int_{\Omega_1} f y \, dx$$

over all domains Ω_1 in D such that

$$\text{(2.3)} \qquad \int_{\Omega_1} dx \geq \alpha$$

for some $\alpha > 0$. A mechanical interpretation of this problem would be the minimization of the total work of the external body force f. If k_1 is the strong material, the volume constraint means that there is a lower bound on the total volume of that material available for the design. At the limit when the weak material k_2 goes to zero, the domain Ω_2 would represent the holes in the domain D.

Let Ω be a measurable domain in $\mathbb{R}^N$ and let $\chi = \chi_\Omega$ of Ω be its characteristic function. We want to relax the shape optimization problem (2.1)-(2.3) to an optimization problem over the set of measurable characteristic functions

$$\text{(2.4)} \qquad X(D) = \left\{ \chi \in L^2(D) : \chi(x) \in \{0, 1\} \text{ a.e. in } D \right\},$$

in $L^2(D)$. Equivalently a function χ is the characteristic function of a domain in D if

$$\text{(2.5)} \qquad \chi(x)\,(1 - \chi(x)) = 0, \quad \text{a.e. in } D.$$

For $\chi \in X(D)$ consider the following variational problem:

$$\text{(2.6)} \qquad \begin{cases} \text{to find } y = y(\chi) \in H_0^1(D) \text{ such that} \\ \int_D [k_1\,\chi + k_2(1-\chi)]\,\nabla y \cdot \nabla\varphi\,dx = \int_D \chi\,f\varphi\,dx, \quad \forall \varphi \in H_0^1(D), \end{cases}$$

with the cost function

$$\text{(2.7)} \qquad J(\chi) = -\int_D \chi\,f\,y(\chi)\,dx$$

to be maximized over all $\chi \in X(D)$ such that

$$\text{(2.8)} \qquad \int_D \chi\,dx \geq \alpha.$$

2.2. Relaxation of the problem.

Many *shape optimization* problems can often be reformulated as optimization problems over the subset of equivalence classes of characteristic functions $X(D)$ in an appropriate function space. The set $X(D)$ is closed and bounded but not convex and we cannot use the weak compactness in $L^2(D)$. The weak compactness can be recovered by further *relaxing* the problem (2.6)-(2.8) from $X(D)$ to its closed convex hull

$$\text{(2.9)} \qquad \overline{\text{co}}\,X(D) = \{\chi \in L^2(D) : \chi(x) \in [0,1] \text{ a.e. in } D\}$$

which is a bounded closed convex set. The elements of $\overline{\text{co}}\,X(D)$ are not necessarily characteristic functions of a domain, that is, identity (2.5) is not necessarily verified. This new formulation allows the "mixing" of the points of Ω_1 and Ω_2 (the mixing of the materials 1 and 2) at the microscale. In general for $k_1 \neq k_2$ an optimal solution of the relaxed problem is not necessarily the characteristic function of a smooth domain Ω_1 in D and the regions Ω_1 and Ω_2 are not necessarily separated by a smooth boundary. The *optimal relaxed solution* in $\overline{\text{co}}\,X(D)$ will be a *composite material*, a *local composition* of the two materials, or some form of probabilistic density of the first material.

For the relaxed problem we have for all χ in $\overline{\text{co}}\,X(D)$

$$\min\{k_1, k_2\} \leq k_1\,\chi(x) + k_2(1-\chi(x)) \leq \max\{k_1, k_2\} \text{ a.e. in } D$$

and by the Lax-Milgram's theorem, the variational equation (2.6) still makes sense and has a unique solution $y(\chi)$ in $H_0^1(D)$ which coincides with the solution of the boundary value problem

$$\text{(2.10)} \qquad \begin{cases} -\text{div}\ \{[k_2 + \chi(k_1 - k_2)]\ \nabla y\} = \chi\,f \text{ in } \mathcal{D}'(D) \\ y = 0 \text{ on } \partial D. \end{cases}$$

To obtain the existence of maximizers χ in $\overline{\text{co}}\,X(D)$, we need a number of technical results. The property that for any sequence $\{\chi_n\}$ in $\overline{\text{co}}\,X(D)$

$$\text{(2.11)} \qquad \chi_n \rightharpoonup \chi \text{ in } L^2(D)\text{–weak} \quad \Rightarrow \quad \chi_n \rightharpoonup \chi \text{ in } L^\infty(D)\text{–weak}\star$$

follows from the following lemma.

Lemma 2.1. *Let D be a measurable subset of $\mathbb{R}^N$ with finite measure, K be a bounded subset of $\mathbb{R}$ and*

$$\mathcal{K} = \{k : D \to \mathbb{R} \,:\, k \text{ is measurable and } k(x) \in K \quad \text{a.e. in } D\}.$$

(i) For any p, $1 \le p < \infty$, and any sequence $\{k_n\} \subset \mathcal{K}$ the following statements are equivalent

a) $\{k_n\}$ converges in $L^\infty(D)$-weak$\star$,
b) $\{k_n\}$ converges in $L^p(D)$-weak,
c) $\{k_n\}$ converges in $\mathcal{D}(D)'$,

where $\mathcal{D}(D)'$ is the space of scalar distributions on D. D is assumed to be a bounded open domain with Lipschitzian boundary. For c) it is assumed that D is an open bounded domain with Lipschitzian boundary.
(ii) If $\{k_n\}$ is a convergent sequence in $L^p(D)$-strong for some p, $1 \le p < \infty$, then it is a convergent sequence in $L^p(D)$-strong for all p, $1 \le p < \infty$.
(iii) The above statements remain true when $\mathcal{K}$ is made up of mappings $k : D \to K$ for some bounded subset $K \subset \mathbb{R}^p$ and a finite integer $p \ge 1$.

If K is convex and compact, then the set $\mathcal{K}$ is closed for all topologies between the $L^\infty(D)$–weak$\star$ topology and the $\mathcal{D}(D)'$ topology.

It is natural to further relax the problem to $\overline{\text{co}}\, X(D)$ and later look at the eventual presence of characteristic functions in the set of solutions. We first establish an existence result in $\overline{\text{co}}\, X(D)$ for this specific problem. Rewrite the functional $J(\chi)$ as the minimum of the energy functional

$$E(\chi, \varphi) = \int_D [k_2 + \chi(k_1 - k_2)]\, \nabla\varphi \cdot \nabla\varphi - 2\,\chi\, f\, \varphi\, dx \tag{2.12}$$

over $H_0^1(D)$

$$J(\chi) = \min_{\varphi \in H_0^1(D)} E(\chi, \varphi).$$

As a result we have the min max problem:

$$\max_{\chi \in M} \min_{\varphi \in H_0^1(D)} E(\chi, \varphi),$$

where

$$M = \{\chi \in \overline{\text{co}}\, X(D) \,:\, \int_D \chi\, dx \ge \alpha\}.$$

For $\alpha = \text{m}(D)$, the maximizer is trivially equal to 1. Thus we assume that $\alpha < \text{m}(D)$.

It is technically advantageous to reformulate the problem by using a Lagrange multiplier $\lambda \ge 0$ to handle the volume constraint:

$$\max_{\chi \in \overline{\text{co}}\, X(D)} \min_{\substack{\varphi \in H_0^1(D) \\ \lambda \ge 0}} G(\chi, \varphi, \lambda) \tag{2.13}$$

where

$$G(\chi, \varphi, \lambda) = E(\chi, \varphi) + \lambda \left[\int_D \chi \, dx - \alpha \right]. \tag{2.14}$$

The existence of saddle points follows from a general result in EKELAND-TEMAM [1, Prop. 2.4, p. 164]. The set $\overline{\text{co}}\, X(D)$ is a nonempty bounded closed convex subset of $L^2(D)$ and the set $H_0^1(D) \times [\mathbb{R}^+ \cup \{0\}]$ is trivially closed and convex. The functional G is concave–convex with the properties:

$$\begin{cases} \forall \chi \in \overline{\text{co}}\, X(D), \\ (\varphi, \lambda) \mapsto G(\chi, \varphi, \lambda) \text{ is convex, continuous, and} \\ \exists \chi_0 \in \overline{\text{co}}\, X(D) \text{ such that } \lim\limits_{\|\varphi\|_{H_0^1} + |\lambda| \to \infty} G(\chi_0, \varphi, \lambda) \to \infty \end{cases} \tag{2.15}$$

$$\begin{cases} \forall \varphi \in H_0^1(D),\ \forall \lambda \geq 0 \\ \chi \mapsto G(\chi, \varphi, \lambda) \text{ is affine and continuous for } L^2(D) - \text{strong}. \end{cases} \tag{2.16}$$

For the first condition and $0 \leq \alpha < \text{m}(D)$ pick $\chi_0(x) = 1$. To check the second condition pick any sequence $\{\chi_n\}$ in $\overline{\text{co}}\, X(D)$ which converges to some χ in $L^2(D)$-strong. Then the sequence also converges in $L^2(D)$-weak and by Lemma 2.1 in $L^\infty(D)$-weak. Hence $G(\chi_n, \varphi, \lambda)$ converges to $G(\chi, \varphi, \lambda)$.

Moreover the set of saddle points $(\hat{\chi}, y, \hat{\lambda})$ is of the form $X \times Y \subset \overline{\text{co}}\, X(D) \times [H_0^1(D) \times [\mathbb{R}^+ \cup \{0\}]$ and is completely characterized by the following variational equation and inequalities (cf. EKELAND-TEMAM [1, Proposition 1.6, p. 157]):

(2.17)

$$\int_D \{[k_2 + (k_1 - k_2)\hat{\chi}] \nabla y \cdot \nabla \varphi - \hat{\chi} f \varphi\} \, dx = 0, \quad \forall \varphi \in H_0^1(D)$$

(2.18)

$$\int_D \left[(k_1 - k_2)|\nabla y|^2 - 2 f y + \hat{\lambda} \right] (\chi - \hat{\chi}) \, dx \leq 0, \quad \forall \chi \in \overline{\text{co}}\, X(D)$$

(2.19)

$$\left[\int_D \hat{\chi} \, dx - \alpha \right] \hat{\lambda} = 0, \quad \int_D \hat{\chi} \, dx - \alpha \geq 0, \quad \hat{\lambda} \geq 0.$$

It is readily seen that for each $\chi \in \overline{\text{co}}\, X(D)$ there exists a unique $y(\chi)$ solution of (2.17) and then

$$\forall \hat{\chi} \in X, \quad \{\hat{\chi}\} \times Y \subset \{\hat{\chi}\} \times [\{y(\hat{\chi}) \times \{\lambda : \lambda \geq 0\}].$$

Therefore $y(\hat{\chi})$ is independent of $\hat{\chi} \in X$, that is $Y = \{y\} \times \Lambda$ and

$$\forall \hat{\chi} \in X, \forall \hat{\lambda} \in \Lambda, \quad y(\hat{\chi}) = y.$$

So we have uniqueness of the y in $H_0^1(D)$.

For each $\hat{\lambda}$, inequality (2.18) is completely equivalent to the following characterization of the maximizer $\hat{\chi} \in X$

$$\hat{\chi}(x) = \begin{cases} 1, & \text{if} \quad (k_1 - k_2)|\nabla y|^2 - 2\,f\,y + \hat{\lambda} > 0 \\ \in [0,1], & \text{if} \quad (k_1 - k_2)|\nabla y|^2 - 2\,f\,y + \hat{\lambda} = 0 \\ 0, & \text{if} \quad (k_1 - k_2)|\nabla y|^2 - 2\,f\,y + \hat{\lambda} < 0. \end{cases} \tag{2.20}$$

Associate with $\lambda \geq 0$ the sets

$$\begin{aligned} D_+(\lambda) &= \{x \in D : (k_1 - k_2)|\nabla y|^2 - 2\,f\,y + \lambda > 0\}, \\ D_0(\lambda) &= \{x \in D : (k_1 - k_2)|\nabla y|^2 - 2\,f\,y + \lambda = 0\}, \\ D_-(\lambda) &= \{x \in D : (k_1 - k_2)|\nabla y|^2 - 2\,f\,y + \lambda < 0\}. \end{aligned} \tag{2.21}$$

In order to complete the characterization of the optimal triplet, we need the following general results.

Lemma 2.2. *Consider a functional*

$$G : X \times Y \to \mathbb{R}$$

for some sets X and Y. Define

$$\begin{aligned} g &= \inf_{x \in X} \sup_{y \in Y} G(x,y) \\ X_0 &= \left\{ x \in X : \sup_{y \in Y} G(x,y) = g \right\}, \\ h &= \sup_{y \in Y} \inf_{x \in X} G(x,y) \\ Y_0 &= \left\{ y \in Y : \inf_{x \in X} G(x,y) = h \right\}. \end{aligned}$$

When $g = h$ the set of saddle points (possibly empty) will be denoted by

$$S = \{(x,y) \in X \times Y : g = G(x,y) = h\}.$$

Then

(i) *In general $h \leq g$ and*

$$\forall (x_0, y_0) \in X_0 \times Y_0, \quad h \leq G(x_0, y_0) \leq g.$$

(ii) *If $h = g$, then*

$$S = X_0 \times Y_0.$$

Associate with an arbitrary solution $(\hat{\chi}, y, \hat{\lambda})$, the characteristic function

$$\chi_{\hat{\lambda}} = \begin{cases} \chi_{D_+(\hat{\lambda})} & D_+(\hat{\lambda}) \neq \varnothing \\ 0, & D_+(\hat{\lambda}) = \varnothing. \end{cases} \tag{2.22}$$

Then from (2.20)

$$\hat{\chi}\left\{(k_1 - k_2)|\nabla y|^2 - 2\,fy + \hat{\lambda}\right\}$$
$$= \chi_{\hat{\lambda}}\left\{(k_1 - k_2)|\nabla y|^2 - 2\,fy + \hat{\lambda}\right\} \quad \text{a.e. in } D$$

and

$$\begin{aligned} G(\hat{\chi}, y, \hat{\lambda}) &= \int_D [\hat{\chi}k_1 + (1-\hat{\chi})k_2]\,|\nabla y|^2 - 2\hat{\chi}fy\,dx + \hat{\lambda}\left[\int_D \hat{\chi}\,dx - \alpha\right] \\ &= \int_D k_2|\nabla y|^2 + \hat{\chi}\left[(k_1 - k_2)|\nabla y|^2 - 2fy + \hat{\lambda}\right]\,dx - \hat{\lambda}\alpha \\ &= \int_D k_2|\nabla y|^2 + \chi_{\hat{\lambda}}\left[(k_1 - k_2)|\nabla y|^2 - 2fy + \hat{\lambda}\right]\,dx - \hat{\lambda}\alpha \\ &= G(\chi_{\hat{\lambda}}, y, \hat{\lambda}). \end{aligned}$$

From Lemma 2.2, $(\chi_{\hat{\lambda}}, y, \hat{\lambda})$ is also a saddle point. So there exists a maximizer $\chi_{\hat{\lambda}} \in X(D)$ which is a characteristic function and necessarily

$$\max_{X(D)} \min_{\substack{\varphi \in H_0^1(D) \\ \lambda \geq 0}} G(\chi, \varphi, \lambda) = \max_{\overline{\text{co}}\,X(D)} \min_{\substack{\varphi \in H_0^1(D) \\ \lambda \geq 0}} G(\chi, \varphi, \lambda).$$

But we can show more than that. If there exists $\hat{\lambda} \in \Lambda$ such that $\hat{\lambda} > 0$, then by construction $\hat{\chi} \geq \chi_{\hat{\lambda}}$ and by (2.19)

$$\alpha = \int_D \hat{\chi}\,dx \geq \int_D \chi_{\hat{\lambda}}\,dx = \alpha.$$

Since $\hat{\chi} = \chi_{\hat{\lambda}} = 1$ a.e. in $D_+(\hat{\lambda})$, then $\hat{\chi} = \chi_{\hat{\lambda}}$. But by construction $\chi_{\hat{\lambda}}$ is independent of $\hat{\chi}$. Thus the maximizer is unique, a characteristic function, and its integral is equal to α.

The case $\Lambda = \{0\}$ is a degenerate one. Set $\varphi = y$ in (2.17) and regroup the terms as follows

$$0 = \int_D [k_2 + (k_1 - k_2)\hat{\chi}]|\nabla y|^2 - f\,y\,\hat{\chi}\,dx$$
$$\int_D [k_2 + \frac{k_1 - k_2}{2}\hat{\chi}]|\nabla y|^2\,dx + \int_D \left\{\frac{k_1 - k_2}{2}\hat{\chi}]|\nabla y|^2 - f\,y\right\}\hat{\chi}\,dx.$$

The integrand of the first integral is positive. From the characterization of $\hat{\chi}$, the integrand of the second one is also positive. Hence they are both zero almost everywhere in D. As a result

$$\mathrm{m}(D_+(0)) = 0 \quad \text{and} \quad \nabla y = 0.$$

Therefore $y = 0$ in D. In particular $D_+(0) = \varnothing$ and from our previous considerations $\chi_\varnothing$, that is $\hat{\chi} = 0$, is a solution. But this can only happen when $\alpha = 0$ since $\int_D \hat{\chi}\, dx \geq \alpha > 0$. So for $\alpha > 0$, the case $\hat{\lambda} = 0$ cannot occur.

In conclusion in all cases, there exists a (unique for $\alpha > 0$) maximizer in $\overline{\mathrm{co}}\, X(D)$ and this maximizer can be chosen (is for $\alpha > 0$) a characteristic function. Therefore

$$\max_{\substack{\chi \in X(D) \\ \int_D \chi\, dx \geq \alpha}} \min_{\varphi \in H_0^1(D)} E(\chi, \varphi) = \max_{\substack{\chi \in \overline{\mathrm{co}}\, X(D) \\ \int_D \chi\, dx \geq \alpha}} \min_{\varphi \in H_0^1(D)} E(\chi, \varphi).$$

Moreover we can also choose that characteristic function such that its integral over D is equal to α. Thus in addition

$$\max_{\substack{\chi \in \overline{\mathrm{co}}\, X(D) \\ \int_D \chi\, dx \geq \alpha}} \min_{\varphi \in H_0^1(D)} E(\chi, \varphi) = \max_{\substack{\chi \in X(D) \\ \int_D \chi\, dx = \alpha}} \min_{\varphi \in H_0^1(D)} E(\chi, \varphi).$$

2.3. Some numerical results.

Here we have considered equation (2.1) over the diamond shaped domain

$$D = \{(x, y) : |x| + |y| < 1\}$$

with $k_1 = 2$ (black in Figures 2.2a and 2.3a), $k_2 = 1$ (white on the figures) and

$$f(x, y) = 56\,(1 - |x| - |y|)^6. \tag{2.24}$$

This function has a sharp peak in $(0, 0)$. Figures 2.2 and 2.3 show the optimal partition as the grid size is reduced. For this example the Lagrange multiplier associated with the problem is strictly positive. The gray zones correspond to the region $D_0(\hat{\lambda}_m)$ where $\hat{\chi} \in [0, 1]$. The presence of this gray zone in the approximated problem is due to the fact that equality for the total area where $\hat{\chi} = 1$ could not be exactly achieved with the chosen triangulation of the domain. Thus the problem had to adjust the value of $\hat{\chi}$ between 0 and 1 in a few triangles in order to achieve equality for the integral of $\hat{\chi}$.

The boundary value problem was approximated by continuous piecewise linear finite elements on each triangle and the function χ by a piecewise constant function on each triangle. The constant on each triangle was constrained to lie between 0 and 1 together with the global constraint on its integral over the whole domain D. The computations have been done on a 6-processor SGI using a standard optimization package which performs well for a large number of optimization variables.

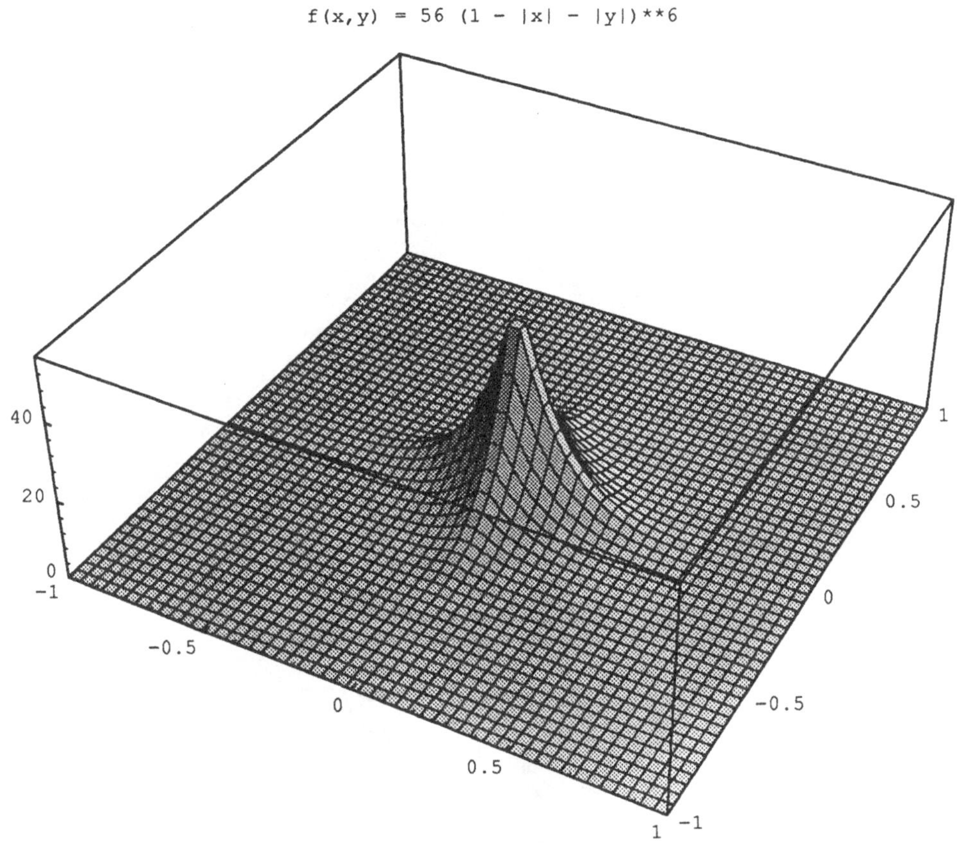

Figure 2.1. Function f

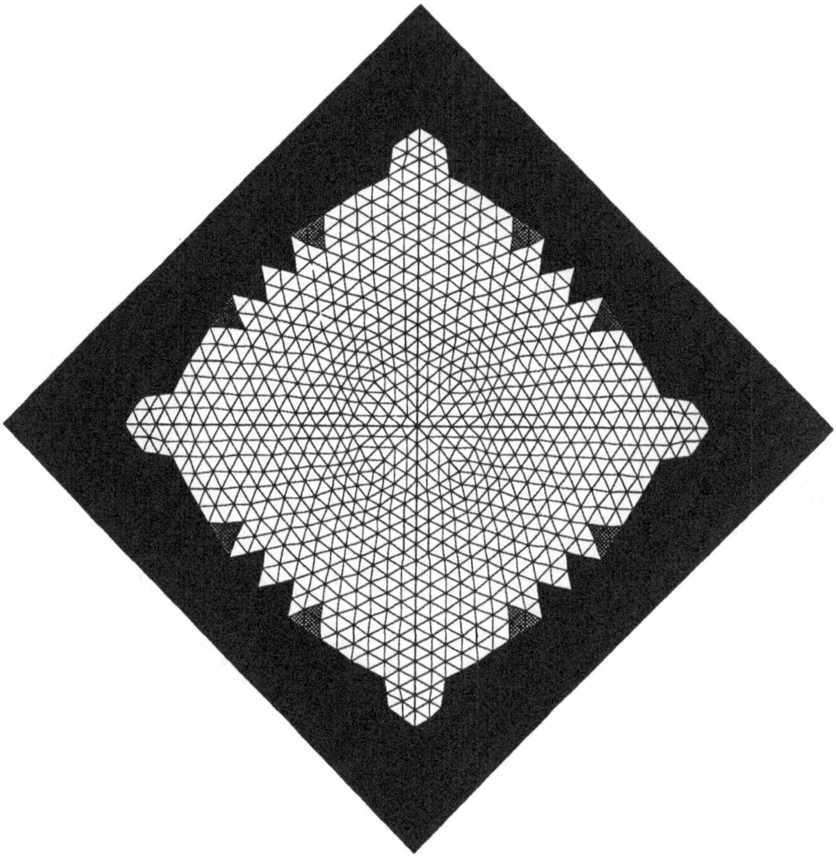

Figure 2.2a. Optimal distribution

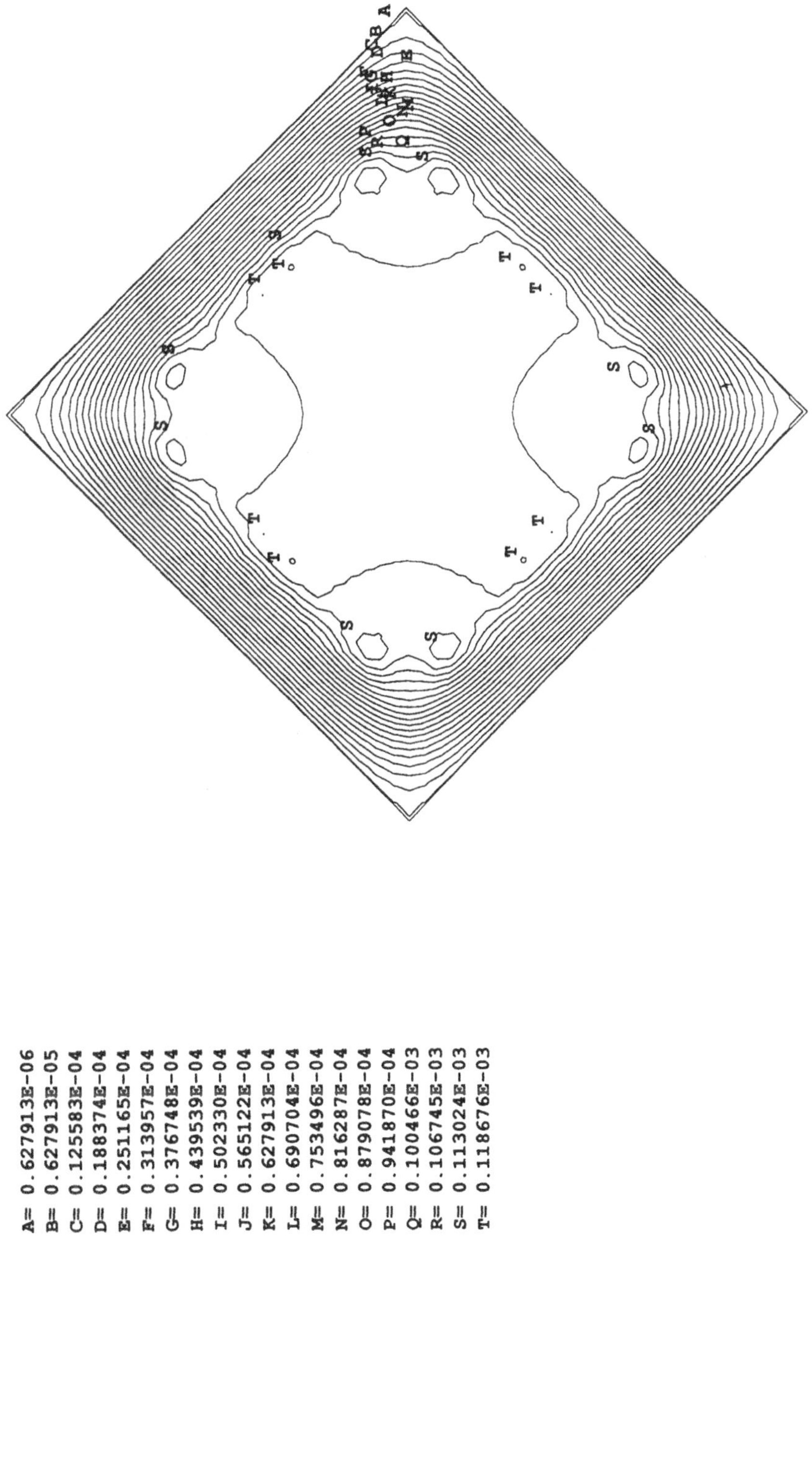
A= 0.627913E-06
B= 0.627913E-05
C= 0.125583E-04
D= 0.188374E-04
E= 0.251165E-04
F= 0.313957E-04
G= 0.376748E-04
H= 0.439539E-04
I= 0.502330E-04
J= 0.565122E-04
K= 0.627913E-04
L= 0.690704E-04
M= 0.753496E-04
N= 0.816287E-04
O= 0.879078E-04
P= 0.941870E-04
Q= 0.100466E-03
R= 0.106745E-03
S= 0.113024E-03
T= 0.118676E-03

Figure 2.2b. Level curves

Figure 2.3a. Optimal distribution

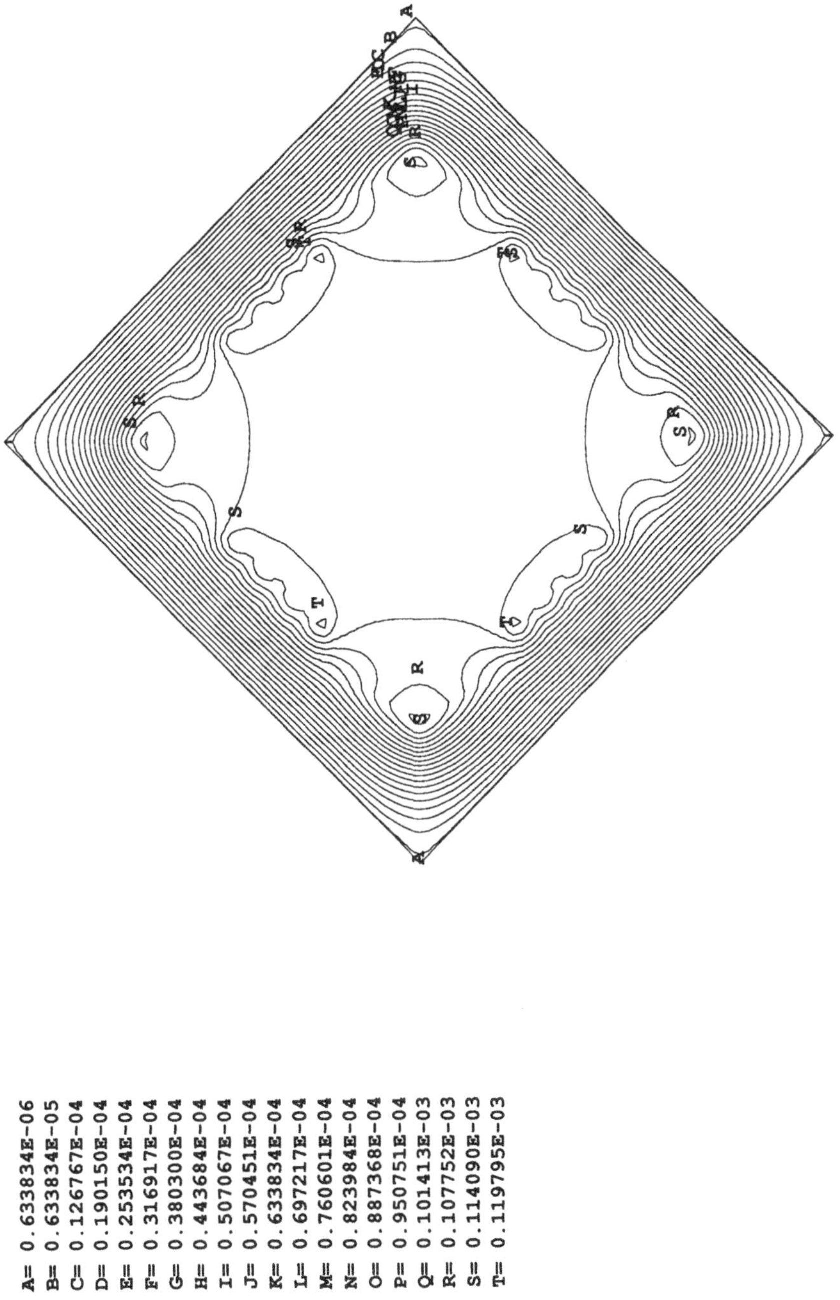
A= 0.633834E-06
B= 0.633834E-05
C= 0.126767E-04
D= 0.190150E-04
E= 0.253534E-04
F= 0.316917E-04
G= 0.380300E-04
H= 0.443684E-04
I= 0.507067E-04
J= 0.570451E-04
K= 0.633834E-04
L= 0.697217E-04
M= 0.760601E-04
N= 0.823984E-04
O= 0.887368E-04
P= 0.950751E-04
Q= 0.101413E-03
R= 0.107752E-03
S= 0.114090E-03
T= 0.119795E-03

Figure 2.3b. Level curves

3. Extensions and conclusions

In this paper we have only considered the simple elliptic equation (2.1). The theory readily extends without conceptual changes to more general elliptic operators of higher orders. It also naturally extends to variational inequalities over a closed convex set $C \subset V$, where V is a fixed Hilbert space of functions on D which are 0 on ∂D. For instance we can consider energy functionals of the general form

$$E(\chi, \varphi) = \int_D k_1 \, \chi \, a_1(x, \varphi, \varphi) + k_2 \, \chi \, a_2(x, \varphi, \varphi) - 2 \, f \, \varphi \, dx$$

where $a_1(x, \varphi, \psi)$ and $a_2(x, \varphi, \psi)$ are suitable continuous bilinear forms on V. Here the problem is of the new form

$$\max_{\chi \in M} \min_{\varphi \in C} E(\chi, \varphi),$$

but the results remain identical.

The same techniques also apply to the vectorial case such as the celebrated optimal thickness of a circular plate of K.-T. Cheng and N. Olhoff [1, 2]. For circular plates the thickness varies between upper bounds $h_{max} > 0$ and $-h_{max}$ and two lower bounds $h_{min} > 0$ and $-h_{min}$, $h_{max} \geq h_{min} > 0$, and the radii between $R_0 > 0$ and $R_1 > R_0$. Here the domain D is chosen as a rectangle

$$D = \{(r, z) \, : \, R_0 < r < R_1, \, -h_{max} < z < h_{max}\}.$$

The material 1 ocupies the region

$$D_{min} = \{(r, z) \, : \, R_0 < r < R_1, \, -h_{min} < z < h_{min}\}$$

and the characteristic function χ corresponds to material 1 in the region $D \backslash D_{min}$, while material 2 occupies the other part of that region. Recent theoretical and numerical results by Y. Hamdouch [1] give also a maximizer which is a characteristic function when the weak material 2 is non zero. When the parameters of the weak material are zero we get a function χ which now takes values between 0 and 1. This corresponds to the microstructure of Cheng and Olhoff. So the characterization of the maximizer χ can be done first and then Homogenization Theory can be used to interpret the meaning of the maximizer at points where it is different from 0 and 1. It is interesting to observe that the fact that $h_{min} > 0$ preserves the coercivity of the associated elliptic operator when the parameters of the weak material go to zero. This is not the case in the problem we have considered in this paper. However if we use $k_0 + k_1\chi + k_2(1 - \chi)$ for $k_0 > 0$ instead of $k_1\chi + k_2(1 - \chi)$ we recover the coercivity.

In Delfour and Zolésio [1] we considered a problem similar to the one in this paper. The difference was that the force f was applied everywhere in D and that the volume constraint was in the opposite direction

$$\int_D \chi \, dx \leq \alpha.$$

Those results can now be sharpened somewhat in the sense that there exists a (unique when $0 \leq \alpha < \mathrm{m}(D)$) maximizer which is a characteristic function.

The results presented are important from the numerical viewpoint. The relaxed problem can be solved and a solution $\hat{\chi} \in \overline{\mathrm{co}}\, M$ can be computed. But the theorem also says that if our computation yields a function $\hat{\chi}$ which is not a characteristic function, then it can always be modified to yield a characteristic function which does not change the state variable y and the value of $E(\hat{\chi}, y)$ at the saddle point.

References

J. Céa and K. Malanowski [1], *An example of a max-min problem in partial differential equations*, SIAM J. Control **8** (1970), 305–316.

K.-T. Cheng and N. Olhoff [1], *Regularized formulation for optimal design of axisymmetric plates*, Internat. J. Solids and Structures **18** (1982), 153–169.

[2], *An investigation concerning optimal design of solid elastic plates*, Internat. J. Solids and Structures **17** (1981), 305–323.

M.C. Delfour and J.P. Zolésio [1], *The optimal design problem of Céa and Malanovski revisited*, Optimal Design and Control (J. Borggaard, J. Burkardt, M. Gunzburger, and J. Peterson, eds.), Birkhäuser Boston Inc., Boston, Basel, Berlin, 1995, pp. 133–150.

[2], *Shape analysis via oriented distance functions*, J. Functional Analysis **123** (1984), 129–201.

I. Ekeland and R. Temam [1], *Analyse convexe et problèmes variationnels*, Dunod, Gauthier-Villars, Bordas, Paris, Bruxelles, Montréal, 1974.

Y. Hamdouch [1], *Optimisation de forme des plaques circulaires*, Mémoire de maîtrise, Université de Montréal, Montréal, mai 1996.

Shape Derivative for the Laplace–Beltrami Equation

F.R. DESAINT AND J.P. ZOLÉSIO

CNRS-Institut Non-Linéaire de Nice, Sophia Antipolis F 06560 Valbonne, France

Abstract. This paper is concerned with the shape tangential sensitivity analysis of the solution to the Laplace-Beltrami boundary value problem with homogeneous Dirichlet boundary conditions. The domain is an open subset ω of a smooth compact manifold of $I\!R^N$. The Speed Method approach is used: the flow transformation of a vector field $V(t,.)$ changes that open subset in ω_t; in general that perturbed set is no more a subset of Γ. The relative boundary γ of ω is smooth enough and $y(\omega_t)$ is the solution in ω_t of the Dirichlet problem with zero boundary value on γ_t. The shape tangential derivative is characterized as being the solution of a similar boundary value problem; that element $y'_\Gamma(\omega; V)$ can be simply defined by the restriction to ω of $\dot{y} - \nabla_\Gamma y.V$ where $\dot{y}$ is the material derivative of y. The study splits in two parts whether the relative boundary γ of ω is empty or not. In both cases the shape derivative depends on the deviatoric part of the second fundamental form of the surface, on the field $V(0)$ through its normal component on ω and on the tangential field $V(0)_\Gamma$ through its normal component on γ.

1. Introduction

1.1. Shape Tangential Derivative. We consider a domain Ω in $I\!R^N$ and its boundary Γ which is a smooth compact manifold with empty boundary and is oriented by the unitary normal field n outgoing to Ω. The solution $y(\Gamma)$ to the elliptic problem associated with the Laplace-Beltrami operator on Γ belongs to the Sobolev space $H^1_*(\Gamma)$ of functions in $H^1(\Gamma)$ with zero mean. For any subset ω of Γ, open in Γ, we denote by γ its relative boundary in Γ and by ν the unitary normal field to γ contained in the linear tangent space to Γ and outgoing to ω. The solution $z(\omega)$ to the Dirichlet Boundary value problem associated with the Laplace-Beltrami operator on ω and Dirichlet homogeneous conditions on γ is in the Sobolev space $H^1_0(\omega)$. We consider a family of perturbed manifolds Γ_t in the form $\Gamma_t = T_t(\Gamma)$ where T_t is a smooth one parameter family of transformations of $I\!R^N$, also we set $\omega_t = T_t(\omega)$ and we consider the perturbed solutions $y(\Gamma_t)$ and $z(\omega_t)$ associated with the perturbed boundary value problems. We emphasis the fact that any such family of transformations can be regarded as being the flow mapping of its "speed vector field" $V(t,x) = \frac{\partial}{\partial t}T_t o(T_t)^{-1}$ and we chose that field V as being the deformation parameter. Following [6], [1] we extend the *shape boundary derivative* of $y(\Gamma)$ at Γ, in the direction V, is defined by $y'_\Gamma(\Gamma, V) = (\frac{d}{dt}(y(\Gamma_t)oT_t(V)))|_{t=0} - \nabla_\Gamma y(\Gamma).V_\Gamma(0)$ We extend that notion to the element $z(\omega)$, introducing here the concept of *shape tangential derivative* of $z(\omega)$ at ω, in the direction of the field V through the shape boundary derivative of the extension by zero to the surface Γ of the element $z(\omega)$. These concepts will be developped in the next section and we shall see in which sense these elements are derivatives. In both situations the shape tangential derivative exists and $z'(\omega; V)$, $y'(\Gamma; V)$ are linear on V. In [1] only the case where γ is empty was considered (for different and easier problems), in that situation no tangential variation is involved in the problem as there is no relative boundary then the terminology was " shape boundary derivative" . The present concept of shape tangential derivative contains as specific case that shape boundary derivative concept but is really a generalization of the *shape derivative* (developped for open domains in $I\!R^N$, see 1.3 bellow) as it characterizes the variation of the solution with respect to the boundary on which a boundary condition is imposed. Following the structure's theorem, cf [6], [8], in the generalized form we establish in the next section, these expressions only depend on the normal component $< V(0), n >$ on ω of the vector field $V(0)$ and on the tangential field $V_\Gamma = V - < V, n > n$ through its normal component $< V_\Gamma(0), \nu >$ on γ. We make use of intrinsic geometry approach and without use of any local coordinates we give a complete analysis of these shape derivatives from which it appears that the shape sensitivity analysis of the solution in both cases deeply depends on the deviatoric part of the second fundamental form of the surface Γ. We underline that the use of flow transformations is the appropriate tool for these tangential problems. An important issue is when the surface does not move but only the domain ω_t moves in Γ. That situation could not be treated by the usual transformations in the form $I_d + \theta$ while in the present setting it suffices to specify the results to tangential vector fields V. Important cases in applications are concerned with constrained families of manifolds and of subdomains. These constraints can be in the form: volume of (Ω_t) and surface of (ω_t) given. Of course there are no obvious manifold structures on such families of manifolds and of subsets. Nevertheless by the choice of vector fields V verifying the linear conditions, $divV(t,.) = 0$ and $div_\Gamma V(t,.) = 0$, these families are stable under the flow transformations and then we can define a candidate for tangential linear space and differential calculus. The shape derivative has to be compared in the usual Banach space setting to the so-called Hadamard semi-derivatives. The same properties occur: that derivative is stable under composition with a smooth enough mapping; when it is linear and continuous on the direction, there exists a linear operator. If that operator continuously depends on the domain (which will be the case for $z'_\Gamma(\omega; V)$) then we get a Frechet derivative when the family of domains is locally isomorphic to an open subset in a Banach space (which is not true for general constrained situations as the three previous examples).

1.2. Material Derivative. The shape sensitivity analysis of eingenvalues has been initiated by J. Hadamard who gave a formal proof of the so called "Hadamard shape derivative" following which the derivative of the first eigenvalue of the usual Dirichlet problem in a bounded smooth open domain of $I\!R^2$ was obtained as an integral on the boundary of the domain of the square of the eigenvalue's normal derivative multiplied by the normal variation of the boundary. It had been shown in [5], [6], [1] that this structure is the general form for any shape derivative of any shape functional. In [3] that structure is also derived for non smooth shape analysis. At this point we must underline here that in all that abondant literature only shape sensitivity analysis of scalar functions was considered, functions such as eigenvalues, ranging in $I\!R$. For such scalar functions depending on a moving domain we reserve classically (from control theory terminology) the name of shape functional. In the present context we are not concerned with scalar functions but with "state variable" (again following the control theory terminology), i.e. with elements such as $z(\omega)$ ranging in a variable Hilbert space depending on the domain ω. Concerning state variable shape sensitivity analysis, the study has been mostly concentrated on the material derivative, see many examples in [1]. That material derivative is usefull but cannot be considered as the "definitive" concept for several reasons, for example the fact that for some transformations which do not move the domain ("do not change the shape") the material derivative is not zero.

1.3. The Shape Derivative. In the classical situation of a moving open domain ω in $I\!R^N$ the situation is easier for the usual Dirichlet problem as the element $z(\omega)$ is obviously extended by zero to an element of some fixed Hilbert space $H_0^1(D)$ where D is any domain containing the moving family of open domains. For general boundary value problems it has been proved in [5], [6] and [8] that as soon as the material derivative $\dot{z}(\omega;V)$ exists say in $H^1(\omega)$ then there exists elements $Z \in C^1([0,\tau[, H^1(D))$ such that $Z(t,.)|_{\omega_t} = z(\omega_t)$. Moreover we have the property that $\frac{\partial}{\partial t}Z(0,.)|_\omega$ is independant on the element Z. That value is then, by definition, the shape derivative $z'(\omega;V)$ of the state variable z at the domain ω in the direction V. The structure's result for the shape boundary derivative states that as soon as the material derivative exists then the shape boundary derivative exists and is given by $\frac{\partial}{\partial t}Z(0,.)|_\omega = \dot{z}|_\omega - [\nabla z.V(0)]|_{|\omega}$. As we adopt the general setting of shape deformation via Speed vector fields, the generalized structure's theorem states that the Eulerian derivative of a shape functional, when it smoothly and linearly depends on the deformation field V (as it is the case for all examples in [2] arising from elasticity, plasticity, fluids dynamics, hyperbolic boundary value problems...) only depends on V through the normal component of the field $V(0)$ on the boundary of the domain. That structure, generalized in the next section for the shape tangential derivative of the state variable, is also true as soon as the material derivative linearly depends on the field V. It turns out that on several examples in which the material derivative is not linear on the field V that structure still occurs. It is the case for example for the solution of unilateral boundary value problem such as contact problems, see [2]. In the tangential situation we consider here, that structure for the shape boundary derivative is preserved in a generalized form as the derivative will depend, again on the vector field $V_\Gamma(0)$ through its normal component $< V_\Gamma(0), \nu >$ on the relative boundary γ but also, at points where the curvature is not zero, on the component $< V(0), n >$. In order to derive these tangential calculus without use of any local coordinates, we make use of the intrinsic calculus developped in [9], [3], [7] and intensive use of the oriented distance function b_Ω. For technical reasons it is usefull to consider that there exists a bounded open domain D containing the family (Γ_t). The right hand side of the tangential problem on Γ_t will be associated with a given element F in $H^s(D)$, $s > \frac{1}{2}$. The Laplace-Beltrami operator on Γ_t will be denoted by Δ_{Γ_t} and z_{Γ_t} will denote the solution of the problem $-\Delta_{\Gamma_t} z_t = f_t$ on Γ_t , $z_t = 0$ on ω, where f_t is the trace of F on Γ_t : $f_t = F|_{\Gamma_t}$. The previous problem is well posed when the relative smooth boundary γ is not empty. In the situation where ω has no boundary in Γ this problem is still well posed without any boundary condition but up to an additive constant for the solution and under the condition that the right hand side f_t has a zero mean value on the surface Γ_t. Now the trace of F on the surface Γ_t cannot be with zero mean value on that moving surface, so the definition of the right hand side f_t should be changed. In order to get a simple example we chose $f_t = F|_{\Gamma_t} - |\Gamma_t|^{-1} \int_{\Gamma_t} F d\Gamma_t$ and then when γ "collapses" in Γ the elliptic problem turns to be $-\Delta_{\Gamma_t} y_t = f_t = F|_{\Gamma_t} - |\Gamma_t|^{-1} \int_{\Gamma_t} F d\Gamma_t$ on Γ_t.

2. Shape Tangential Derivative

2.1. Tangential Calculus. Let D be a smooth domain in $I\!R^N$ and Γ an oriented C^2 compact manifold included in the hold all D. For simplicity we assume that Γ is the boundary of a smooth domain $\Omega \subset D$. Then Γ is a compact manifold of codimension 1 in $I\!R^N$, without boundary. We consider an open subset ω of Γ. Its relative boundary in Γ will be denoted by γ; γ is a manifold of codimension 2 in $I\!R^N$ and is also assumed to be of class C^2. In this context, we shall study the shape sensitivity analysis of the solution of the boundary value problem associated with the Laplace-Beltrami operator respectively in ω with homogeneous Dirichlet boundary conditions on γ and in the whole surface Γ without any boundary condi-

tion. The Laplace-Beltrami operator is defined in [6], [1], [9], [4] by use of intrinsic geometry and without any local coordinates f: the tangential gradient of a scalar function f defined over Γ, $\nabla_\Gamma f$, is the restriction to Γ of $\nabla F - < \nabla F, \nabla b_\Omega > \nabla b_\Omega$, where F is any smooth extension of f to a neighbourhood of Γ and where b_Ω is the oriented distance function to the open set Ω defined by $b_\Omega = d_\Omega$ in $D - \Omega$ and $b_\Omega = -d_{D-\Omega}$ in Ω, ∇b_Ω being an extension to a neighbourhood of Γ of the normal field on Γ. The tangential divergence of a field e defined on Γ is given by the restriction to Γ of $divE - < DE.\nabla b_\Omega , \nabla b_\Omega >$, where E is any smooth extension of e to the neighbourhood of Γ. Finally the Laplace-Beltrami operator $\Delta_\Gamma \varphi$ is defined by the usual chain rule $div_\Gamma(\nabla_\Gamma \varphi)$. The Sobolev spaces associated with the two boundary value problems are respectively $H^1_o(\omega)$ and $H^1_*(\Gamma)$ defined as follows:
$H^1_*(\Gamma) = \{\varphi \in H^1(\Gamma) / \int_\Gamma \varphi d\Gamma = 0\}$ and $H^1_0(\omega) = \{\varphi \in H^1(\omega) / \varphi = 0 \text{ on } \gamma \}$.
The state variables are defined by:

$$(1) \qquad y(\Gamma) \in H^1_*(\Gamma), \text{verifying } -\Delta_\Gamma y(\Gamma) = f \text{ on } \Gamma.$$

$$(2) \qquad z(\omega) \in H^1_0(\omega), \text{verifying } z(\omega) = 0 \text{ on } \gamma \; -\Delta_\Gamma z(\omega) = f \text{ in } \omega.$$

While the problem (1) is a specific situation of (2) we shall be obliged in that situation to distinguish the case where ω is empty (or in fact with zero capacity). In that case the usual Poincaré inequality does not exist and the solution will be defined up to an additive constant as soon as the right hand side f will be with zero mean value over Γ. We make use of the usual identification of the quotient Hilbert space $\frac{H^1(\Gamma)}{R}$ with $H^1_*(\Gamma)$. Then when the relative smooth boundary ω "collapses " in Γ that space should be changed for $H^1_0(\omega)$ in that study.

2.2. Flow Mapping. A smooth enough non autonomeous field V is given,
$V \in C^0([0,\tau[, C^k(\bar{D}; I\!R^N))$ with $V.\vec{n}_D = 0$ on ∂D, $\vec{n}_D$ being the unitary normal field on ∂D. We call $T_t(V)$ the flow transformation associated with the field V, $T_t(x) = x + \int_0^t V(\sigma, T_\sigma(V)(x))d\sigma$. From the condition on the boundary of D we get that the boundary ∂D is globally invariant under the transformation $T_t(V)$. That transformation, we shall simply denote by T_t, maps D onto itself. The inverse mapping is itself the flow mapping associated with the vector field V^t defined by $V^t(s,x) = -V(t-s,x)$. We get $T_t \in C^1([0,\tau], C^2(D, I\!R^N))$ and $(T_t)^{-1} \in C^0([0,\tau], C^k(D, I\!R^N))$. The usual semi group property holds: $T_{t+s}(V) = T_s(V_t)oT_t(V)$ where V_t is the field $V_t(\sigma, .) = V(t+\sigma, .)$. In case of autonomeous vector field V the fields V^t, V_t coincide and we have a group. In all situations the sensitivity analysis through the semi group property is done at $t = 0$. If the field V satisfies $< V(t,.), n(.) >= 0$ on Γ then the surface Γ is globally invariant under the transformation T_t. We consider the one parameter family of surfaces $\Gamma_t = T_t(V)(\Gamma)$ with $\Gamma = \Gamma_0$, and the one parameter family of subsets $\omega_t = T_t(V)(\omega) \subset \Gamma_t$. Notice that Γ_t is also the C^k boundary of Ω_t, where $\Omega_t = T_t(V)(\Omega)$. Both Γ_t and Ω_t are defined with the transformation T_t : $T_t(\Gamma) = \Gamma_t$ is a C^k surface in D, being the boundary of the bounded open domain $\Omega_t = T_t(\Omega)$. We also have $\omega_t = T_t(\omega)$ and $\gamma_t = T_t(\gamma)$.

Let Γ be a manifold with regularity C^2, we assume that $\Gamma = \partial\Omega$, where Ω is a bounded domain in $I\!R^N$. n is the unitary normal field on Γ, outgoing to Ω. There exists a neighbourhood $\mathcal{U}$ of Γ in $I\!R^N$ such that the projection mapping P onto Γ is well defined; then for any element φ in $C^1(\Gamma; I\!R)$ the tangential gradient $\nabla_\Gamma \varphi$ is the restriction to Γ of $\nabla(\varphi o P)$ (see [1], [7]). Similarly the tangential divergence of a smooth tangent field E, $E \in C^1(\Gamma; I\!R^N)$ $E(x).n(x) = 0$ is defined by $div_\Gamma E = div(EoP)|_\Gamma$ (the restriction to Γ of $div(EoP)$). For a given element

φ in $L^2(\Gamma)$, the tangential gradient $\nabla_\Gamma\varphi$ is the distribution on Γ defined by $\left\langle \nabla_\Gamma\varphi, E\right\rangle = - \int_\Gamma \varphi div_\Gamma(E) d\Gamma$ where $\left\langle , \right\rangle$ is the duality brackets between $\mathcal{D}'(\Gamma; T\Gamma)$ and $\mathcal{D}(\Gamma; T\Gamma)$ where $T\Gamma$ is the tangent fiber bundle ($E \in \mathcal{D}(\Gamma; \mathbb{R}^N)$ and $E.n = 0$, i.e. $E(x)$ is in the linear tangent space $T_x\Gamma$ to Γ at x). We consider the Hilbert spaces: $L^2_*(\Gamma) = \{\varphi \in L^2(\Gamma)/\bar{\varphi} = \int_\Gamma \varphi d\Gamma = 0\}$ and $H^1_*(\Gamma) = \{\varphi \in L^2_*(\Gamma)/\nabla_\Gamma\varphi \in L^2(\Gamma; T\Gamma)\}$ The Laplace Beltrami operator is defined by $\Delta_\Gamma\varphi = div_\Gamma(\nabla_\Gamma\varphi)$ or similarly $\Delta_\Gamma\varphi = (\Delta(\varphi oP))|_\Gamma$ (the restriction to Γ of $\Delta(\varphi oP)$, Δ being the Laplace operator in $\mathbb{R}^N$).

For any g in $L^2_*(\Gamma)$, the problem $-\Delta_\Gamma y = g$ on Γ possesses a unique solution y in $H^1_*(\Gamma)$. y is the solution of the following weak problem:
$y \in H^1_*(\Gamma), \forall\varphi \in H^1_*(\Gamma), \int_\Gamma \nabla_\Gamma y.\nabla_\Gamma\varphi d\Gamma = \int_\Gamma g\varphi d\Gamma$.
The operator $-\Delta_\Gamma$ is an isomorphism from $H^2(\Gamma) \bigcap H^1_*(\Gamma)$ onto $L^2_*(\Gamma)$. In fact the sobolev space $H^1_*(\Gamma)$ is isomorphic to $\frac{H^1(\Gamma)}{\mathbb{R}}$ (the linear space of functions defined up to an additive constant): the dual space of $\frac{H^1(\Gamma)}{\mathbb{R}}$ is the linear space $\{\varphi \in H^{-1}(\Gamma), \left\langle \varphi, 1\right\rangle = 0\}$ so that the right hand side g must be taken with zero mean on Γ. As f is given in $H^s(D), s > \frac{1}{2}$, its restriction to each surface Γ is an element of $L^2(\Gamma)$ and we shall denote by $\bar{f}_\Gamma = f - \frac{1}{meas(\Gamma)} \int_\Gamma f d\Gamma$.
We notice here that for φ in $H^1_*(\Gamma)$, we have $\int_\Gamma \bar{f}_\Gamma\varphi d\Gamma = \int_\Gamma f\varphi d\Gamma$ where f is the restriction of f to Γ. For simplicity we shall denote by $\bar{f}$ for $\bar{f}_\Gamma$ and finally the problem can be written:
$y \in H^1_*(\Gamma), \forall\varphi \in H^1_*(\Gamma), \int_\Gamma \nabla_\Gamma y.\nabla_\Gamma\varphi d\Gamma = \int_\Gamma f\varphi d\Gamma$.

The usual change of variable on Γ_t can be written as:
$\int_{\Gamma_t} \varphi d\Gamma_t = \int_\Gamma \varphi oT_t\omega(t) d\Gamma$
where $\omega(t)$ is given by $||M(DT_t).n||$, $M(DT_t)$ being the cofactor's matrix of DT_t, $M(DT_t) = detDT_t({}^*DT_t)^{-1}$ (for short we shall write ${}^*DT_t^{-1}$) (cf [4]).

2.3. Shape Tangential Derivative Structure. We assume from now the field V smooth enough in the variable (t, x) so that the mapping $(t \to T_t(V)^{-1}) \in C^1([0, \tau[, W^{1,\infty}(D)^N)$, this is the case for example when V is autonomeous. We shall make use of extension operator P_0 continuously and linearly defined from the boundary Sobolev space $H^1(\Gamma)$ in $H^{\frac{3}{2}}(D) \cap H^1_0(D)$ verifying $P_0.\varphi|_\gamma = \varphi$ where $\varphi \in H^1(\Gamma)$ is such that $\frac{\partial}{\partial n}(P_0.\varphi)|_\Gamma = 0$. To derive the existence of this operator, it is sufficient to solve fourth order elliptic boundary value problems in the domain Ω and in the complementary domain $D - \Omega$ with the two previous boundary conditions on Γ.

2.3.1. *The case without boundary.*

Proposition 2.1. *Let P_0 be a simultaneous continuous extension mapping in*

$$\mathcal{L}(H^{1+\epsilon}(\Gamma), H^{\frac{3}{2}+\epsilon}(D) \cap H^1_0(D)) \ , \ 0 < \epsilon < \epsilon_o < \frac{1}{2},$$

verifying, $\forall\varphi \in H^{1+\epsilon}(\Gamma)$, $P_0.\varphi|_\Gamma = \varphi$ *and* $\frac{\partial}{\partial n}(P_0.\varphi) = 0$ *on* Γ.
We consider the element $Y(t) = [\ P_0.(\ y(\Gamma_t)oT_t(V)\ )\ ]o(T_t(V))^{-1}$.
Assume:
i) The material derivative $\dot{y}(\Gamma; V) = [\frac{d}{dt}(\ y(\Gamma_t)oT_t(V))]_{t=0}$ *exists in* $H^1(\Gamma)$ *for all surfaces* Γ *in* D.
ii) The following regularity:

$$(t \to y(\Gamma_t)oT_t(V)\ ) \in C^0([0, \tau],\ H^{1+\epsilon}(\Gamma)\ ).$$

Then Y belongs to $C^1([0,\tau[, H^{\frac{1}{2}+\epsilon}(D))$,($\tau \in I\!R^*_+$). Moreover the restriction of $Y(t)$ to the surface Γ_t is equal to $y(\Gamma_t)$ and $(\frac{\partial}{\partial t}Y(0))|_\Gamma$, the restriction to Γ of the derivative $\frac{\partial}{\partial t}Y(0,.)$, is an element of $L^2(\Gamma)$ which does not depend on the choice of the smooth extension Y.

Proof. The first part is just the use of the chain rule derivative as the derivative with respect to t of $(y^t = y(\Gamma_t)oT_t(V))$ is precisely the material derivative. Then $f(t) = P_o.y^t$ belongs to

$$C^1([o,\tau[, H^1(D)) \cap C^0([0,\tau[, H^{\frac{3}{2}+\epsilon}(D)).$$

Then from [1] (prop 2.38 page 71) we get $f(t)oT_t(V)^{-1} \in C^1([o,\tau[, H^{\frac{1}{2}+\epsilon}(D))$. We turn to the second part, let Y_1, Y_2 be two such smooth extensions of y. Being given an element $\varphi \in C^\infty_{comp}(D)$ we consider the derivative with respect to t at $t = 0$ of the integral

$$\frac{\partial}{\partial t}\Big[\int_{T_t(V)(\Gamma)} (Y_2(t) - Y_1(t))\, \varphi\, d\Gamma_t \Big]_{t=0}$$

$$= \frac{\partial}{\partial t}\Big[\int_\Gamma (Y_2(t)oT_t(V) - Y_1(t)oT_t(V))\, \varphi oT_t(V)\, det(DT_t(V))\, \|(DT_t(V)^*)^{-1}.n\|\, d\Gamma_t \Big]_{t=0}.$$

This derivative vanishes as, on Γ, we immediately get: $Y_2(t)oT_t(V) = Y_1(t)oT_t(V) = y(\Gamma_t)oT_t(V)$. A direct calculation of the first derivative gives, see [1],

$$\int_\Gamma [\, (\frac{\partial}{\partial t}Y_2(0,.) - \frac{\partial}{\partial t}Y_1(0,.) + \frac{\partial}{\partial n}((Y_2(0) - Y_1(0)) < V(0), n >)\, \varphi$$
$$+ < \nabla\varphi, n > (Y_2(0) - Y_1(0)) < V(0), n > + \varphi\,(Y_2(0) - Y_1(0))\, div_\Gamma V\,] d\Gamma$$

From the properties of the extensions Y_1, Y_2, and from the surjectivity of the trace operator on Γ (from which the normal derivative on Γ of the element φ can be chosen equal to zero, i.e. $(\frac{\partial}{\partial n}\varphi)|_\Gamma = 0$, $\varphi|_\Gamma = \psi$), we get

$$\forall \psi \in L^2(\Gamma)\,,\ \int_\Gamma (\frac{\partial}{\partial t}Y_2(0,.) - \frac{\partial}{\partial t}Y_1(0,.)\ \psi\ d\Gamma = 0.$$

Definition 2.1. *The shape boundary derivative $y'_\Gamma(\Gamma; V)$ is the element $(\frac{\partial}{\partial t}Y(0))|_\Gamma \in L^2(\Gamma)$ where Y is any smooth extension of y verifying : $0 < \epsilon < \frac{1}{2}$, $Y \in C^1([0,\tau[, H^{\frac{1}{2}+\epsilon}(D))$, $Y(0,.)|_\Gamma = y(\Gamma)$ and $\frac{\partial}{\partial n}Y(0) = 0$ on Γ. Under this context we shall say that $y(\Gamma_t)$ is shape differentiable in $L^2(\Gamma)$.*

Proposition 2.2. *The shape boundary derivative is given by*

(3) $$y'_\Gamma(\Gamma; V) = \dot{y}(\Gamma; V) - < \nabla_\Gamma y(\Gamma), V(0) >$$

Proposition 2.3. *Assume that the mapping $\dot{y}(\Gamma) = (V \longrightarrow \dot{y}(\Gamma; V))$ is linear and continuous,*

$$\dot{y}(\Gamma) \in \mathcal{L}(\, C^2(D, I\!R^N) \cap H^1_0(D, I\!R^N)\,,\ H^1(\Gamma)\,).$$

Then $y'_\Gamma(\Gamma; V)$ linearly depends on $V(0)|_\Gamma$ through its normal component $< V(0), n >$.

Proof. By a classical argument developped in [5], [1], [6] that mapping depends in fact only on the value of the field V at $t = 0$. From now V stands for $V(0)$. We consider the kernel $\mathcal{N}_0$ of the linear mapping $y'_\Gamma = \dot{y}(\Gamma) - \nabla_\Gamma y^* = (\, V \longrightarrow \dot{y}(\Gamma)(V) - < \nabla_\Gamma y\,, V >)$ which is closed in the Banach space $E = C^0([0,\tau[, C^2(D, R^N) \cap H^1_0(D, R^N))$ and we consider the classical factorization of the linear mapping through the quotient space $Q = E\,/\,\mathcal{N}$, where $\mathcal{N} = \{\, V \in E \mid < V, n > = 0\, \}$. We denote by r_n the usual canonical surjective mapping from E onto Q. We shall prove that the following inclusion holds: $\mathcal{N} \subset \mathcal{N}_0$, then by an usual argument there exists a linear mapping $\bar{y}' \in \mathcal{L}(Q, H^1(\Gamma))$ such that $y'_\Gamma = \bar{y}' o r_n$. We consider the linear mapping $\mathcal{T}_n \in \mathcal{L}(E, C^1(\Gamma))$ defined by $\mathcal{T}_n(V) = < V\,,\, n >$ and we denote by R its range in $C^1(\Gamma)$. It can be easily verified that the mapping $\mathcal{T}_n$ induces an isomorphism $i \in \mathcal{L}(Q, R)$ between the Banach spaces Q and R. Finally we get $y'_\Gamma = (\bar{y}')o(i^{-1})o\mathcal{T}_n$ but $\mathcal{T}_n(V) =< V, n >$. We turn now to the previous inclusion, let be given a fixed element $\varphi \in C^\infty(D)$ and consider the derivative with respect to t at $t = 0$ of the following integral:

$$a(t) = \int_{\omega_t} \varphi\, y_t\, d\Gamma_t = \int_{\omega_t} \varphi Y(t,.)\, d\Gamma_t = \int_{\omega} \varphi o T_t\; Y(t) o T_t\; j(t)\; d\gamma_t$$

Where j is the "tangential jacobian" in the change of variable: $j(t) = det(DT_t)\|(^*DT_t)^{-1}.n\|$. Assuming that φ is taken with $\frac{\partial}{\partial n}\varphi = 0$ on Γ (which does not imply and is neither a consequence of $\frac{\partial}{\partial n_t}\varphi = 0$ on Γ_t); then using the classical shape calculus, see[1], we get $a'(0) = \int_\omega (\frac{\partial}{\partial t}Y(0)\,\varphi + Y(0)\varphi H < V, n >)\; d\Gamma$. By classical argument $V \in \mathcal{N}$ implies that the surface does not move: $\Gamma_t = \Gamma$, so that the state variable is constant with respect to $t:\; V \in \mathcal{N}$ implies $y_t = y$. Then we get that $V \in \mathcal{N}$ implies $a(t) = a(0)$ and then $a'(0) = 0$. From the previous expression of $a'(0)$, which holds for any $\varphi \in L^2(\Gamma)$ we get: $V \in \mathcal{N}$ implies $y'_\Gamma(\Gamma; V) = 0$, that is the desired inclusion.

2.3.2. *Shape Tangential Derivative.* It is the case when the variable z also depends on the relative boundary γ. We specify here the study to the specific case under consideration: z depends on ω through the homogeneous Dirichlet boundary condition. As the state variable $z(\omega)$ ranges in the Sobolev space $H^1_0(\omega)$ we simply adapt the present context by considering the usual extension by zero of the element z to the whole surface Γ : $\tilde{z}(\omega) = z(\omega)$ a. e. in ω, $\tilde{z}(\omega) = 0$ in $\Gamma \setminus \omega$. The boundary γ of ω being smooth and then with zero capacity (in the surface) that element is well defined in $H^1(\Gamma)$. Then we consider smooth extensions in the form

$$Z(t) = (\;P_0.(\,\tilde{z}(\omega_t)oT_t(V)\,)\;o(T_t(V))^{-1}\;).$$

If the material derivative $\dot{z}(\omega; V) = [\frac{d}{dt}(\,z(\omega_t)oT_t(V))]_{t=0}$ exists in $H^1(\omega)$, for all subset ω open in Γ and for all surface Γ, then the material derivative $\dot{\tilde{z}}\;(\Gamma; V)$ exists in $H^1_0(\Gamma)$ and is itself $\tilde{\dot{z}}(\omega; V)$, the extension by zero to the whole surface Γ of the material derivative $\dot{z}(\omega; V)$. Then from the previous results there exists the shape boundary derivative of the state variable $\zeta(\Gamma_t) = \tilde{z}(\omega_t)$ given by

$$\zeta'_\Gamma(\Gamma; V) = \frac{\partial}{\partial t}Z(0,.)|_\Gamma \;\in\; H^1(\Gamma)$$

and we get $\zeta'_\Gamma(\Gamma; V) = \dot{\tilde{z}}\;(\Gamma; V) - < \nabla_\Gamma \tilde{z}(\omega)\,,\, V(0) >$.
Then we define the shape tangential derivative

$$z'_\Gamma(\omega; V) = \zeta'_\Gamma(\Gamma; V)|_\omega \;\in\; H^1_0(\omega)$$

as being the restriction to the open subset ω of $\zeta'_\Gamma(\Gamma; V)$ which is the shape boundary derivative of $\tilde{z}\;(\omega_t)$. We get the similar structure:

Proposition 2.4. *If the material derivative $\dot{z}(\omega; V)$ exists in $H^1_0(\omega)$ and is linear and continuous on the field V and if $(t \to z(\omega_t)oT_t\,) \in C^0([0, \tau[, H^{1+\epsilon}(\omega)\,)$, then the shape tangential derivative $\zeta'_\Gamma(\Gamma; V)$ exists in $L^2(\Gamma)$ and depends on V through the elements $v =< V(0)\,,\, n >$ $\in C^1(\omega)$
, and $w =< V(0)\,,\, \nu > \in C^1(\gamma)$. More precisely there exists an element $\tilde{\zeta'_\Gamma}(\omega;\, .) \in \mathcal{L}(C^1(\bar{\omega}) \times C^1(\gamma)\,,\, L^2(\Gamma)\;)$ such that*

$$\zeta'_\Gamma(\omega; V) = \tilde{\zeta'_\Gamma}(\omega;\, (V.n\,,\, V_\Gamma.\nu)\;) \tag{4}$$

Proof. We remark that if the material derivative $\dot{z}$ is linear and continuous then so is the material derivative $\dot{\tilde{z}}$ just in the same sense as previously for the state variable y. The proof is similar to the previous one for the shape boundary derivative. The set $\mathcal{N}$ is now defined as

$$\mathcal{N} = \{ V \in E \mid < V, n >= 0 \ < V, \nu >= 0 \}.$$

And the set

$$R = \{ (v, w) \in C^1(\Gamma) \times C^1(\gamma) \mid v =< V, n >, \ w =< V, \nu > , V \in E \}.$$

In order to complete that proof which is similar to the previous one we have only to show that $V \in \mathcal{N}$ implies that $Z'_\Gamma(\omega; V) = 0$. We proceed as previously with

$$b(t) = \int_{\omega_t} \varphi \, z_t \, d\Gamma_t = \int_{\omega_t} \varphi \, Z(t, .) \, d\Gamma_t = \int_\omega \varphi o T_t \; Z(t) o T_t \; j(t) \; d\Gamma.$$

Assuming that φ is taken such that $\frac{\partial}{\partial n}\varphi = 0$ on Γ, then using the classical shape calculus, see[1], we get $b'(0) = \int_\omega (\frac{\partial}{\partial t} Z(0) \, \varphi + Z(0)\varphi H < V, n >) \; d\Gamma + \int_\gamma Z(0)\varphi < V_\Gamma , \nu > d\gamma$. By classical argument $V \in \mathcal{N}$ implies that the set does not move: $\omega_t = \omega$, so that the state variable is constant with respect to t : $V \in \mathcal{N}$ implies $z_t = z$. Then we get that $V \in \mathcal{N}$ implies $b(t) = b(0)$, and then $b'(0) = 0$. From the previous expression of $b'(0)$, which holds for any $\varphi \in L^2(\Gamma)$ we get: $V \in \mathcal{N}$ implies $z'_\Gamma(\omega; V) = 0$, that is the desired inclusion. The quotient linear space is then define as

$$\mathcal{Q} = C^2(\bar{D}, R^N)/\mathcal{N}$$

the linear mapping $\zeta'_\Gamma(\Gamma; V)$ induces an element $\bar{\zeta}'_\Gamma(\Gamma; V) \in \mathcal{L}(\mathcal{Q}, L^2(\omega))$ such that

$$z'_\Gamma(\omega, .) = \bar{z}'_\Gamma(\omega, .) \, o \, r_{n,\nu}$$

where $r_{n,\nu}$ is the surjective mapping from E onto $\mathcal{Q}$. The trace mapping $\tau_{n,\nu} \in \mathcal{L}(E, C^1(\bar{\omega}) \times C^1(\gamma))$ induces an injective linear mapping $i = \bar{\tau}_{n,\nu} \in \mathcal{L}(\mathcal{Q}, R)$ where R is the range of $\tau_{n,\nu}$. Finally we get $z'_\Gamma(\omega, .) = \tilde{z}'_\Gamma(\omega, .) o \tau_{n,\nu}$ where $\tilde{z}'_\Gamma(\omega, .) = \bar{z}'_\Gamma(\omega, .) \, o \, \bar{\tau}^{-1}_{n,\nu} \in \mathcal{L}(C^1(\bar{\omega}) \times C^1(\gamma), L^2(\Gamma))$.
Such derivatives are usefull in practice for example to derive the Shape gradient of functionals governed by tangential problems. In the case of Energy shape functional such as "compliance" in structural mechanics, such examples can be founded in [10].

2.4. Shape Boundary Derivative of Vector Fields. We define the shape boundary derivative of a vector field $u_t \in C^1(\Gamma_t)$ as being the vector field with each component being the shape boundary derivative of the corresponding component of u, u being the element in $C^1(\Gamma)$ such that $u_{t|_{t=0}} = u$.

Definition 2.2.

$$(u_t)'_\Gamma = \dot{u} - D_\Gamma u.V(0)$$

An important example is the tangential field obtained as a gradient: let $\Phi \in \mathcal{D}(IR^N)$, the shape boundary derivative or shape tangential derivative of $\nabla_{\Gamma_t}\Phi$ is given by:

$$(\nabla_{\Gamma_t}\Phi)'_\Gamma = \widehat{\nabla_{\Gamma_t}\Phi}^{\cdot} - D_\Gamma(\nabla_\Gamma \Phi).V(0)$$

where $\widehat{\nabla_{\Gamma_t}\Phi}^{\cdot}$ is the material derivative of $\nabla_{\Gamma_t}\Phi$, $(\widehat{\nabla_{\Gamma_t}\Phi}^{\cdot})_i = (\widehat{\nabla_{\Gamma_t}\Phi_i}^{\cdot})$.
We know from [1] that the material derivative $\dot{n}$ of n is given by:

(5)
$$\dot{n}(V) = -({}^*DV.n)_\Gamma = -{}^*DV.n + < \epsilon(V)n, n > n = n'_\Gamma(V) + D_\Gamma n.V_\Gamma = -\nabla_\Gamma(V.n) + D_\Gamma n.V_\Gamma.$$

Lemma 2.1. *The shape boundary derivative n'_Γ of n is given by:*

$$n'_\Gamma(V) = \dot{n}_{|\Gamma} - D_\Gamma n.V_\Gamma = -\nabla_\Gamma(V.n). \tag{6}$$

Proof. Let ξ be a test function given in $\mathcal{D}(I\!R^N)$, $n_{t,i}$ corresponding to the ith component of the normal n, we have the following derivatives computed at $t = 0$:

$$\frac{\partial}{\partial t}\Big(\int_{\Gamma_t} \xi\, n_{t,i} d\Gamma_t\Big)|_{t=0} = \int_\Gamma \Big(\xi\, n'_i + n_i \frac{\partial \xi}{\partial n} + H\, \xi\, n_i V.n\Big) d\Gamma. \tag{7}$$

¿From Stokes' formula we have

$$\frac{\partial}{\partial t}\Big(\int_{\Gamma_t} \xi\, n_{t,i} d\Gamma_t\Big)|_{t=0} = \frac{\partial}{\partial t}\Big(\int_{\Omega_t} \partial_{x_i}\xi\, dx\Big)|_{t=0} = \int_\Gamma \partial_{x_i}\xi\, V.n\, d\Gamma \tag{8}$$

and

$$\int_\Gamma n'_i \xi d\Gamma = \int_\Gamma \Big(\partial_{x_i}\xi V.n - n_i \frac{\partial \xi}{\partial n} V.n - H\, \xi n_i V.n\Big) d\Gamma, \tag{9}$$

from the surjectivity of the restriction mapping, the element ξ can be extended to D with the condition $\frac{\partial \xi}{\partial n} = 0$ on Γ. Then

$$\int_\Gamma n'_\Gamma\, \xi d\Gamma = \int_\Gamma \Big(\nabla_\Gamma \xi - H \xi n\Big) V.n d\Gamma.$$

And the conclusion derives by the use of tangential by part integration formula on Γ that we recall here for convenience:

$$\int_\Gamma \nabla_\Gamma \varphi.e\, d\Gamma = \int_\Gamma (-\varphi div_\Gamma e + H\, \varphi\, e.n)\, d\Gamma$$

As we understand through these results, the shape boundary derivative study is governed by the material derivative one. On the considered example we shall then begin with it and for doing this we need to study the continuity of the solution on the parameter t.

3. Material Derivative

3.1. Shape Continuity. In order to simplify we shall concentrate the study on the case where γ is empty. In a final section we shall indicate the modifications introduced by the presence of the relative boundary γ. The element y_t being in $H^1(\Gamma_t)$, the transported solution $y_t o T_t$ is in the fixed Sobolev space $H^1(\Gamma)$. We consider the continuity of $y_t o T_t$ in the fixed space $H^1(\Gamma)$. In fact, in order to preserve the zero mean value property, we consider the element $j(t) y_t o T_t$. We shall denote by ${}^*DV$ the transposed of the jacobian matrix DV.

Proposition 3.1. *Let be given $V \in C^0([0,\tau[, C^2(\bar{D}, I\!R^N))$ and $F \in H^{\frac{1}{2}+\epsilon}(D)$. Let $y(\Gamma_t)$ denotes the solution of the problem (1) in $\Gamma_t = T_t(V)(\Gamma)$. Then $j(t) y_t o T_t \longrightarrow y(\Gamma)$ in $H^1_*(\Gamma)$ as $t \to 0$.*

Proof. We have

$$\int_{\Gamma_t} \nabla_{\Gamma_t} y_t . \nabla \varphi_t d\Gamma_t = \int_{\Gamma_t} f \varphi_t d\Gamma_t, \qquad \forall \varphi_t \in H^1_*(\Gamma_t) = \int_{\Gamma_t} \bar{f}_{\Gamma_t} \varphi_t d\Gamma_t, \qquad \forall \varphi_t \in H^1(\Gamma_t)$$

and

$$\int_\Gamma \nabla_\Gamma y . \nabla \varphi d\Gamma = \int_\Gamma f \varphi d\Gamma, \qquad \forall \varphi \in H^1_*(\Gamma) = \int_\Gamma \bar{f}_\Gamma \varphi d\Gamma, \qquad \forall \varphi \in H^1(\Gamma)$$

but $\varphi_t \in H^1_*(\Gamma_t)$ implies that $\psi = \varphi_t o T_t \in H^1(\Gamma)$ then:

$$\int_\Gamma (\nabla_{\Gamma_t} y_t) o T_t . (\nabla \varphi_t) o T_t d\Gamma = \int_\Gamma j f o T_t \varphi_t o T_t d\Gamma, \qquad \forall \varphi_t \in H^1_*(\Gamma_t)$$

and we get by substraction of these two equalities, using (8) and the notations introduced at proposition 3.3

$$(10)\qquad \int_\Gamma \Big\{ < DT_t^{-1}\nabla_\Gamma(j(t)y_to T_t - y), \nabla\psi > + < (DT_t^{-1} - Id)\nabla_\Gamma y, \nabla\psi > +$$

$$< ({}^*DT_t^{-1} - Id)\nabla(j(t)YoT_t), {}^* DT_t^{-1}\nabla\psi >$$
$$- < ({}^*DT_t^{-1} - Id)\nabla(j(t)YoT_t), n_toT_t >< n_toT_t, {}^* DT_t^{-1}\nabla\psi >$$

$$- < Id\nabla(j(t)YoT_t), (n_toT_t - n) >< n, {}^* DT_t^{-1}\nabla\psi > - < Id\nabla(j(t)YoT_t), n_toT_t > \times$$

$$< (n_toT_t - n), {}^*DT_t^{-1}\nabla\psi > - < \frac{j(t)}{j(t)}^* DT_t^{-1}YoT_t\nabla(j(t)-1), {}^* DT_t^{-1}\nabla\psi >$$

$$+ < \frac{1}{j(t)}^* DT_t^{-1}YoT_t\nabla(j(t)-1), n_toT_t >< \omega(t)n_toT_t, {}^* DT_t^{-1}\nabla\psi > \Big\} d\Gamma$$

$$= \int_\Gamma \{j(t)foT_t - f\}(\psi - c)d\Gamma, \ \forall\psi \in H^{\frac{3}{2}}(D)$$

$$= \int_\Gamma \{j(t)\bar{f}_{\Gamma_t}oT_t - \bar{f}_\Gamma\}\psi d\Gamma, \ \forall\psi \in H^{\frac{3}{2}}(D)$$

where $c = \frac{1}{|\Gamma|}\int_\Gamma \psi d\Gamma$. When $t \searrow o$, all the terms in (10) go to zero and we get $j(t)y_toT_t \longrightarrow y$, weakly in $H^1_*(\Gamma)$. But we can write $j(t)y_toT_t = (j(t) - j(0))y_toT_t + j(0)y_toT_t$ where $j(0) = 1$. Then we have the weak continuity of $t \longrightarrow y_t$. To get the strong continuity, we show that $\|j(t)y_toT_t\|_{H^1(\Gamma)} \longrightarrow \|y\|_{H^1(\Gamma)}$ as $t \searrow 0$.

3.2. Existence of the Material Derivative $\dot{y}$. The material derivative $\dot{y}$ is classically defined in mechanics of continuous media as being a pointwise limit. Here following [6] we consider the analogous derivative in $H^1(\Gamma)$-norm, see also [1].
From the strong continuity of the transported element $y_toT_t \in H^1(\Gamma)$, we consider the following derivative:

Definition 3.1. *Let us denote by $\dot{y}$ the material derivative*
$\dot{y} = \lim_{t\searrow 0}\{\frac{y_toT_t - y}{t}\}$.
This limit must be understood as the limit in the $\frac{H^1(\Gamma)}{I\!R}$ norm of the quotient:
$\|\dot{y} - \{\frac{y_toT_t - y}{t}\}\| \longrightarrow 0,\ t \searrow 0$.

Theorem 3.1. *Let D be a bounded domain in $I\!R^N$, Ω a smooth domain included in D, Γ its boundary, $V \in C^0([0,\tau[, C^2(\bar{D}, I\!R^N) \cap H^1_0(D, I\!R^N))$ a smooth vector field and $T_t(V)$ its flow transformation; it is a mapping from $\bar{D}$ onto $\bar{D}$. Let F be given in $H^{\frac{1}{2}+\epsilon}(D)$.*
Let $y_t \in H^1_*(\Gamma_t)$ be the solution of problem (1) formulated on Γ_t, where $\Gamma_t = T_t(V)(\Gamma)$.
The mapping $t \longrightarrow y_t$ is strongly differentiable in $\frac{H^1(\Gamma)}{I\!R}$ and $\dot{y}$ is the unique solution of the following problem:

(11)
$$-\Delta_\Gamma\dot{y} + 2div_\Gamma(\epsilon_\Gamma(V(0)).\nabla_\Gamma y) = \frac{\partial F}{\partial n}V(0).n + div_\Gamma(fV(0)) - \frac{1}{|\Gamma|}\int_\Gamma(\frac{\partial F}{\partial n} + Hf)V(0).nd\Gamma + c,$$

where the constant is given by $c = \frac{1}{|\Gamma|}\int_\Gamma H < D_\Gamma(V(0)).\nabla_\Gamma y, n > d\Gamma$, $\epsilon_\Gamma(V) = \frac{D_\Gamma V + {}^* D_\Gamma V}{2}$ *is the tangential deformation tensor on* Γ *and* H *is the mean curvature on* Γ.

The material derivative $\dot{y}$ depends on both the tangential component $V_\Gamma(0)$ and the normal component $V(0).n$ of the field $V(0)$. The term $\Delta_\Gamma \dot{y}$ has zero mean value on Γ so that the right hand side associated with the equation (2) has zero mean value too, so that $\dot{y}$ is well defined up to an additive constant which is fixed by the zero mean value property.

3.3. Equation Satisfied by the Transported Element y^t. We set $y^t = j(t)y_t oT_t$ element of $H^1_*(\Gamma)$. It is convenient to introduce the following quotient:
$\dot{\theta}(V) = \lim_{t \searrow 0} \frac{1}{t}(y^t - y)$ where $V = (\frac{\partial T_t}{\partial t})oT_t^{-1}$. We characterize the problem whose solution is the transported element y^t.

Proposition 3.2. *y^t is the solution of the weak-problem (10) below.*

Proof. The weak formulation of the problem is

$$\int_{\Gamma_t} \nabla_{\Gamma_t} y_t . \nabla_{\Gamma_t} \varphi d\Gamma_t = \int_{\Gamma_t} f(\varphi - c) d\Gamma_t, \qquad \forall \varphi \in H^1(\Gamma_t),$$

with $c = \frac{1}{|\Gamma_t|}\int_{\Gamma_t} \varphi d\Gamma_t$.
Or we have equivalently:

$$\int_{\Gamma_t} \nabla_{\Gamma_t} y_t . \nabla \varphi d\Gamma_t = \int_{\Gamma_t} \bar{f}_{\Gamma_t} \varphi d\Gamma_t, \qquad \forall \varphi \in H^1(\Gamma_t).$$

Let $\varphi \in H^1_*(\Gamma_t)$ (or $H^1(\Gamma_t)$) be defined by $\varphi = \psi o T_t^{-1}$, where now ψ is in $H^1(\Gamma)$. Any element in $H^1(\Gamma)$ can be looked as the trace on Γ of some ψ in $H^{\frac{3}{2}}(D)$ with $\frac{\partial \psi}{\partial n} = 0$ on Γ; we have $\frac{\partial \varphi}{\partial n_t} \neq 0$ on Γ_t.
Notice that $(\nabla \varphi) o T_t|_\Gamma = {}^* DT_t^{-1} \nabla \psi|_\Gamma = {}^* DT_t^{-1} \nabla_\Gamma \psi$ gives:

$$(12) \qquad \int_\Gamma < (\nabla_{\Gamma_t} y_t) oT_t, {}^* DT_t^{-1} \nabla \psi > j(t) d\Gamma = \int_\Gamma j(t) \bar{f}_{\Gamma_t} oT_t \psi d\Gamma, \quad \forall \psi \in H^{\frac{3}{2}}(D).$$

The following lemma is a lemma of transport by help of which we replace the weak formulation of the initial perturbed problem on the fixed manifold.

Lemma 3.1. *a)* $\varphi \in H^1(\Gamma_t) \iff (\varphi - c) \in H^1_*(\Gamma_t)$, *where* $c = \frac{1}{|\Gamma|}\int_\Gamma \varphi d\Gamma$
b) $\varphi \in H^1_*(\Gamma_t) \iff \psi = j(t)\varphi o T_t \in H^1_*(\Gamma)$
c) $\bar{f}_{\Gamma_t} \in L^2_*(\Gamma_t) \iff j(t) \bar{f}_{\Gamma_t} o T_t \in L^2_*(\Gamma)$,

we have

$$(13) \quad \int_\Gamma < (\nabla_{\Gamma_t} y_t) oT_t, {}^* DT_t^{-1} \nabla \psi > j(t) d\Gamma = \int_\Gamma j(t) f oT_t (\psi - c) d\Gamma, \qquad \forall \psi \in H^{\frac{3}{2}}(D)$$

where $c = \frac{1}{|\Gamma|}\int_\Gamma \psi d\Gamma$.

Proposition 3.3. *For each t, let $Y(t) \in H^{\frac{3}{2}}(D)$ be an extension of y_t (i.e. $Y_{|\Gamma_t} = y_t$). We have:* $(\nabla_{\Gamma_t} y_t) oT_t = {}^* DT_t^{-1} \nabla (Y oT_t) - < {}^* DT_t^{-1} \nabla (Y oT_t), n_t oT_t > n_t oT_t$.

Using the proposition 3.2 and the Lemma 3.1, we can write:

$$\int_\Gamma \{< {}^*DT_t^{-1}\nabla(YoT_t), {}^*DT_t^{-1}\nabla\psi > - < {}^*DT_t^{-1}\nabla(YoT_t), n_toT_t >< n_toT_t, {}^*DT_t^{-1}\nabla\psi >\} j(t)d\Gamma$$

$$= \int_\Gamma j(t)\bar{f}_{\Gamma_t}oT_t(\psi - c)d\Gamma, \forall\psi \in H^{\frac{3}{2}}(D)$$

$$= \int_\Gamma j(t)\bar{f}_{\Gamma_t}oT_t\psi, \qquad \forall\psi \in H^{\frac{3}{2}}(D).$$

Using proposition 3.3, we can write:

$$(14) \qquad (\nabla_{\Gamma_t}y_t)\,oT_t = \frac{1}{j(t)}\nabla_\Gamma y^t + \frac{1}{j(t)}\{({}^*DT_t^{-1} - Id)\nabla(j(t)YoT_t)$$

$$- < ({}^*DT_t^{-1} - Id)\nabla(j(t)YoT_t), n_toT_t > n_toT_t$$

$$- < Id\nabla(j(t)YoT_t), n_toT_t - n > n - < Id\nabla(j(t)YoT_t), n_toT_t > (n_toT_t - n)\}$$

$$-{}^*DT_t^{-1}YoT_t\frac{\nabla(j(t)-1)}{j(t} + < {}^*DT_t^{-1}YoT_t\frac{\nabla(j(t)-1)}{j(t)}, n_toT_t > n_toT_t,$$

with $Y_{|\Gamma_t}$ solution of $-div_{\Gamma_t}(\nabla_{\Gamma_t}y_t) = f$ on Γ_t ; $j(t)YoT_{t|\Gamma} = y^t$.
Then we have the weak-formulation of the problem associated with y^t:

$$(15) \qquad \int_\Gamma \{< \nabla_\Gamma y^t, {}^*DT_t^{-1}\nabla\psi > + < ({}^*DT_t^{-1} - Id)\nabla(j(t)YoT_t), {}^*DT_t^{-1}\nabla\psi >$$

$$- < ({}^*DT_t^{-1} - Id)\nabla(j(t)YoT_t), n_toT_t >< n_toT_t, {}^*DT_t^{-1}\nabla\psi >$$

$$- < Id\nabla(j(t)YoT_t), n_toT_t - n >< n, {}^*DT_t^{-1}\nabla\psi >$$

$$- < Id\nabla(j(t)YoT_t), n_toT_t >< (n_toT_t - n), {}^*DT_t^{-1}\nabla\psi >$$

$$- < j(t){}^*DT_t^{-1}YoT_t\frac{\nabla(j(t)-1)}{j(t)}, {}^*DT_t^{-1}\nabla\psi >$$

$$+ < {}^*DT_t^{-1}YoT_t\frac{\nabla(j(t)-1)}{j(t)}, n_toT_t >< j(t)n_toT_t, {}^*DT_t^{-1}\nabla\psi > \Big\} d\Gamma$$

$$= \int_\Gamma j(t)\bar{f}_{\Gamma_t}oT_t(\psi - c)d\Gamma, \quad \forall\psi \in H^{\frac{3}{2}}(D).$$

We substract the term $\int_\Gamma < \nabla_\Gamma y, \nabla\psi > d\Gamma$ (where y is the solution of the initial problem (1) in Γ) to the first term of the left hand side of the above equality (15) and we divide the final expression by t:

$$\frac{1}{t}\int_\Gamma < \nabla_\Gamma y^t, {}^*DT_t^{-1}\nabla\psi > d\Gamma - \frac{1}{t}\int_\Gamma < \nabla_\Gamma y, \nabla\psi > d\Gamma$$

$$(16) \qquad = \int_\Gamma < DT_t^{-1}(\nabla_\Gamma(\frac{y^t - y}{t})), \nabla\psi > d\Gamma + \int_\Gamma < (\frac{DT_t^{-1} - Id}{t})\nabla_\Gamma y, \nabla\psi > d\Gamma,$$

this enables us to make appear $\dot{y}$, taking the limit of this term, as t goes to zero.

We have to exhibe the whole equation whose solution is $\theta^t = \frac{y^t - y}{t}$ and then to use this equation in order to prove the existence of $\dot\theta$ (which is directly linked to the existence and the characterization of $\dot y$), that is to say that θ^t is bounded in $H^1(\Gamma)$.

Remark 3.1. *An obvious technic to show the existence and the characterization of* $\dot y = \frac{\partial}{\partial t}(y_t o T_t)|_{t=0}$ *would be the use of the implicit function theorem. This way, we would directly get the result concerning the material derivative if the right hand side* $f|_\Gamma$ *of the equation is supposed more regular than* $L^2(\Gamma)$. *Here,* $f|_\Gamma$ *belongs to* $L^2(\Gamma)$ *does not imply the strong convergence in* $H^{-1}(\Gamma)$ *of the quotient* $\frac{f o T_t - f}{t}$, $t \searrow 0$, *as it is required to apply the implicit function theorem. In [1, page 73] there are counter examples for which one cannot expect the mapping* $t \longrightarrow f o T_t$ *to be strongly differentiable in* $H^{-1}(\Gamma)$ *for any* f *in* $L^2(\Gamma)$.

3.4. Boundedness of θ^t in $H^1(\Gamma)$. The above equation (15) satisfied by y^t enables us to make appear θ^t, and then, taking the limit of all the terms of the above equation, as t goes to zero, we will get the caracterization of $\dot\theta$ and then the characterization of the material derivative $\dot y$.
Using (16), we can rewrite (15) in a different way where θ^t explicitly appears:

$$\int_\Gamma \Big\{ < DT_t^{-1}\nabla_\Gamma\theta^t, \nabla\psi > + < (\frac{DT_t^{-1} - Id}{t})\nabla_\Gamma y, \nabla\psi >$$

$$+ < (\frac{{}^*DT_t^{-1} - Id}{t})\nabla(j(t)Y o T_t), {}^*DT_t^{-1}\nabla\psi >$$

$$- < (\frac{{}^*DT_t^{-1} - Id}{t})\nabla(j(t)Y o T_t), n_t o T_t >< n_t o T_t, {}^*DT_t^{-1}\nabla\psi >$$

$$- < Id\nabla(j(t)Y o T_t), (\frac{n_t o T_t - n}{t}) >< n, {}^*DT_t^{-1}\nabla\psi >$$

$$- < Id\nabla(j(t)Y o T_t), n_t o T_t >< (\frac{n_t o T_t - n}{t}), {}^* DT_t^{-1}\nabla\psi >$$

$$- < \frac{j(t)}{j(t)} {}^*DT_t^{-1} Y o T_t \frac{\nabla(j(t) - 1)}{t}, {}^* DT_t^{-1}\nabla\psi >$$

$$+ \text{ij}(t)^{-1} {}^*DT_t^{-1} Y o T_t \frac{\nabla(j(t)-1)}{t}, n_t T_t >< j(t) n_t o T_t, {}^*DT_t^{-1}\nabla\psi > \Big\} d\Gamma$$

$$= \int_\Gamma \frac{j(t) f o T_t - f}{t}(\psi - c) d\Gamma, \ \forall\psi \in H^{\frac{3}{2}}(D).$$

Proposition 3.4. *Let* $h > 0$ *be a given real number. The application (* $t \longrightarrow DT_t^{-1}$ *) is differentiable on* $[0, h]$ *and we have* $\forall t \in [o, h]$, $\exists\, \alpha \in]0, 1[$ *such that* $DT_t^{-1} = Id - tDV(\alpha t)$.

Proof. The differentiability of the above application is obvious as soon as

$$T_t(V) \in C^1([0, \tau[, C^2(\bar D, I\!R^N))$$

Let us take $\psi = \theta^t$ in (12), then we have:

$$(17) \quad \left| \int_\Gamma < DT_t^{-1}\nabla_\Gamma\theta^t, \nabla_\Gamma\theta^t > d\Gamma \right| = \left| ||\nabla_\Gamma\theta^t||^2_{L^2(\Gamma)} - t\int_\Gamma < DV(\theta t)\nabla_\Gamma\theta^t, \nabla_\Gamma\theta^t > d\Gamma \right|$$

$$\geq ||\nabla_\Gamma\theta^t||^2_{L^2(\Gamma)} \left| 1 - t||V||_{W^{1,\infty}} \right|,$$

$$\geq \alpha||\nabla_\Gamma\theta^t||^2_{L^2(\Gamma)}, \text{ as soon as } t \leq \frac{1-\alpha}{||V||_{W^{1,\infty}}}, \alpha \in I\!R_+$$

and the whole equation (15) gives:

$$\alpha\|\nabla_\Gamma\theta^t\|^2_{L^2(\Gamma)} \leq \int_\Gamma \Big| \Big\{ \frac{j(t)\bar{f}_{\Gamma_t}oT_t - \bar{f}_\Gamma}{t}.\theta^t - < (\frac{DT_t^{-1} - Id}{t})\nabla_\Gamma y, \nabla_\Gamma\theta^t > \tag{18}$$

$$- < (\frac{{}^*DT_t^{-1} - Id}{t})\nabla(j(t)YoT_t),{}^* DT_t{}^{-1}\nabla_\Gamma\theta^t >$$

$$+ < (\frac{{}^*DT_t{}^{-1} - Id}{t})\nabla(j(t)YoT_t), n_t oT_t >< n_t oT_t,{}^* DT_t{}^{-1}\nabla_\Gamma\theta^t >$$

$$+ < Id\nabla(j(t)YoT_t), (\frac{n_t oT_t - n}{t}) >< n,{}^* DT_t{}^{-1}\nabla_\Gamma\theta^t >$$

$$+ < Id\nabla(j(t)YoT_t), n_t oT_t >< (\frac{n_t oT_t - n}{t}),{}^* DT_t{}^{-1}\nabla_\Gamma\theta^t >$$

$$+ < \frac{1}{j(t)}{}^* DT_t{}^{-1} j(t)YoT_t \frac{\nabla(j(t) - 1)}{t},{}^* DT_t{}^{-1}\nabla_\Gamma z^t >$$

$$- < \frac{1}{j(t)}{}^* DT_t{}^{-1}YoT_t \frac{\nabla(j(t) - 1)}{t}, n_t oT_t >< j(t)n_t oT_t,{}^* DT_t{}^{-1}\nabla_\Gamma\theta^t > \Big\} \Big| d\Gamma$$

(18) gives the boundedness of $\nabla_\Gamma\theta^t$ in $L^2(\Gamma)$, as long as T_t is $C^1([0,\tau[, C^2(\bar{D}, I\!R^N))$, f is in $H^{\frac{1}{2}+\epsilon}(D)$, Γ is of class C^2 (so that the normal fields n_t and n are always defined and bounded in $L^2(\Gamma)$) and y is in $H^1_*(\Gamma)$.
Then there exists a subsequence $(t_k) \searrow 0$ such that

$$\frac{y^{t_k} - y}{t_k} = \theta^{t_k} \underset{t_k\searrow 0}{\longrightarrow} \dot{\theta}, \text{ weakly in } H^1_*(\Gamma). \tag{19}$$

Now, to characterize the equation satisfied by $\dot{\theta}$, we have to go into the limit in the terms of (12). First, we need three results of convergence:

Proposition 3.5. *Let Γ be of class C^k, $T_t(V) \in C^1([0,\tau[, C^2(\bar{D}, I\!R^N)), k \geq 1$, then the mapping $t \longmapsto n_t oT_t$ is strongly differentiable in $L^\infty(\Gamma)$ and we have* $\frac{n_t oT_t - n}{t} \underset{t\searrow 0}{\longrightarrow} \dot{n} = -{}^*DV.n + < DV.n, n > n.$

Proof. we have $n_t oT_t = \|{}^*DT_t^{-1}.n\|^{-1}_{I\!R^3}{}^*DT_t{}^{-1}.n$ (cf. [1], p. 139). Then we differentiate this expression and the result becomes obvious.

Proposition 3.6. *Let $\delta > 0$ be given; the mapping $(t \in [0,\delta[\longmapsto j(t) \in C^{k-1}(I\!R^N), k \geq 1)$ is differentiable and $j'(0) = divV(0) - < DV(0).n, n >_{I\!R^N} = div_\Gamma V(0)$.*

Proof. We have $j(t) = detDT_t\|{}^*DT_t^{-1}.n\|_{I\!R^N}$ which is differentiable as

$$T_t(V) \in C^1([0,\tau[, C^2(\bar{D}, I\!R^N))$$

and we have $\frac{\partial det(DT_t)}{\partial t}|_{t=0} = divV(0)$, $\frac{\partial\|{}^*DT_t^{-1}.n\|}{\partial t}|_{t=0} = - < DV(0).n, n >$. Then we get the result of proposition 3.6.

Proposition 3.7. *Let $f \in H^{\frac{1}{2}+\epsilon}(D), V \in C([0,\epsilon[, \mathcal{D}^k(I\!R^N, I\!R^N))$ be given, $k \geq 1$, then the mapping $(t \longmapsto foT_t)$ is weakly differentiable in $H^{-\frac{1}{2}}(D)$: $\frac{foT_t - f}{t} \underset{t\searrow 0}{\longrightarrow} \nabla f.V$, weakly in $H^{-\frac{1}{2}}(D)$. Furthermore, the restriction to Γ of f, $f|_\Gamma$ is an element of $H^\epsilon(\Gamma) \subset L^2(\Gamma)$ and then* $\frac{foT_t - f}{t} \underset{t\searrow 0}{\longrightarrow} \nabla f.V$, *weakly in $H^{-1}(\Gamma)$.*

Remark 3.2. *From proposition 3.6, the remark 3.1 becomes obvious: the quotient* $\frac{foT_t - f}{t} \xrightarrow[t\searrow 0]{} \nabla f.V$ *weakly in* $H^{-1}(\Gamma)$ *but not strongly so that the implicit function theorem cannot be applied here in order to derive the existence and the characterization of the material derivative.*

Proof. cf. [1], p. 72-73.
Using these propositions 3.5, 3.6, 3.7 we have the following convergences as $t \searrow 0$:

$$\frac{1}{t}\int_\Gamma \Big\{ < \nabla_\Gamma \theta^t, {}^* DT_t^{-1} \nabla\psi > - < \nabla_\Gamma y, \nabla\psi > \Big\} d\Gamma \xrightarrow[t\searrow 0]{} T_1,$$

$$\int_\Gamma \Big\{ < (\tfrac{{}^*DT_t^{-1} - Id}{t}) \nabla(j(t) Y o T_t), {}^* DT_t^{-1}\nabla\psi > - < (\tfrac{{}^*DT_t^{-1} - Id}{t}) \nabla(j(t) Y o T_t), n_t o T_t >< n_t o T_t, {}^* DT_t^{-1}\nabla\psi > \Big\} d\Gamma \xrightarrow[t\searrow 0]{} T_2,$$

$$-\int_\Gamma \Big\{ < \nabla(j(t) Y o T_t), (\tfrac{n_t o T_t - n}{t}) >< n, {}^* DT_t^{-1}\nabla\psi > - < \nabla(j(t) Y o T_t), n_t o T_t >< (\tfrac{n_t o T_t - n}{t}), {}^* DT_t^{-1}\nabla\psi > \Big\} d\Gamma \xrightarrow[t\searrow 0]{} T_3,$$

$$\int_\Gamma \Big\{ < -{}^*DT_t^{-1} Y o T_t \tfrac{\nabla(j(t)-1)}{t}, {}^* DT_t^{-1}\nabla\psi > + < \tfrac{1}{j(t)} {}^*DT_t^{-1} Y o T_t \tfrac{\nabla(j(t)-1)}{t}, n_t o T_t >< j(t) n_t o T_t, {}^* DT_t^{-1}\nabla\psi > \Big\} d\Gamma \xrightarrow[t\searrow 0]{} T_4,$$

$$\int_\Gamma \Big\{ \frac{j(t) - j(0)}{t} . f o T_t + \frac{f o T_t - f}{t} \Big\} . (\psi - c) d\Gamma \xrightarrow[t\searrow 0]{} T_5$$

with

$T_1 = \int_\Gamma \Big\{ < \nabla_\Gamma \dot\theta, \nabla\psi > - < D_\Gamma V \nabla_\Gamma y, \nabla\psi > \Big\} d\Gamma.$

$T_2 = \int_\Gamma < -({}^*DV\nabla Y)_\Gamma, \nabla\psi > d\Gamma = \int_\Gamma < -{}^*D_\Gamma V \nabla Y, \nabla_\Gamma\psi > d\Gamma,$

$T_3 = -\int_\Gamma \Big\{ < \nabla Y, \dot n >< n, \nabla\psi > d\Gamma + < \nabla Y, n >< \dot n, \nabla\psi > \Big\} d\Gamma,$

$T_4 = \int_\Gamma < (-Y\nabla j'(0))_\Gamma, \nabla\psi > d\Gamma = -\int_\Gamma < Y\nabla_\Gamma j'(0), \nabla_\Gamma\psi > d\Gamma,$

$T_5 = \int_\Gamma \{\nabla F V + f div_\Gamma V\}\{\psi - c\} d\Gamma$, with $c = \frac{1}{|\Gamma|}\int_\Gamma \psi d\Gamma$ and $\psi \in H^1(\Gamma)$,

$\quad = \int_\Gamma \{\overline{\nabla F.V + f div_\Gamma V}\}\psi d\Gamma$ but now $\psi \in H^1_*(\Gamma)$

where the integral should be understood as the duality bracket between $H^{-1}(\Gamma)$ and $H^1(\Gamma)$. As $\forall\psi \in H^1(\Gamma)$, $\nabla_\Gamma(\psi + c) = \nabla_\Gamma\psi$ then the weak formulation (12) holds for test functions ψ with zero mean value property.

Remark 3.3. *In the expression of* T_3, *we need to characterize the material derivative* $\dot n$ *of the normal n:* $\dot n = -{}^*DV.n + < DV.n, n >$; *we easily get* $< \dot n, n >= 0$ *which implies that* $< \dot n, \nabla\psi >=< \dot n, \nabla_\Gamma\psi >$.

θ^t is weakly convergent in $H^1_*(\Gamma)$ and $\dot\theta$ is then the solution of the weak equation:

$$T_1 + T_2 + T_3 + T_4 = T_5, \ \forall\psi \in H^1_*(\Gamma) \tag{20}$$

which can be written, after an integration by parts on Γ:

$$-\Delta_\Gamma\dot\theta + div_\Gamma(D_\Gamma V\nabla_\Gamma y) + div_\Gamma({}^*D_\Gamma V\nabla_\Gamma y) + div_\Gamma(y\nabla_\Gamma j'(0)) \tag{21}$$

$$= \nabla F.V + F div_\Gamma V - \frac{1}{|\Gamma|}\int_\Gamma(\nabla F.V + F div_\Gamma V) d\Gamma + c = \overline{\nabla F.V + F div_\Gamma V} + c$$

with $j'(0) = div_\Gamma V$. As

$$\nabla F.V + F.div_\Gamma V = \frac{\partial F}{\partial n}V.n + \nabla_\Gamma f.V_\Gamma + F div_\Gamma V = \frac{\partial F}{\partial n}V.n + div_\Gamma(FV)$$

and

$$\int_\Gamma (\nabla F.V + F.div_\Gamma V)d\Gamma = \int_\Gamma \frac{\partial F}{\partial n}V.nd\Gamma + \int_\Gamma div_\Gamma(FV)d\Gamma$$

then after an integration by parts we get $\frac{\partial F}{\partial n}V.nd\Gamma + \int_\Gamma div_\Gamma(FV)d\Gamma = \int_\Gamma HF < V, n >$.
The determination of the constant c is done taking the mean of the equation (21):
$\int_\Gamma -\Delta_\Gamma \dot{\theta} d\Gamma = 0 = -2\int_\Gamma(\epsilon_\Gamma(V).\nabla_\Gamma y)d\Gamma + |\Gamma|c$
and after an integration by parts on Γ we get:
$c = \frac{1}{|\Gamma|}\int_\Gamma 2H < \epsilon_\Gamma(V).\nabla_\Gamma y, n > d\Gamma = \frac{1}{|\Gamma|}\int_\Gamma D_\Gamma(V).\nabla_\Gamma y, n > d\Gamma$.
Finally we have proved the following:

Proposition 3.8. *the mapping $t \to \theta^t$ is weakly differentiable in $H^1_*(\Gamma)$ and we have :*

$$-\Delta_\Gamma \dot{\theta} + 2div_\Gamma(\epsilon_\Gamma(V).\nabla_\Gamma y) + div_\Gamma(\nabla_\Gamma(y.div_\Gamma V)) \tag{22}$$

$$= \frac{\partial F}{\partial n}V.n + div_\Gamma(FV) - \frac{1}{|\Gamma|}\int_\Gamma (\frac{\partial F}{\partial n} + HF)V.nd\Gamma + c,$$

where $c = \frac{1}{|\Gamma|}\int_\Gamma < D_\Gamma(V).\nabla_\Gamma y, n > d\Gamma$.

it can be verified that $\dot{\theta}$ has zero mean value. In order to prove the theorem 3.1 we notice that $\dot{\theta} = j'(0)y + j(0)\dot{y}$, but $j(0) = 1$ and $j'(0) = div_\Gamma V$ so that in the previous proposition we get:
$\Delta_\Gamma \dot{\theta} = \Delta_\Gamma \dot{y} + \Delta_\Gamma(y.div_\Gamma V(0))$, but

$$\Delta_\Gamma(y.div_\Gamma V(0)) = div_\Gamma(div_\Gamma V(0)\nabla_\Gamma y) + div_\Gamma(y.\nabla_\Gamma(div_\Gamma V(0)))$$

and we get the equation for $\dot{y}$ in theorem 3.1. To verify that this convergence is strong, it is enough to verify classically that we do have from (15) the convergence of the following norms: $||\theta^t||^2 \longrightarrow ||\dot{\theta}||^2$, $t \searrow 0$.

3.5. Smoothness Result. In fact, for a given right hand side member $f \in L^2_*(\Gamma)$ and a smooth manifold Γ, the solution y to the Laplace Beltrami equation is more regular:

Theorem 3.2. *Let $f \in L^2_*(\Gamma)$, Γ be of class C^2. Then $y \in H^2 \cap H^1_*(\Gamma)$.*

Proof. Let us take $V \in \mathcal{D}(\Gamma; I\!R^N)$ an autonomeous vector field which, at each point $x \in \Gamma$, is tangent: $V(x).n(x) = 0, \quad \forall x \in \Gamma$.
Then we know that Γ is globaly invariant under the flow transformation $T_t(V) : T_t(V)(\Gamma) = \Gamma$.
As a result, we have $y_t = y(\Gamma_t) = y, \forall t$ and then $\dot{y} = \nabla_\Gamma y.V$ on Γ, so that for any such field V we get that $\nabla_\Gamma y.V$ lies in $H^1_0(\Gamma)$.$But\Gamma$ is supposed to be a smooth compact manifold then it derives that $\nabla_\Gamma y(\Gamma)$ itself is in $H^1(\Gamma; T\Gamma)$.

4. Calculus of the Shape Tangential Derivative

4.1. The Oriented Distance Function. The existence of the material derivative $\dot{y}$ allows us to derive the characterization of the shape boundary derivative y'_Γ which is defined in $L^2(\Gamma)$ as we have shown that $\dot{y}$ is in $H^1_*(\Gamma)$, V being in $C^0([0,\tau[, C^2(\bar{D}, I\!R^N))$ and F regular enough, say $F \in H^{\frac{3}{2}+\epsilon}(D)$, in order to get its shape derivative F'_Γ well defined in $L^2(\Gamma)$. Following [7] we consider the oriented distance function b_Ω defined as follows:

$$b_\Omega(x) = \begin{cases} d_\Gamma(x) & if x \in I\!R^N \setminus \overline{\Omega} \\ -d_\Gamma(x) & if x \in \Omega. \end{cases}$$

This function has many interesting properties in relation with the topology of Ω (cf [7]) and we recall some of them. For each $k \geq 2$, b_Ω is C^k in a neighbourhood $\mathcal{U}$ of Γ if and only if the domain Ω is C^k itself. The gradient of b_Ω is an extension to that neighbourhood $\mathcal{U}$ of the normal vector field n defined on Γ. The hessian matrix $D^2 b$ (second order derivatives of b, Γ having a C^2 boundary) contains the curvatures and the main directions of curvatures (respectively in its eigenvalues and eigenfunctions). Γ being compact it is convenient to consider $\mathcal{U}$ in tubular form:$\mathcal{U}_h(\Gamma) = \{x \in D \mid |b_\Omega(x)| < h\}$. The projection mapping defined from $\mathcal{U}$ onto Γ is differentiable and we have:

$$(23) \qquad DP^* = DP = Id - \nabla b.\nabla b_\Omega{}^* - b_\Omega D^2 b_\Omega; \qquad DP.\tau = \tau; \qquad DP.n = 0.$$

Also the tangential operators are easily expressed with that projection mapping,for example fom [7]:

Lemma 4.1. *Φ being an element in $H^1(\mathcal{U})$, that is defined in a neighbourhood $\mathcal{U}$ of Γ and such that: $\Phi = \varphi oP, \varphi \in H^1(\Gamma)$, we have then $\nabla\Phi = (\nabla_\Gamma\varphi)oP$.*

The proof is based on the remark that $\nabla\Phi =^* DP.(\nabla_\Gamma\varphi)oP$.
An important issue is the characterization of the following shape boundary derivatives,

$$(\nabla_{\Gamma_t}[\Phi|_{\Gamma_t}])'_\Gamma, \quad (\Delta_{\Gamma_t}[\Phi|_{\Gamma_t}])'_\Gamma, \text{ with } \Phi \in \mathcal{D}(I\!R^N)$$

4.2. Basic Shape Tangential Derivatives of Test Functions.

Proposition 4.1. *Let $\Phi \in \mathcal{D}(I\!R^N)$ in the form $\Phi|_\Gamma = \varphi op$. The shape boundary derivative of its tangential gradient verifies:*

(24)

$$\widehat{(\nabla_{\Gamma_t}\phi)}_{|_\Gamma} = -D^2 b.\nabla_\Gamma\varphi V.n + \nabla_\Gamma\varphi.\nabla_\Gamma(V.n)\vec{n} - \nabla_\Gamma\varphi.(D^2 b.V_\Gamma)\vec{n} + ({}^*D_\Gamma(\nabla_\Gamma\varphi).V_\Gamma)_\Gamma \quad on \quad \Gamma.$$

Proof. In view of definition 2.2 by definition we have, n_t being the normal to Γ_t,

$$\nabla_{\Gamma_t}\Phi = \nabla\Phi|_{\Gamma_t} - \nabla\Phi.n_t n_t.$$

Then by a direct calculus we get the following expression:

Lemma 4.2. *The material derivative on Γ of $\nabla_{\Gamma_t}\Phi$ is given by:*

$$\widehat{\nabla_{\Gamma_t}\Phi}|_\Gamma = -({}^*DV.\nabla\Phi)_\Gamma + \nabla_\Gamma(\dot{\Phi}) - \nabla\Phi.\dot{n}n - \nabla\Phi.n\dot{n}, \qquad on\,\Gamma.$$

Lemma 4.3. *φ being defined on Γ we consider $\Phi = \varphi oP$, then the material derivative of Φ is given by:*

$$\dot{\phi}(V) =^* DP(\nabla_\Gamma\varphi)oP.V \qquad and \qquad \dot{\phi}(V)|_\Gamma = \nabla_\Gamma\varphi.V_\Gamma.$$

We have the following decomposition on Γ for the transposed Jacobian matrix DV :

Lemma 4.4.

$$^*DV = ^* D_\Gamma V_\Gamma + V.n^* D_\Gamma n + \nabla_\Gamma(V.n).^*n + n.^*(DV.n).$$

Lemma 4.5. *n is the normal to Γ, b the oriented distance function and τ a tangential vector,* $D_\Gamma n = D^2 b|_\Gamma$, $\quad D_\Gamma n.\tau = D^2 b.\tau$, $\quad D_\Gamma n.n = -\frac{\partial}{\partial n}(\nabla b) = 0.$

In the expression of lemma 4.2, we use lemmas 4.1, 4.3 and 2.1 and we get

(25)

$$\widehat{\nabla_{\Gamma_t}\dot{\Phi}} = -(^*DV.(\nabla_\Gamma\varphi)oP)_\Gamma + (\nabla(^*DP.(\nabla_\Gamma\varphi)oP.V))_\Gamma - \nabla\Phi.(-\nabla_\Gamma(V.n) + D_\Gamma n.V_\Gamma)n..$$

in the third term of the right hand side of (25) we have, in view of Lemma 4.1 :

$$(26) \qquad -\nabla\Phi.(-\nabla_\Gamma(V.n) + D_\Gamma n.V_\Gamma)n = -\nabla_\Gamma\varphi(-\nabla_\Gamma(V.n) + D^2 b.V_\Gamma)n.$$

From (25) and (26) we conclude the proof of Proposition 4.1 with Lemma 4.4 and the following result:

Lemma 4.6. $\nabla_\Gamma(\dot{\widehat{\varphi o p}}) = - < V, n > D^2 b.\nabla_\Gamma\varphi +^* D_\Gamma(\nabla_\Gamma\varphi).V_\Gamma \;+\; (^*DV.\nabla_\Gamma\varphi)_\Gamma$ *on* Γ

Proof. In the second term of the right hand side of (25), we consider the term $\mathcal{A} = (\nabla(^*DP.(\nabla_\Gamma\varphi)oP.V))$. We shall use the following identity:
Let E and V be respectively a matrix and a vector. The following identity holds:

$$\nabla_\Gamma(E.V) = {}^*D_\Gamma E.V + {}^*D_\Gamma V.E.$$

Then $\mathcal{A} = \mathcal{A}_1 + \mathcal{A}_2$, with

$$\mathcal{A}_1 = {}^*D(^*DP.(\nabla_\Gamma\varphi)oP).V_\Gamma$$
$$\mathcal{A}_2 = {}^*DV.(^*DP.(\nabla_\Gamma\varphi)oP)$$

we have:

$$(27) \qquad {}^*D(^*DP.(\nabla_\Gamma\varphi)oP)_{ij} = \partial_i[(^*DP)_{jk}](\nabla_\Gamma\varphi)oP)_k + (^*DP)_{jk}\partial_i[(\nabla_\Gamma\varphi)oP)_k]$$

but as

$$\partial_i[(^*DP)_{jk} = 0 - \partial_i(\partial_j b\partial_k b) - \partial_i b\partial^2_{jk}b - b\partial^3_{ijk}b \,(\text{ and } b = 0 \text{ on } \Gamma)$$

and

$$(28) \qquad \partial_i[(\nabla_\Gamma\varphi)oP)_k] = \partial_i[(\nabla_\Gamma\varphi)_k oP] = \partial_i[(\nabla_\Gamma\varphi.e_k)oP] = (DP)_{il}(\partial_l(\nabla_\Gamma\varphi.e_k))oP$$

then on Γ we get:

$$(29) \qquad (\mathcal{A}_1)_i = \{\partial_i(DP_{jk})(\nabla_\Gamma)_k + (DP)_{jk}\partial_i(\nabla_\Gamma\varphi)_k\}V_j,$$

$$(\mathcal{A}_1)_i = \{-(\partial_i(\partial_j b\partial_k b) + \partial_i b\partial^2_{jk}b)(\nabla_\Gamma\varphi)_k + (\delta_{jk} - \partial_j b\partial_k b - b\partial^2_{jk}b)\partial_i(\nabla_\Gamma\varphi)_k\}V_j$$

and finally in a similar way, we get for the ith component of $\mathcal{A}_2$:

$$(30) \qquad (\mathcal{A}_2)_i = \partial_i V_j\{(\delta_{jk} - \partial_j b\partial_k b - b\partial^2_{jk}b)(\nabla_\Gamma\varphi)_k\}.$$

Then we can write the final expressions for the tangential components $(\mathcal{A}_1)_\Gamma$ and $(\mathcal{A}_2)_\Gamma$ of $\mathcal{A}_1$ and $\mathcal{A}_2$:

$$(\mathcal{A}_1)_\Gamma = -V.nD^2 b.\nabla_\Gamma\varphi +^* D_\Gamma(\nabla_\Gamma\varphi).V_\Gamma \qquad (\mathcal{A}_2)_\Gamma = (^*DV.\nabla_\Gamma\varphi)_\Gamma$$

amd the Lemma 4.6 is proved. Notice that :
${}^*D_\Gamma(\nabla_\Gamma\varphi) = (\nabla_\Gamma((\nabla_\Gamma\varphi)_1), ..., \nabla_\Gamma((\nabla_\Gamma\varphi)_N))$ and then $(D_\Gamma(\nabla_\Gamma\varphi).V)_i = \nabla_\Gamma((\nabla_\Gamma\varphi)_i).V.$

Proposition 4.2. *With $\Phi = \varphi op$ we get:*

(31) $$(\nabla_{\Gamma_t}\phi)'_\Gamma = -D^2b.\nabla_\Gamma\varphi V.n + \nabla_\Gamma\varphi.\nabla_\Gamma(V.n)\vec{n}.$$

Proof. The shape derivative $(\nabla_{\Gamma_t}\Phi)'_\Gamma$, defined by $(\nabla_{\Gamma_t}\Phi)'_\Gamma = \dot{\widehat{\nabla_{\Gamma_t}\Phi}} - D_\Gamma(\nabla_\Gamma\varphi).V_\Gamma$ is equal to:

(32)
$$(\nabla_{\Gamma_t}\Phi)'_\Gamma = -D^2b.\nabla_\Gamma\varphi V.n + \nabla_\Gamma\varphi.(\nabla_\Gamma(V.n) - D^2b.V_\Gamma)\vec{n} + \{{}^*D_\Gamma(\nabla_\Gamma\varphi) - D_\Gamma(\nabla_\Gamma\varphi)\}.V_\Gamma.$$

This expression is not satisfying as according to the structure's theorem it should only depend on the normal component of the vector field V. In that direction we conclude making use of the following result :

Lemma 4.7. $\{ {}^*D_\Gamma(\nabla_\Gamma\varphi) - D_\Gamma(\nabla_\Gamma\varphi)\}.V_\Gamma = < \nabla_\Gamma\varphi, D^2b.V_\Gamma > n.$

Proof. The vector ${}^*D_\Gamma(\nabla_\Gamma\varphi).V_\Gamma$ has no normal component. Conversely, $D_\Gamma(\nabla_\Gamma\varphi).V_\Gamma$ has a normal and a tangential components. If τ design a tangential vector, we have:
$< D_\Gamma(\nabla_\Gamma\varphi).V_\Gamma, n > + < \nabla_\Gamma\varphi, D^2b.V_\Gamma >= 0$
and
$< \{{}^*D_\Gamma(\nabla_\Gamma\varphi) - D_\Gamma(\nabla_\Gamma\varphi)\}.V_\Gamma, \tau >= 0,$
the Lemma is then proved.

Proposition 4.3. *Let $\Phi \in \mathcal{D}(I\!R^N)$ be such that $\Phi|_\Gamma = \varphi$. We have:*

(33) $$(\Delta_{\Gamma_t}[\Phi|_{\Gamma_t}])'_\Gamma = -div_\Gamma\{(2D^2b.\nabla_\Gamma\varphi - H\nabla_\Gamma\varphi)V.n\} - H\Delta_\Gamma\varphi V.n$$

For shortness we shall simply write $(\Delta_{\Gamma_t}\Phi)'_\Gamma$ for $(\Delta_{\Gamma_t}[\Phi|_{\Gamma_t}])'_\Gamma$, the shape boundary derivative of the Laplace-Beltrami operator of the restriction of Φ to the moving boundary .
Proof. In order to characterize the shape derivative of $\Delta_{\Gamma_t}\Phi$, we use the by parts integration formula:

(34) $$\int_{\Gamma_t} \Delta_{\Gamma_t}\Phi.\Psi d\Gamma_t = \int_{\Gamma_t} -\nabla_{\Gamma_t}\Phi.\nabla_{\Gamma_t}\Psi d\Gamma_t, \quad \forall\Psi \in \mathcal{D}(I\!R^N)$$

then using the derivative of an integral defined on Γ_t we get, $\forall\Psi \in \mathcal{D}(I\!R^N)$:

$$\int_\Gamma \left\{(\Delta_{\Gamma_t}\Phi)'_\Gamma.\Psi d\Gamma_t + \Delta_\Gamma\varphi\frac{\partial\Psi}{\partial n}V.n + H\Delta_\Gamma\varphi V.n\Psi\right\}d\Gamma$$
$$= -\int_\Gamma \left\{(\nabla_{\Gamma_t}\Phi)'_\Gamma.\nabla_\Gamma\Psi + \nabla_\Gamma\varphi.(\nabla_{\Gamma_t}\Psi)'_\Gamma + H\nabla_\Gamma\varphi.\nabla_\Gamma\Psi V.n\right\}d\Gamma,$$

using proposition 4.2, Ψ being chosen such that $\frac{\partial\Psi}{\partial n} = 0$ on Γ, we have, $\forall\Psi \in \mathcal{D}(I\!R^N)$:

$$\int_\Gamma \left\{(\Delta_{\Gamma_t}\Phi)'_\Gamma + H\Delta_\Gamma\varphi V.n\right\}\Psi d\Gamma$$
$$= \int_\Gamma \left\{ < D^2b.\nabla_\Gamma\varphi, \nabla_\Gamma\Psi > + < \nabla_\Gamma\varphi, D^2b.\nabla_\Gamma\Psi > -H\nabla_\Gamma\varphi.\nabla_\Gamma\Psi\right\}V.nd\Gamma.$$

Regrouping these terms we get (33) and proposition 4.3 is proved.

4.3. The Result.

Theorem 4.1. *Let y_t be the solution of the problem (1) in Γ_t with the right hand side F in $H^{\frac{3}{2}+\epsilon}(D)$. Then the shape boundary derivative y'_Γ of y_t exists in $L^2(\Gamma)$ and is the unique element in $\frac{H^1(\Gamma)}{I\!R}$ solution of the following equation:*

$$-\Delta_\Gamma y'_\Gamma = -2div_\Gamma\Big\{(D^2b - \frac{1}{2}HId)\nabla_\Gamma y V.n\Big\} + \overline{(\frac{\partial F}{\partial n} + HF)V.n}$$

where y is the solution of the problem (1) in Γ, H is the main curvature of Γ, D^2b is the curvatures' matrix and $(D^2b - \frac{1}{2}HId)$ is its deviatoric part.

Proof.
The weak formulation of the problem associated with y_t leads to

$$\int_\Gamma y'_\Gamma \Delta_\Gamma \varphi d\Gamma = \int_\Gamma -(\frac{\partial F}{\partial n} + HF)V.n\varphi d\Gamma - \int_\Gamma \Big\{y(\Delta_{\Gamma_t}\Phi)'_\Gamma + Hy\Delta_\Gamma\varphi V.n\Big\}d\Gamma, \qquad \forall\varphi \in \mathcal{D}(I\!R^N)$$

that is, using proposition 4.3:

$$\int_\Gamma y'_\Gamma \Delta_\Gamma \varphi d\Gamma = \int_\Gamma \Big\{-(\frac{\partial F}{\partial n} + HF) < V, n > \varphi - Hy\Delta_\Gamma\varphi < V, n > \Big\}d\Gamma + \int_\Gamma Hy\Delta_\Gamma\varphi < V, n > d\Gamma$$

$$+ \int_\Gamma ydiv_\Gamma\Big\{(2D^2b - HId)\nabla_\Gamma\varphi V.n\Big\}d\Gamma, \qquad \forall\varphi \in \mathcal{D}(I\!R^N)$$

and finally, by use of integration by parts we get the result of theorem 4.1.

5. The Problem with Dirichlet Conditions

We turn to the case where ω is an open subset of Γ with a relative boundary in Γ denoted by γ; γ is a manifold of codimension 2 in $I\!R^N$ and is also of class C^2. Then we consider the Laplace Beltrami equation, with a right-hand side F with the same regularity as previously. On γ we consider the Dirichlet boundary condition:

(35) $$-\Delta_\Gamma z = f \text{ in } \omega, \qquad z = 0 \text{ on } \gamma$$

and it has a unique solution z in $H_0^1(\omega)$. Here, f denotes the restriction of F to ω.
We shall denote by ν the normal field, outgoing to ω and contained in Γ; n still denotes the normal field outgoing to Ω ($\Gamma = \partial\Omega$). Then we have $< \nu, n >= 0$. $< V, \nu >$ will denote the component of V following the unitary vector ν and similarly, $< \nabla_\Gamma z, \nu >=< \nabla z, \nu >= \frac{\partial z}{\partial \nu}$ is the component of $\nabla_\Gamma z$ following ν. We still have $T_t(\Gamma) = \Gamma_t$ and then $T_t(\omega) = \omega_t$, $T_t(\gamma) = \gamma_t$. Notice that the space $H_0^1(\omega)$ is transported into the space $H_0^1(\omega_t)$. The operator Δ_Γ on ω will denote the restriction to ω of Δ_Γ. Let z_toT_t be the transport of the solution z_t of the problem (35) in $H_0^1(\omega_t)$, the mapping $t \to z_toT_t$ is strongly continuous in $H_0^1(\omega)$. It is also strongly differentiable in $H_0^1(\omega)$ and the material derivative $\dot{z}$ of z_t is given by the previous equation (11). This mapping is also differentiable and we have:

Theorem 5.1. *Let z_t be the solution in $H_0^1(\omega_t)$ of the problem (35) formulated in ω_t with F given in $H^{\frac{3}{2}+\epsilon}(D)$. Then the Shape tangential derivative $z'_\Gamma(\omega, V)$ of z_t exists in $L^2(\Gamma)$ and is the unique solution of the following equation in ω:*

(36)
$$-\Delta_\Gamma z'_\Gamma(\omega, V) = -2div_\Gamma\Big\{(D^2b - \frac{1}{2}HId)\nabla_\Gamma z < V(0), n >\Big\} + (\frac{\partial F}{\partial n} + HF) < V(0), n >$$

And boundary condition :

$$z'_\Gamma(\omega, V) = -\frac{\partial z}{\partial \nu} < V, \nu > \quad on \ \ \gamma,$$

where H is the main curvature of Γ, D^2b is the curvatures' matrix and $(D^2b - \frac{1}{2}HId)$ denotes the deviatoric part of the curvature.

Remark 5.1. *The derivative depends on both components $< V(0), n >$ and $< V(0)_\Gamma, \nu >$ of the speed vector field V. If V let both Γ (and resp. ω) globaly invariant (that is $< V, n >= 0$ and -resp. $< V, \nu >= 0$) then the shape tangential derivative $z'_\Gamma(\omega, V)$ is zero.*

Proof. By the use of the same technic (used to characterize the material derivative $\dot{y}(\Gamma, V)$) we would derive a similar characterization for the material derivatives $\dot{Z}(\Gamma, V)$,$\dot{z}(\omega, V) = \dot{Z}(\Gamma, V)|_\omega$, and we would verify that this last element defines a linear and continuous mapping in $\mathcal{L}(E, L^2(\omega))$. Then from the structure theorem we derive the existence of the shape tangential derivative in the form $z'_\Gamma(\omega, V) = \tilde{z}'_\Gamma(\omega, (V.n, V_\Gamma.\nu)\ )$. By the introduction of

$$a_\omega(v) = \tilde{z}'_\Gamma(\omega, (v, 0)\ )\ ,\ \ a_\gamma(w) = \tilde{z}'_\Gamma(\omega, (0, w)\ )$$

we get $z'_\Gamma(\omega, V) = a_\omega(V.n) + a_\gamma(V_\Gamma.\nu)$. The first term $a_\omega(V.n)$ is similar to the previous calculus of the shape boundary derivative $y'_\Gamma(\Gamma, V)$, i.e. the equation solved by $z'_\Gamma(\omega, V)$ in ω is the same that the one solved by $y'_\Gamma(\Gamma, V)$ in the surface Γ. The characterization of the second term $a_\gamma(V_\Gamma.\nu)$ will furnish the boundary condition on γ. For this we consider a tangent vector field to Γ : $V \in E$ with $V.n = 0$. We recall the differentiability of a functionnal depending on the boundary γ_t :

Lemma 5.1. *ψ being a state variable depending on the manifold ω_t, we set $J(\gamma_t) = \int_{\gamma_t} \psi(\omega_t) d\gamma_t$. Then for any vector field V we have:*

(37)
$$dJ(\gamma; V) = \int_\gamma [\ \psi'_\Gamma(\omega; V) + \nabla_\Gamma\psi.V_\Gamma + \psi\,\theta'(0)\ ]d\gamma$$

where θ is the associated "jacobian" in the change of variable.

In dimension 3, if λ is a parametrization of the path γ, the boundary of ω in Γ (it is a closed path) then $\theta(t)(\lambda(s)) = (\ \frac{|(T_t o\lambda)'|}{|\lambda'|}\ )o\lambda^{-1}$. This θ is continuously differentiable on t as the relative boundary γ is of class C^2. Let $\Phi \in \mathcal{D}(I\!\!R^N)$, we have

$$0 = \int_{T_t(\gamma)} z_t\,\Phi\, d\gamma_t = \int_\gamma z_t o T_t\,\Phi o T_t\,\theta(t) d\gamma.$$

From Lemma 5.1 we get:

$$\int_\gamma [z'_\Gamma\Phi + \nabla_\Gamma z.V_\Gamma\ \Phi]\, d\gamma\ = 0$$

and then we derive

$$z'_\Gamma(\omega, V)|_\gamma = -\frac{\partial}{\partial \nu} z(\omega) < V_\Gamma, \nu > .$$

Then the theorem 5.1 is proved.

References

1) J.Sokolowski and J.P. Zolésio, "Introduction to Shape Optimization", Computational Mathematics, vol. 16, Springer Verlag, 1992.

2) F. Desaint and J.P. Zolésio, "Shape Derivative of the First Eigenvalue of the Laplace Beltrami Equation", Boundary Control and Variation, Lecture Note in Pure and Applied Mathematics,163, Chapter 7, Marcel Dekker, inc. 1994 .

3) M.C. Delfour and J.P.Zolésio, "Structure of Shape Derivatives for Nonsmooth Domains", Journal of Functional Analysis, vol. 104, No 1, February 15th, 1992.

4) M.C. Delfour and J.P.Zolésio, "A Boundary Differential Equation for Thin Shells", J. D.E., vol. 119, No 2, July 1st, 1995.

5) J.P. Zolésio, " Identification de Domaines par Déformation", Thèse de doctorat d'état, University of Nice, France, 1979.

6) J.P. Zolésio, "The Material Derivative (or Speed) Method for Shape Optimization", in "Optimization of distributed parameter structures", vol.II, E.J. Haug and J. Céa, Eds, pp.1089 - 1151, Sijhoff and Nordhoff, Alphen aan den Rijn,1981.

7) M.C. Delfour and J.P. Zolésio, " Shape Analysis Via Oriented Distance Functions", J.F.A. **123** (1994), 129-201.

8) J.P. Zolésio, " Introduction to Shape Optimization Problems and Free Boundary Problems", Kluwer Academic Publishers, NATO ASI Series C, Mathematical and Physical Sciences, Vol. 380, M. Delfour ed., p.p. 397-457, 1992.

9) M.C. Delfour and J.P. Zolésio, " Tangential Differential Equations for Dynamical Thin/Shallow shells", J.D.E, 1996.

10) M.C. Delfour and J.P. Zolésio, " On the Design and Control of Systems Governed by Differential Equations on Manifolds ", Control and Cybernetics, Polish Acad. Sc., Warsaw, 1996.

F. R. DESAINT and J. -P. ZOLÉSIO CNRS-INLN, 1361 Route des Lucioles 06560 Valbonne, France.

An Energy Principle for a Free Boundary Problem for Navier–Stokes Equations

Raja DZIRI* and J.P. ZOLÉSIO*

Abstract

We study the existence of an optimal domain for an energy shape minimization problem. By showing that the problem considered can be relaxed to measurable sets we avoid regularity hypotheses. The study leads, in the nonsmooth case, to a distributed optimality condition, which can be expressed, in the smooth case, by an extra-equation satisfied on the boundary of the optimal domain.

1 Introduction

We deal with a steady free boundary problem for a Navier-Stokes flow in three dimensions. Since the model is not variational, the free boundary cannot be derived through a variational principle but via the energy minimization of the given system with respect to the shape of the domain occupied by the fluid, that is a control problem. That energy will take the following form :

$$e_0(\Omega) = e_0(\Omega, u_\Omega) = \int_\Omega (\frac{1}{2}|u_\Omega|^2 + \rho g x_3 + k|\varepsilon(u_\Omega)|^2)dx.$$

To obtain an existence result, we have to add the surface energy $\theta P_D(\Omega)$, where θ is the surface tension and $P_D(\Omega)$ is the perimeter of Ω in D. We denote by $e_\theta(\Omega)$ the sum $e_0(\Omega) + \theta P_D(\Omega)$.
The minimization of $e_\theta(\,.\,)$ on the family of measurable subsets Ω having a given measure is a control problem with quadratic cost $e_0(\,.\,)$ and a highly nonlinear control Ω. Effectively, $e_0(\Omega)$ is defined via the solution u_Ω of the steady Navier-Stokes problem : $-k\Delta u + Du.u + \nabla p = f$ and $divu = 0$ in Ω with the boundary condition $\vec{u}.\vec{n} = 0$ and the associated "natural" condition $(\varepsilon(u).n)_\Gamma = \varepsilon(u).n - < \varepsilon(u)n, n > n$ on Γ, where $(\,.\,)_\Gamma$ denotes the tangential component.
In order to derive existence results, we make the usual physical assumption that in the outer domain, $\Omega^c = D \backslash \overline{\Omega}$, there is another fluid having the same rheology but eventually with arbitrarily small viscosity and/or density. The two fluids will be assumed to be immiscible. Let us denote by $K = k_1\chi_\Omega + k_2\chi_{\Omega^c}$ the function characterizing the viscosity parameters of the two fluids occupying respectively the domains Ω and Ω^c. To be able to manage that condition together with $\vec{u}.\vec{n} = 0$ on Γ and $divu = 0$ in Ω, $divu = 0$ in Ω^c when Ω is a nonsmooth measurable subset of D, we need to relax the definition of the problem. The main point is to transfer the pointwise condition $u.n = 0$ on Γ onto a distributed integral condition. That condition turns out to be the $L^2(D)^3$-orthogonality to all the functions in the form: $\chi_\Omega\nabla p + \chi_{\Omega^c}\nabla q$, with p, $q \in H_0^1(D)$. Using that relaxed formulation and changing that orthogonality condition into a $(H^{-1}(D, I\!R^3), H_0^1(D, I\!R^3))$ "orthogonality", we derive

École des Mines de Paris, Centre de Mathématiques Appliquées, B.P. 207, 06904 Sophia Antipolis, France.

As is usual in control theory, the Eulerian derivative $de_0(\Omega; V)$ will be characterized through an "adjoint" problem which turns out to be associated to a linearization of the problem in the neighborhood of the solution u_Ω and having a forcing term that derives from the chosen energy equation. If $(\mathbb{W}, P)$ is the solution of that linear problem, the necessary optimality condition for the optimality of $e_\theta(\Omega)$ leads to a new kind of free boundary extra condition which is quadratic in the variables u and $\mathbb{W}$. This new condition generalizes the Bernoulli condition through a similar energy principle (cf. [6]). The Bernoulli-like problem is called a "self-adjoint problem". Meanwhile, the steady Navier-Stokes wave problem under consideration is not self-adjoint in that sense that the free boundary condition deriving from the energy principle leads to a coupled system involving one more linear equation. The derived condition is a nonlocal one :

$$\left[\frac{1}{2}\nu|\varepsilon(u)|^2 + < -\nu\nabla_\Gamma(\mathrm{div}_\Gamma u) + \nabla p, \mathbb{W} > -\nu\varepsilon(u)..\varepsilon(\mathbb{W}) - \nabla_\Gamma P.u + \rho g x_3\right]_\Gamma + \theta H = cst \text{ on } \Gamma$$

where $\nu = 2K$, $[\,.\,]_\Gamma$ designates the jump at the boundary Γ, P (resp. p) is the pressure associated to the adjoint problem (resp. to the state equations), H is the mean curvature of Γ.

2 Notation and Relaxed Problem

Let Ω be a measurable subset contained in the universe $D(\subset \mathbb{R}^3)$.
We introduce the following continuous forms :

$$a_0(.,.) : [H_0^1(D, \mathbb{R}^3)]^2 \longrightarrow \mathbb{R}; \ (u,v) \longrightarrow \int_D K\, Du..Dv dx$$

(where K is a symmetric matrix in $L^\infty(D; \mathbb{R}^9)$).

$$a_1(.;.,.) : [H_0^1(D, \mathbb{R}^3)]^3 \longrightarrow \mathbb{R}; \ (u,v,w) \longrightarrow \int_D < Dv.u, w > dx$$

and

$$b(.,.) : H_0^1(D, \mathbb{R}^3) \times \mathcal{E}(\Omega) \longrightarrow \mathbb{R}\,; (v, L) \longrightarrow < L, v >_{H^{-1}(D,\mathbb{R}^3), H_0^1(D,\mathbb{R}^3)}$$

with $\mathcal{E}(\Omega) = \overline{E(\Omega)}^{H^{-1}(D,\mathbb{R}^3)}$ and $E(\Omega) = \{l = \chi_\Omega \nabla p + \chi_{\Omega^c}\nabla q; \ p, q \in H_0^1(D)\}$.
Throughout this paper, we assume that

$$k_2\, Id_{\mathbb{R}^3} \le K \le k_1\, Id_{\mathbb{R}^3}; \quad 0 < k_2 \le k_1. \tag{1}$$

Let $B \in \mathcal{L}(H_0^1(D, \mathbb{R}^3), \mathcal{E}(\Omega)')$; $\mathcal{E}(\Omega)'$ is the dual space of $\mathcal{E}(\Omega)$, defined by :

$$< Bv, l >_{\mathcal{E}(\Omega)', \mathcal{E}(\Omega)} = b(v, L) \ \ \forall l \in \mathcal{E}(\Omega), \ \ \forall v \in H_0^1(D, \mathbb{R}^3).$$

We denote by B^* its adjoint operator.

We consider the following variational problem called (Q):
Given $f \in L^2(D; \mathbb{R}^3)$, find $u \in H_0^1(D, \mathbb{R}^3)$ and $L \in \mathcal{E}(\Omega)$ such that

$$a(u; u, v) + b(v, L) = (f, v) \ \ \forall v \in H_0^1(D, \mathbb{R}^3) \tag{2}$$

$$b(u, l) = 0 \ \ \forall l \in \mathcal{E}(\Omega) \tag{3}$$

where $a(u;v,w) = a_0(u,v) + a_1(u;v,w)$.
For $u \in H_0^1(D,\mathbb{R}^3)$, we define a linear operator $\mathcal{A}(u)$ in $\mathcal{L}(H_0^1(D,\mathbb{R}^3), H^{-1}(D,\mathbb{R}^3))$ by

$$< \mathcal{A}(u)v, w >= a(u;v,w) \ \ \forall v, w \in H_0^1(D,\mathbb{R}^3).$$

We associate with problem (Q) the following problem (P) : Find $u \in \text{Ker}B = X(\Omega)$ such that

$$a(u;u,v) = (f,v) \ \ \forall v \in X(\Omega). \tag{4}$$

Proposition 2.1 *Problem (P) has at least one solution in $X(\Omega)$.*

Proof *The following hypotheses are satisfied :*

- *for all $v \in X(\Omega)$,*
$$a(v;v,v) \geq k_2 \parallel v \parallel^2_{H_0^1(D,\mathbb{R}^3)} . \tag{5}$$
Indeed, it suffices to see that the elements of $X(\Omega)$ are of divergence-free.
Let u in $X(\Omega)$. For all $p \in H_0^1(D)$, ∇p belongs to $\mathcal{E}(\Omega)$.
So, $< \nabla p, u >= \int_D p \ divu dx = 0 \ \ \forall p \in H_0^1(D)$.
Thanks to the density of $H_0^1(D)$ in $L^2(D)$, we obtain $div\, u = 0$ a.e. in D. Then, $a(u;v,v) = 0 \ \forall \ u, v \in X(\Omega)$. Hence $a(v;v,v) = a_0(v,v)$. As the matrix K verifies (1), we obtain inequality (5).

- *the linear space $X(\Omega)$ is separable and, for all $v \in X(\Omega)$, the mapping $u \longrightarrow a(u;u,v)$ is sequentially weakly continuous on $X(\Omega)$ because the injection of $H_0^1(D)$ in $L^r(D)$, $r \in [1,6[$, is compact.*

According to [3], Problem (P) has at least one solution in $X(\Omega)$.

Proposition 2.2 *Problem (Q) has at least one solution $(u,L) \in H_0^1(D,\mathbb{R}^3) \times \mathcal{E}(\Omega)$.*

Proof *The bilinear form b(.,.) satisfies the "inf-sup condition" :*

$$\inf_{l\in\mathcal{E}(\Omega)} \sup_{v\in H_0^1(D,\mathbb{R}^3)} \left\{ \frac{b(v,l)}{\parallel v \parallel_{H_0^1(D,\mathbb{R}^3)} \parallel l \parallel_{\mathcal{E}(\Omega)}} \right\} \geq \eta \tag{6}$$

where $\eta > 0$ is a constant. Indeed, for $l \in \mathcal{E}(\Omega)$, we have

$$\sup_{v\in H_0^1(D,\mathbb{R}^3)} \frac{b(v,l)}{\parallel v \parallel_{H_0^1(D,\mathbb{R}^3)}} = \sup_{v\in H_0^1(D,\mathbb{R}^3)} \frac{< l, v >_{H^{-1}(D,\mathbb{R}^3), H_0^1(D,\mathbb{R}^3)}}{\parallel v \parallel_{H_0^1(D,\mathbb{R}^3)}} = \parallel l \parallel_{H^{-1}(D,\mathbb{R}^3)}$$

Thus, (6) happen with η=1.
Let us recall the definition of the polar set of $X(\Omega)$, $X^0 = \{l \in H^{-1}(D,\mathbb{R}^3), < l, v >= 0 \ \ \forall v \in X(\Omega)\}$.
Now, we shall prove that B^* is an isomorphism from $\mathcal{E}(\Omega)$ onto X^0. It is obvious that B^* is one-to-one from $\mathcal{E}(\Omega)$ onto its range $R(B^*)$ and that its inverse is continuous. From the Closed Range Theorem of Banach, we deduce that $R(B^*) = X^0$. Then, B^* is an isomorphism from $\mathcal{E}(\Omega)$ onto X^0. In that particular case X^0 coincides with $\mathcal{E}(\Omega)$.
Finally let u be a solution of Problem (P). Then, $\mathcal{A}(u)u - f$ is in X^0 . We know that there exists a unique $L \in \mathcal{E}(\Omega)$ such that $\mathcal{A}(u)u - f = -B^*L$ or equivalently in a variational form :

$$a(u;u,v) + b(v,L) = (f,v) \ \ \forall v \in H_0^1(D,\mathbb{R}^3).$$

Proposition 2.3 *Let C=c(D) $\| \mathcal{I} \|$, $\mathcal{I}$ being the canonical injection $H_0^1(D, \mathbb{R}^3) \hookrightarrow L^4(D; \mathbb{R}^3)$ and c(D) the Poincaré's constant.*
Assume

$$k_2^2 - C \| f \|_{L^2(D;\mathbb{R}^3)} > 0 \tag{7}$$

Then problem (Q) has a unique solution.

Proof *Let u_i (i=1,2) be two solutions of Problem (P) and $u = u_1 - u_2$. We have,*

$$a_0(u, v) + a_1(u; u_1, v) + a_1(u_2; u, v) = 0 \ \forall v \in X(\Omega).$$

For $v = u$, we get $a_0(u, u) = -a_1(u; u_1, u)$. Then,

$$\begin{aligned} k_2 \| u \|^2_{H_0^1(D,\mathbb{R}^3)} &\leq \| u \|^2_{L^4(D,\mathbb{R}^3)} \| u_1 \|_{H_0^1(D,\mathbb{R}^3)} \\ k_2 \| u \|^2_{H_0^1(D,\mathbb{R}^3)} &\leq \frac{C(D)}{k_2} \| f \|_{L^2(D;\mathbb{R}^3)} \| u \|^2_{L^4(D,\mathbb{R}^3)} \end{aligned}$$

Using the canonical injection $H_0^1(D, \mathbb{R}^3) \hookrightarrow L^4(D; \mathbb{R}^3)$, we obtain

$$(k_2^2 - C \| f \|_{L^2(D;\mathbb{R}^3)}) \| u \|^2_{H_0^1(D,\mathbb{R}^3)} \leq 0$$

Hypothesis (7) gives : $\| u \|_{H_0^1(D,\mathbb{R}^3)} = 0$ and the uniqueness is proved for Problem (P). We deduce (as in the proof of proposition 2.2) the existence of a unique distribution L in $\mathcal{E}(\Omega)$ such that (u,L) is the unique solution of Problem (Q).

3 Continuity Process

In this section, we study the continuity of Problem (P) with respect to the domain when the matrix K is reduced to the scalar function $k_1\chi_\Omega + k_2\chi_{\Omega^c}$. First, we shall recall an important property of bounded perimeter sets.

3.1 Bounded Perimeter Sets

We denote by $BPS(D)$ the family of bounded perimeter sets of D :

$$BPS(D) = \{\Omega \subset D; measurable|$$

$$\sup \left\{ \int_\Omega div(g)dx | g \in C_c^\infty(D; \mathbb{R}^3), \max_D \| g(x) \| \leq 1 \right\} < \infty.$$

It is immediate that $\{\chi_\Omega | \Omega \in BPS(D)\}$ is contained in $BV(D)$ and the norm of χ_Ω is given by : $\| \chi_\Omega \|_{BV(D)} = meas\, \Omega + \| \nabla \chi_\Omega \|_{M^0(D)}$.
The perimeter of $\Omega \in BPS(D)$ (relative to D))is given by

$$P_D(\Omega) = \sup \left\{ \int_\Omega div\, g dx | g \in C_c^\infty(D; \mathbb{R}^3), \max_D \| g(x) \| \leq 1, \right\}. \tag{8}$$

We have the following compactness result

Lemma 3.1 *Let $\{\Omega_n\}$ be a sequence in $BPS(D)$ such that*

$$P_D(\Omega_n) \leq c.$$

Then, there exists a subsequence $\{\Omega_{n_k}\}$ and $\Omega \in BPS(D)$ such that

$$\chi_{\Omega_{n_k}} \longrightarrow \chi_\Omega \text{ in } L^1(D)(\text{ denoted } \Omega_{n_k} \overset{char(D)}{\longmapsto} \Omega).$$

Moreover, for any g in $C_c(D; I\!R^3)$, we have

$$< \nabla\chi_{\Omega_{n_k}}, g > \longrightarrow < \nabla\chi_\Omega, g > \text{ and } P_D(\Omega) \leq \liminf P_D(\Omega_{n_k}).$$

Remark 3.1 *Note that the perimeter for any measurable subset Ω of D could be defined by (8) as an element of $I\!R^+ \cup \{\infty\}$. For more details see for example [1].*

3.2 Kuratowski Limit

Let us denote by $F^\perp$ the orthogonal of a closed subspace F of $H_0^1(D, I\!R^3)$.
We will often use the characterization of elements in the linear space of divergence free functions (denoted X_0) by means of the curl operator (see [3]).

Lemma 3.2 *Every function v of X_0 has the following form :*

$$v = curl\Phi$$

where $\Phi \in H^2(D, I\!R^3)$ with div Φ=0 is the unique solution of

$$\begin{aligned} (-\Delta\Phi, curl\ w) &= (curl\ v, curl\ w)\ \forall w \in X_0, \\ \Phi.n_D &= 0 \text{ on } \partial D \end{aligned} \tag{9}$$

where n_D is the unit normal vector field on ∂D, directed outward to D.

Let $\Omega_n, n \in I\!N^*$, be a sequence of sets in BPS(D) such that $P_D(\Omega_n) \leq M$ independently of n. We introduce the following continuous forms

$$\begin{aligned} c_{\mu_n} &: X_0 \longrightarrow H_0^1(D, I\!R^3);\ v = \text{curl}\Phi \longmapsto \text{curl}(\mu_n\Phi) \\ b_0(.,.) &: H_0^1(D, I\!R^3) \times L_0^2(D) \longrightarrow I\!R;\ (\varphi, p) \longmapsto -\int_D p \text{ div } \varphi dx =< \nabla p, \varphi >, \end{aligned}$$

where $\mu_n \in C_c^\infty(D)$ such that the support of μ_n satisfies $(\text{spt}\,\mu_n) \cap \partial^*\Omega_n = \emptyset$, $\partial^*\Omega_n$ is the reduced boundary of Ω_n. We know (see for example [2]) that $P_D(\Omega_n) = \mathcal{H}^2(\partial^*\Omega_n)$ $(< +\infty)$, $\mathcal{H}^2$ is the Hausdorff 2-dimensional measure in $I\!R^3$. It is important to notice that using the generalized Gauss-Green formula for finite perimeter sets, we obtain that $c_{\mu_n}(X_0) \subset X_n$.
For any given g in $H^{-1}(D, I\!R^3)$, we consider the following problem :
Find $(\varphi_n, p_n) \in H_0^1(D, I\!R^3) \times L_0^2(D)$ such that

$$\begin{aligned} ((\varphi_n, P_{X_n}\psi)) + \frac{1}{n}((\varphi_n, c_{\mu_n} P_{X_n^\perp \cap X_0}\psi)) + b_0(\psi, p_n) &= < g, \psi >\ \forall \psi \in H_0^1(D, I\!R^3), \\ b_n(\varphi_n, l_n) &= 0\ \forall l_n \in \mathcal{E}(\Omega_n). \end{aligned} \tag{10}$$

Lemma 3.3 *Problem (10) is well-posed.*

Proof *First, it is easy to see that*

$$\forall \varphi \in X_n, \ \sup_{\psi \in X_0} ((\varphi, \psi)) + \frac{1}{n}((\varphi, c_{\mu_n} P_{X_n^\perp} \psi)) \geq \| \varphi \|^2_{H_0^1(D, \mathbb{R}^3)} .$$

It remains to prove that

$$\forall \psi \in X_0, \ \psi \neq 0, \ \sup_{\varphi \in X_n} ((\varphi, \psi)) + \frac{1}{n}((\varphi, c_{\mu_n} P_{X_n^\perp} \psi)) > 0.$$

The worst situation would be when $P_{X_n}\psi = -\frac{1}{n} c_{\mu_n} P_{X_n^\perp}\psi$. *It would imply that* $\tilde{\psi} = P_{X_n}\psi + \frac{1}{n} P_{X_n^\perp}\psi = (1 - c_{\mu_n})\frac{1}{n} P_{X_n^\perp}\psi = c_{1-\mu_n} P_{X_n^\perp}\tilde{\psi}$.

$$\begin{aligned} \| \tilde{\psi} \|_{H_0^1} &\leq \| rot \|_{\mathcal{L}(H^2, H^1)} \| (1 - \mu_n)\Phi_{2,n} \|_{H^2} \leq c_0(D) \| 1 - \mu_n \|_{H^2} \| \Phi_{2,n} \|_{H^2} \\ &\leq c_1(D) \| 1 - \mu_n \|_{H^2} \| \tilde{\psi} \|_{H_0^1} . \end{aligned}$$

With a suitable choice of μ_n, *the previous inequality gives :* $\psi = 0$. *Moreover, the following "inf-sup" conditions :*

$$\sup_{\varphi \in H_0^1(D, \mathbb{R}^3)} \frac{b_n(\varphi, l)}{\| \varphi \|} = \| l \|_{H^{-1}(D, \mathbb{R}^3)} \quad and \quad \sup_{\psi \in H_0^1(D, \mathbb{R}^3)} \frac{b_0(\psi, p)}{|p|_{L^2} \| \psi \|} \geq \beta_0 > 0$$

are obviously satisfied for respectively all $l \in \mathcal{E}(\Omega_n)$ *and all* $p \in L_0^2(D)$. *Then problem (10) is uniquely solvable.*

Theorem 3.1 *Let* Ω_n *be a bounded sequence in* $BPS(D)$.
Assume that $\Omega_n \overset{char(D)}{\longrightarrow} \Omega$. *Then, the linear space* $X(\Omega)$ *is contained in the Kuratowski Limit of* $X(\Omega_n)$.

Proof *Let* v *in* $X(\Omega)$. *From lemma 3.3, there exists a unique pair* $(v_n, p_n) \in X(\Omega_n) \times L_0^2(D)$ *satisfying*

$$((v_n, \psi)) + \frac{1}{n}((v_n, c_{\mu_n} P_{X_n^\perp \cap X_0} \psi)) + b_0(\psi, p_n) = ((v, \psi)) \ \ \forall \psi \in H_0^1(D, \mathbb{R}^3). \tag{11}$$

With $\psi = v_n - v$, *we obtain*
$\| v_n - v \| \leq \frac{c_0}{n} \| v \|$ *(*c_0 *is a constant). Then,* $v_n \longrightarrow v$ *in* $H_0^1(D, \mathbb{R}^3)$*-strong.*

3.3 The continuity result

The fact that $X(\Omega_n)$ is contained in the Kuratowski limit of $X(\Omega)$ when $\Omega_n \overset{char(D)}{\longmapsto} \Omega$ allows us to characterize the weak limit u_Ω as the unique solution of problem (P) relative to Ω.

Theorem 3.2 *Let* Ω_n *be a sequence in* $BPS(D)$. *Assume that* $\Omega_n \overset{char(D)}{\longmapsto} \Omega$. *Then, there exists a subsequence* $\{u_{\Omega_{n_k}}\}$ *such that* $u_{\Omega_{n_k}} \rightharpoonup u_\Omega$ *weakly in* $H_0^1(D, \mathbb{R}^3)$.

Proof *Let* $v \in X(\Omega)$. *There exists a sequence* $\{v_n \in X(\Omega_n)\}_{n \in \mathbb{N}^*}$ *such that*

$$v_n \longrightarrow v \quad strongly \ in \ H_0^1(D, \mathbb{R}^3).$$

As $\chi_{\Omega_n} \to \chi_\Omega$ *in* $L^2(D)$*, we can extract a susbsequence (still denoted* χ_{Ω_n}*) converging to* χ_Ω *almost everywhere in D. Then,* $K(\Omega_n)Dv_n \to K(\Omega)Dv$ *in* $L^2(D, \mathbb{R}^3)$*-strong. This follows from the boundedness of* $K(\Omega_n)$ *in* $L^\infty(D)$ *and the strong convergence of* v_n *in* $H_0^1(D, \mathbb{R}^3)$*. We know that*

$$a_n(u_n; u_n, v_n) - \int_D f v_n dx = 0 \ \forall n \in \mathbb{N}^*.$$

Then, there exists a subsequence such that

$$\begin{array}{ccc} u_{n_k} & \rightharpoonup & u \text{ in } H_0^1(D, \mathbb{R}^3) - weak \\ & and & \\ \chi_{\Omega_{n_k}} Dv_{n_k} & \longrightarrow & \chi_\Omega Dv \text{ in } L^2(D) - strong. \end{array}$$

By passing to the limit, we get

$$a(u; u, v) = \int_D f v dx.$$

This proves the continuity of the solution of Problem (P) with respect to the domain.

3.4 An existence result

The minimization problem we consider is the following

$$\min\{e_\sigma(\Omega) | \Omega \subset D, \text{measurable}, \text{meas}\Omega = m_0\} \tag{12}$$

$$\text{where} \quad e_\sigma(\Omega) = \frac{1}{2}|u_\Omega|^2_{L^2(D)} + \int_D K(\Omega)\varepsilon(u_\Omega)..\varepsilon(u_\Omega)dx + \int_D \rho g x_3 dx + \sigma P_D(\Omega).$$

We denote by $e_0(\Omega)$ the term $e_\sigma(\Omega) - \sigma P_D(\Omega)$ and by $e_1(\Omega)$ the term

$$\frac{1}{2}|u_\Omega|^2_{L^2(D)} + \int_D K(\Omega)\varepsilon(u_\Omega)..\varepsilon(u_\Omega)dx.$$

Proposition 3.1 *There exists, at least, a measurable subset* Ω *in* $BPS(D)$ *that is a solution of the minimization problem (12).*

Proof *Let* Ω_n *be a minimizing sequence. The sequence* $\{e_\sigma(\Omega_n)\}_n$ *being bounded, there exists a constant c such that :*

$$P_D(\Omega_n) \leq c.$$

Therefore, referring to lemma 3.1, $\exists\, \Omega$ *and a subsequence* $\{\Omega_{n_k}\}$ *in* $BPS(D)$ *such that* $\Omega_{n_k} \overset{char(D)}{\longmapsto} \Omega$*,* $\text{meas}\Omega = m_0$ *and* $P_D(\Omega) \leq \liminf P_D(\Omega_{n_k})$*.*

¿From theorem 3.2, there exists a sequence $u_{\Omega_{n_k}}$ *weakly convergent (in* $H_0^1(D, \mathbb{R}^3)$*) to* u_Ω*. Then, using the lower semi-continuity of the norm in* $H_0^1(D, \mathbb{R}^3)$ *and the perimeter, we deduce that* Ω *is a solution of the problem (12) considered .*

4 Differentiability Result

In this section we shall study the derivability of the flow u with respect to the domain. T_t is a given transformation defined in $\overline{D}$.

4.1 Material derivative

We apply the *Speed Method* cf.[4] : Given a family of vector fields $V \in C^0([0,\tau); C^2(\overline{D}, \mathbb{R}^3))$ satisfying :

$$V(t,x).n_D(x) = 0 \text{ for a.e. } x \in \partial D.$$

and

$$\exists c > 0, \forall x, y \in \mathbb{R}^3, \| V(.,y) - V(.,x) \|_{C^0([0,\tau);\mathbb{R}^3)} \leq c|y-x|$$

We know from [4] that there exists an interval I, $0 \in I$, and a family of one-to-one transformations $\{T_t(V), t \in I\}$ mapping $\overline{D}$ onto $\overline{D}$
satisfying :

$$V = \frac{\partial}{\partial t} T_t(V) \circ T_t^{-1}(V).$$

As we deal with incompressible fluids, transformations $\{T_t\}$ that we use shall satisfy $\det DT_t$=1.

Definition 4.1 *For any transformation T_t, $t \in I$, we define the following mappings :*

1. $H_0^1(D, \mathbb{R}^3) \longrightarrow H_0^1(D, \mathbb{R}^3); \quad w \longmapsto DT_t w \circ T_t^{-1}$
2. $\mathcal{E}(\Omega_t) \longrightarrow \mathcal{E}(\Omega); \quad l_t \longmapsto l_t \star T_t = l^t$ *defined by*

$$< l^t, w >=< l_t, DT_t w \circ T_t^{-1} > \quad \forall w \in H_0^1(D, \mathbb{R}^3).$$

Remark 4.1 *It is easy to see that these mappings are in fact isomorphisms.*

For all $t \in I = [0,\tau)$ and $v \in H_0^1(D, \mathbb{R}^3)$, we have

$$\int_D K(\Omega_t) Du_t..Dv dx + \int_D < Du_t.u_t, v > dx + < L_t, v >= \int_D fv dx \tag{13}$$

$$< l_t, u_t >_{H^{-1}(D,\mathbb{R}^3), H_0^1(D,\mathbb{R}^3)} = 0 \quad \forall l_t \in \mathcal{E}(\Omega_t). \tag{14}$$

We apply the change of coordinates defined by the transformation T_t $(t \in I)$:

$$\left\{ \begin{array}{ll} \int_D K(\Omega) D(u_t \circ T_t).DT_t^{-1}..D(v \circ T_t).DT_t^{-1} dx & + \\ \int_D < D(u_t \circ T_t).DT_t^{-1}.(u_t \circ T_t), v \circ T_t > dx & + \\ + < L^t, DT_t^{-1} v \circ T_t >= \int_D f \circ T_t.v \circ T_t dx & \end{array} \right. \tag{15}$$

$$< l^t, DT_t^{-1}(u_t \circ T_t) >= 0 \tag{16}$$

Set $w = DT_t^{-1}(v \circ T_t)$ and $u^t = DT_t^{-1}(u_t \circ T_t)$, we obtain :

$$\left\{ \begin{array}{ll} \int_D K(\Omega) D[DT_t u^t].DT_t^{-1}..D[DT_t w].DT_t^{-1} dx & + \\ \int_D < D[DT_t u^t].u^t, DT_t w > dx & + \\ + < L^t, w >= \int_D f \circ T_t.DT_t w dx \quad \forall w \in H_0^1(D, \mathbb{R}^3) & \end{array} \right. \tag{17}$$

$$< l_t \star T_t, u^t >= 0 \quad \forall l_t \in \mathcal{E}(\Omega_t) \tag{18}$$

Remark 4.2 *Equation (18) can be replaced by $< l, u^t >= 0 \quad \forall l \in \mathcal{E}(\Omega)$.*

We prove the differentiability at the origin of the mapping

$$t \longmapsto u_t \circ T_t.$$

First let us recall a weak form of the Implicit Function Theorem (and its proof), se also [5] :

Theorem 4.1 *Let F and G be two Banach spaces and*

$$\Phi : I \times F \longrightarrow G, \quad I \text{ is an open set in } \mathbb{R}.$$

We assume :

$$\left\{ \begin{array}{l} \forall g' \in G' (\text{ dual of} G) \\ s \longmapsto < \Phi(s,f), g' >_{G\times G'} \text{ is continuously differentiable} \\ \frac{\partial}{\partial s}\Phi(s,f) \text{ denotes its weak derivative.} \end{array} \right. \tag{19}$$

$$(s,f) \longmapsto \frac{\partial}{\partial s}\Phi(s,f) \text{ is continuous from } I \times F \text{ into } G - weak. \tag{20}$$

There exists a function U such that

$$U \in Lip(I,F) \tag{21}$$

$$\Phi(s, U(s)) = 0 \quad \forall s \in I$$

The mapping $f \longrightarrow \Phi(s,f)$ is differentiable and

$$(s,f) \longmapsto \frac{\partial}{\partial f}\Phi(s,f) \text{ is continuous.} \tag{22}$$

At $(s_0, U(s_0))$,

$$\frac{\partial}{\partial f}\Phi(s_0, U(s_0)) \text{ is an isomorphism from } F \text{ onto } G.$$

Then, the mapping $s \longmapsto u(s)$ is differentiable in F-weak on $s = s_0$ and

$$U'(s_0) = -\frac{\partial}{\partial f}\Phi(s_0, U(s_0))^{-1} . \frac{\partial}{\partial s}\Phi(s_0, U(s_0))$$

Proof *Set $t = U(s_0 + \varepsilon) - U(s_0) \in F$. It is obvious that t goes to 0 as ε goes to 0 and that $\Phi(s_0 + \varepsilon, U(s_0) + t) = \Phi(s_0, U(s_0)) = 0$. For any $g' \in G'$, the mapping*

$$(s,f) \longmapsto < \Phi(s,f), g' >$$

is continuously differentiable. Then, for $k > 0$, there exists $r > 0$ such that

$$|\Phi(s_0 + \varepsilon, U(s_0) + t) - \Phi(s_0, U(s_0)) - \varepsilon\frac{\partial}{\partial s}\Phi(s_0, U(s_0)) - \frac{\partial}{\partial f}\Phi(s_0, U(s_0)).t, g' > | \leq k(|\varepsilon| + \parallel t \parallel)$$

which is equivalent to

$$| < \varepsilon T^{-1} . \frac{\partial}{\partial s}\Phi(s_0, U(s_0)) + t, T^* g' > | \leq k(|\varepsilon| + \parallel t \parallel)$$

where $T = \frac{\partial}{\partial f}\Phi(s_0, U(s_0))$.
Since u is Lipschitz-continuous, one can find a constant K such that $\parallel t \parallel \leq K\varepsilon$. Hence, $\forall k > 0$, $\exists r > 0$ such that

$$| < T^{-1} . \frac{\partial}{\partial s}\Phi(s_0, U(s_0)) + \frac{t}{\varepsilon}, T^* g' > | \leq k(1 + K) \quad for \ |\varepsilon| \leq r. \tag{23}$$

Besides, $T^*(G') = F'$. *Thus,*

$$\frac{t}{\varepsilon} \rightharpoonup -T^{-1}.\frac{\partial}{\partial s}\Phi(s_0, U(s_0)) \ (as\ \varepsilon \to 0).$$

in F*-weak.*

We apply the previous theorem to show the existence of the material derivative of u_Ω solution of problem (P). First, we derive the following result

Proposition 4.1 *The mapping*

$$t \longmapsto (u^t, L^t)$$

is weakly differentiable at the origin.

Proof *We apply theorem 4.1 with* $I = [0,\tau)$, $F = X(\Omega)$, $G = X(\Omega)'$ *and*

$$\begin{aligned} < \Phi(s,v), w >_{X',X} &= \textstyle\int_D K(\Omega) D[DT_s v].DT_s^{-1}..D[DT_s w].DT_s^{-1} dx + \\ &+ \textstyle\int_D < D[DT_s v].v, DT_s w > dx - \int_D f \circ T_s.DT_s w dx. \end{aligned} \tag{24}$$

Since $T \in C^1([0,\tau); C^2(\overline{D}, I\!R^3))$, *condition (19) and (20) are obviously satisfied and we have*

$$\begin{aligned} < \frac{\partial}{\partial s}\Phi(s,v), w > &= \int_D K(\Omega)(D[DV(s)v].DT_s^{-1}..D[DT_s w].DT_s^{-1} \\ &- D[DT_s v].DV(s)..D[DT_s w].DT_s^{-1} + D[DT_s v].DT_s^{-1}..D[DV(s)w].DT_s^{-1} \\ &- D[DT_s v].DT_s^{-1}..D[DT_s w].DV(s))dx + \int_D < D[DV(s)v].v, DT_s w > dx \\ &+ \int_D < D[DT_s v].v, DV(s)w > dx - \int_D f \circ T_s.DV(s) w dx \\ &- < Df.V(s), DT_s w >_{H^{-1}(D,I\!R^3), H_0^1(D,I\!R^3)} . \end{aligned} \tag{25}$$

The mapping $t\ (\in I) \longmapsto u^t = DT_t^{-1}(u_t \circ T_t)$ *satisfies* $\Phi(t, u^t) = 0$. *To prove condition (21), we need to compute the difference between equation (4) and equation (17) taken at* $t \in (0,\tau)$. *The first term intervening is*

$$\begin{aligned} &\int_D K(\Omega) D[DT_t u^t] DT_t^{-1}..D[DT_t w] DT_t^{-1} dx - \int_D K(\Omega) Du..Dw dx = \\ &= \int_D K(\Omega) D[DT_t u^t - u] DT_t^{-1}..D[DT_t w] DT_t^{-1} dx \\ &+ \int_D K(\Omega) Du.(DT_t^{-1} - I)..D[DT_t w] DT_t^{-1} dx \\ &+ \int_D K(\Omega) Du..D[(DT_t - I)w] DT_t^{-1} dx + \int_D K(\Omega) Du..Dw(DT_t^{-1} - I) dx \\ &= \int_D K(\Omega) D(u^t - u)..Dw dx + \int_D K(\Omega) D(u^t - u)(DT_t^{-1} - I)..Dw dx \\ &+ \int_D K(\Omega) D(u^t - u) DT_t^{-1}..Dw(DT_t^{-1} - I) dx \\ &+ \int_D K(\Omega) D[(DT_t - I)(u^t - u) DT_t^{-1}..D[DT_t w] DT_t^{-1} dx \\ &+ \int_D K(\Omega) D(u^t - u) DT_t^{-1}..D[(DT_t - I)w] DT_t^{-1} dx \end{aligned} \tag{26}$$

$$+\int_D K(\Omega)D[DT_tu]DT_t^{-1}..D[DT_tw]DT_t^{-1}dx$$
$$+\int_D K(\Omega)Du(DT_t^{-1}-I)..D[DT_tw]DT_t^{-1}dx$$
$$+\int_D K(\Omega)Du..D[(DT_t-I)w]DT_t^{-1}dx+\int_D K(\Omega)Du..Dw(DT_t^{-1}-I)dx$$

The second one is :

$$\int_D < D[DT_tu^t].u^t, DT_tw > dx - \int_D < Du.u, w > dx = \tag{27}$$
$$= \int_D < D[DT_tu^t].u^t - Du.u, DT_tw > dx + \int_D < Du.u, (DT_t - I)w > dx$$
$$= \int_D < D[(DT_t - I)u^t].u^t + D(u^t-u).u^t + Du(u^t-u), DT_tw > dx +$$
$$+ \int_D < Du.u, (DT_t - I)w > dx$$
$$= \int_D < D(u^t-u).u^t + Du(u^t-u), w > dx$$
$$+ \int_D < D(u^t-u).u^t + Du(u^t-u), (DT_t - I)w > dx$$
$$+ \int_D < D[(DT_t - I)u^t].u^t, DT_tw > dx + \int_D < Du.u, (DT_t - I)w > dx$$

Therefore, we obtain :

$$\int_D K(\Omega)D(u^t-u)..Dwdx + \int_D < D(u^t-u)u^t + Du(u^t-u), w > dx =$$
$$= \int_D (f\circ T_t - f).DT_twdx + \int_D f.(DT_t - I)wdx + \ldots$$

Using the fact that $t \longmapsto f\circ T_t$ *is weakly differentiable in the space* $H^{-1}(D, I\!R^3)$ *([4]), the regularity of the transformations* $\{T_t\}$ *and condition (7), we deduce (21). On the other hand, the mapping* $v \longmapsto \Phi(s,v)$ *is clearly differentiable and*

$$(s,v) \longmapsto \frac{\partial}{\partial v}\Phi(s,v) \quad \textit{is continuous.}$$

Indeed, for all $(s,\overline{v}) \in I \times X(\Omega)$,

$$< \frac{\partial}{\partial v}\Phi(s,\overline{v})v, w >_{X',X} \;=\; \int_D K(\Omega)D[DT_sv].DT_s^{-1}..D[DT_sw].DT_s^{-1}dx$$
$$+ \int_D < D[DT_s\overline{v}]v, DT_sw > dx + \int_D < D[DT_sv]\overline{v}, DT_sw > dx \quad \forall v,w \textit{ in } X(\Omega).$$

Finally, for any $F \in X(\Omega)'$, *there exists a unique function* $\overline{u} \in X(\Omega)$ *(under hypothesis(7)) such that :* $\forall w \in X(\Omega)$

$$< \frac{\partial}{\partial v}\Phi(0,u)\overline{u}, w > = < F, w >_{X',X}$$

or equivalently

$$\int_D K(\Omega)D\overline{u}..Dwdx + \int_D < D\overline{u}.u + Du.\overline{u}, w > dx = < F, w >_{X',X}$$

As a consequence, the mapping $t \longmapsto u^t$ is weakly differentiable in $X(\Omega)$ and its derivative (at the origin) is

$$\tilde{u} = -\frac{\partial}{\partial v}\Phi(0,u)^{-1}.\frac{\partial}{\partial t}\Phi(0,u)$$

More explicitly, it means that, for all $v \in X(\Omega)$, $\tilde{u}$ satisfies

$$\int_D K(\Omega)D\tilde{u}..Dvdx + \int_D < Du.\tilde{u} + D\tilde{u}.u, v > dx =$$
$$- \int_D K(\Omega)(D[DV(0)u]..Dv - Du.DV(0)..Dv + Du..D[DV(0)v] - Du..Dv.DV(0))dx$$
$$- \int_D < D[DV(0)u].u, v > dx - \int_D < Du.u, DV(0)v > dx$$
$$+ \int_D f.DV(0)vdx + < Df.V(0), v >_{H^{-1}(D,I\!R^3),H_0^1(D,I\!R^3)} .$$

Besides, from equation (17), we deduce that the mapping $t \longmapsto L^t$ is weakly differentiable in $H^{-1}(D, I\!R^3)$.

Remark 4.3 *We cannot consider the classical Implicit function theorem for it needs more regularity especially the strong differentiability of the mapping $t \longmapsto f \circ T_t$ in $H^{-1}(D, I\!R^3)$. It is proved (cf. [4]) that for any $F \in H^s(D)$, $s \geq 1$,*

$$\frac{F \circ T_t - F}{t} \longrightarrow \nabla F.V(0) \ \ (t \to 0) \ \ \text{strongly in } H^{s-1}(D).$$

If $s - 1 < 0$, the convergence hold only in H^{s-1}-weak.

Corollary 4.1 *The solution of Problem (2)-(3) is weakly differentiable in $H_0^1(D, I\!R^3)$ and the weak material derivative $\dot{u}$ satisfies :*

$$\int_D K(\Omega)D\dot{u}..Dvdx + \int_D < D\dot{u}.u, v > dx + \int_D < Du.\dot{u}, v > dx = \int_D K(\Omega)Du.DV(0)..Dvdx$$
$$- \int_D K(\Omega)Du..D[DV(0)v]dx + \int_D K(\Omega)Du..Dv.DV(0)dx + \int_D < Du.DV(0)u, v > dx$$
$$- \int_D < Du.u, DV(0)v > dx + \int_D f.DV(0)vdx + < Df.V(0), v >_{H^{-1},H_0^1} \ \ \forall v \in X(\Omega).$$

5 Optimality Condition

Let Ω be a solution of the considered minimization Problem (12). Our aim is to derive the necessary optimality condition satisfied by the associated flow.
A distributed necessary optimality condition will be presented when Ω is nonsmooth. In the smooth case, the same condition can be expressed as a boundary equation. In both cases, we need to introduce an "adjoint-Problem".

5.1 Nonsmooth case

Let $(I\!\!U, L) \in H_0^1(D, I\!R^3) \times \mathcal{E}(\Omega)$ be the solution of the following adjoint-Problem :
$\forall \ (v, l) \in H_0^1(D, I\!R^3) \times \mathcal{E}(\Omega)$,

$$\int_D K(\Omega)DI\!\!U..Dvdx + \int_D < Dv.u, I\!\!U > dx + \int_D < Du.v, I\!\!U > dx + < L, v >= \tag{28}$$
$$\int_D uvdx + 2\int_D K(\Omega)\varepsilon(u)..\varepsilon(v)dx$$

$$< l, I\!\!U >= 0. \tag{29}$$

Our attention is turned to $e_1(\Omega)$ and its Eulerian derivative. The remaining terms of $e_\sigma(\Omega)$ are easy to compute.

$$de_1(\Omega;V) = \int_D u\dot{u}dx + 2\int_D K(\Omega)\varepsilon(\dot{u})..\varepsilon(u)dx - \int_D K(\Omega)\varepsilon(u)..S(0)(u)dx, \tag{30}$$

where $S(0)(u) = Du.DV(0) +^* (Du.DV(0))$.
Thanks to equation (28), we can express differently the right hand side term of (30) :

$$\begin{aligned}
de_1(\Omega;V) &= \int_D K(\Omega)D\mathbb{U}..D\dot{u}dx + \int_D < D\dot{u}.u,\mathbb{U} > dx + \int_D < Du.\dot{u},\mathbb{U} > dx + < L,\dot{u} > \\
&- \int_D K(\Omega)\varepsilon(u)..S(0)(u)dx \\
&= < L, DV(0)u > + \int_D K(\Omega)Du.DV(0)..D\mathbb{U}dx - \int_D K(\Omega)Du..D[DV(0)\mathbb{U}]dx \\
&+ \int_D K(\Omega)Du..D\mathbb{U}DV(0)dx - \int_D < Du.u, DV(0)\mathbb{U} > dx + \int_D < Du.DV(0)u,\mathbb{U} > dx \\
&- \int_D K(\Omega)\varepsilon(u)..S(0)(u)dx + \int_D f.DV(0)\mathbb{U}dx + < Df.V(0),\mathbb{U} >
\end{aligned}$$

The "Eulerian semi-derivative" of $P_D(\Omega)$ at Ω in the direction V is defined by (cf. [7]) :

$$\begin{aligned}
d_-P_D(\Omega;V) &= \liminf_{t>0,t\longmapsto 0} t^{-1}(P_D(\Omega_t) - P_D(\Omega)) \\
&= \inf\{\liminf_{n\to\infty} t_n^{-1}(P_D(\Omega_{t_n}) - P_D(\Omega))|\,\{t_n\} \in I\!R^N,\ t_n > 0,\ t_n \longmapsto 0\}
\end{aligned}$$

where $\Omega_t = T_t(\Omega)$.

Proposition 5.1 *Let Ω be an optimal solution in $BPS(D)$ of problem (12), then for any admissible field $V = (V_1, V_2, V_3)$, we have*

$$\begin{aligned}
&\int_D K(\Omega)Du.DV(0)..D\mathbb{U}dx + \int_D K(\Omega)Du..D\mathbb{U}DV(0)dx + \int_D < Du.DV(0)u,\mathbb{U} > dx \\
&+ < Df.V(0),\mathbb{U} > + < L, DV(0)u > - < L, DV(0)\mathbb{U} > - \int_D K(\Omega)\varepsilon(u)..S(0)(u)dx + \\
&\qquad \int_D gV_3(0)dx + \sigma\, d_-P_D(\Omega;V) \geq 0
\end{aligned} \tag{31}$$

Remark 5.1 *We note that in a neighborhood of an optimal domain Ω we have*

$$\begin{aligned}
e_0(\Omega_t) - e_0(\Omega) + \sigma(P_D(\Omega_t) - P_D(\Omega)) &\geq 0 \\
tde_0(\Omega;V) + to(t) + \sigma(P_D(\Omega_t) - P_D(\Omega)) &\geq 0,\ (o(t) \longrightarrow 0;\ as\ t \searrow 0).
\end{aligned}$$

Hence, $d_-P_D(\Omega;V) > -\infty$.
On the other-hand,

$$\begin{aligned}
P_D(\Omega_t) &= \sup\{\int_\Omega div(DT_t^{-1}goT_t)dx|g \in C_c^\infty(D, I\!R^3),\ \max_{x\in D} \| g(x) \| \leq 1\} \\
&\leq \sup_g\{\int_\Omega div((DT_t^{-1} - I)goT_t)dx\} + \sup_g\{\int_\Omega div(goT_t)dx\},\ 0 < \theta < 1.
\end{aligned}$$

Then,

$$\begin{aligned}
P_D(\Omega_t) - P_D(\Omega) &\leq \sup_g\{\int_\Omega div(DT_t^{-1} - I)goT_tdx\} \\
t^{-1}(P_D(\Omega_t) - P_D(\Omega)) &\leq \sup_g\{\int_\Omega div(-DV(\theta\, t)goT_t)dx\}
\end{aligned}$$

This proves that

$$d^-P_D(\Omega;V) = \limsup_{t>0,t\to 0} t^{-1}(P_D(\Omega_t) - P_D(\Omega)) \le \limsup_{t>0,t\to 0} \sup_g \{\int_\Omega div(-DV(\theta\, t)goT_t)dx\}.$$

In general, in the nonsmooth case these $d^-P_D(\Omega;V)$ and $d_-P_D(\Omega;V)$ don't coincide and the perimeter is not derivable.

5.2 Smooth case

We assume in this section that Ω and f are sufficiently smooth.
To simplify, we shall denote by Ω_i, $i = 1,2$, respectively Ω and $D \setminus \overline{\Omega}$.
The adjoint state $(\mathbb{U}, L) \in H_0^1(D, \mathbb{R}^3) \times \mathcal{E}(\Omega)$ is defined as the unique solution of : $\forall\ (v,l) \in H_0^1(D, \mathbb{R}^3) \times \mathcal{E}(\Omega)$,

$$\sum_{i=1}^{2} 2k_i \int_{\Omega_i} \varepsilon(\mathbb{U})..\varepsilon(v)dx + \int_{\Omega_i} < Du.v + Dv.u, \mathbb{U} > dx + < L, v >=$$

$$\int_{\Omega_i} uvdx + 2k_i \int_{\Omega_i} \varepsilon(u)..\varepsilon(v)dx \tag{32}$$

$$< l, \mathbb{U} >= 0 \tag{33}$$

Equation (33) is equivalent to

$$\begin{cases} \text{div}\,\mathbb{U} &= 0 \text{ in } D \\ \mathbb{U}.n &= 0 \text{ on } \Gamma = \partial\Omega \end{cases}$$

The velocity $u_i = u_{|\Omega_i}$ and its correspondent pressure p_i satisfy (i=1,2) :

$$-k_i\Delta u_i + Du_i.u_i + \nabla p_i = f \ \ in\ \ \Omega_i \tag{34}$$

$$divu_i = 0 \ \ in\ \ \Omega_i \tag{35}$$

$$u_i.n = 0 \ \ on\ \ \Gamma \tag{36}$$

$$(\varepsilon(u_i).n)_\Gamma = 0 \ \ on\ \ \Gamma \tag{37}$$

From classical regularity results for elliptic problems, we deduce that
$(u_i, p_i) \in H^2(\Omega_i, \mathbb{R}^3) \times H^1(\Omega_i)$. The expressions of the multipliers L and L are given by

$$L = \chi_\Omega \nabla p_1 + \chi_{\Omega^c} \nabla p_2 \quad \text{and} \quad L = \chi_\Omega \nabla P_1 + \chi_{\Omega^c} \nabla P_2.$$

Let $v \in H_0^1(D, \mathbb{R}^3)$, then

$$\int_{\Omega_i} (-k_i\Delta u_i + Du_i.u_i + \nabla p_i)vdx = \int_{\Omega_i} f.vdx.$$

But,
$$-\int_{\Omega_i} \Delta u_i.vdx = 2\left(\int_{\Omega_i} \varepsilon(u_i)..\varepsilon(v)dx - \int_\Gamma < \varepsilon(u_i)n, v > d\Gamma\right)$$

So the weak form of (34) can be written as follows

$$2k_i \int_{\Omega_i} \varepsilon(u_i)..\varepsilon(v)dx + \int_{\Omega_i} Du_i.u_ivdx + \int_{\Omega_i} \nabla p_ivdx - 2\int_\Gamma k_i < \varepsilon(u_i).\vec{n}, v > d\Gamma = \int_{\Omega_i} f.vdx,\ i = 1,2.$$

We obtain by addition

$$2\int_D K(\Omega)\varepsilon(u)..\varepsilon(v)dx \;+\; \int_D < Du.u, v > dx + \sum_{i=1,2}\int_{\Omega_i} \nabla p_i v dx$$
$$-\; 2\int_\Gamma < (k_1\varepsilon(u_1) - k_2\varepsilon(u_2)).n, v > d\Gamma = \int_D f.vdx$$

Since, this is true for all $v \in H_0^1(D, I\!R^3)$ and taking (37) into account, we deduce that

$$[K < \varepsilon(u).n, n >]_\Gamma =< (k_1\varepsilon(u_1) - k_2\varepsilon(u_2)).n, n >= 0 \quad \text{on } \Gamma, \tag{38}$$

where $[\ ]_\Gamma$ designates the jump at the interface Γ.
The same regularity and continuity properties are satisfied by the adjoint state.
Before the computation of the optimality condition, we need to characterize the shape derivatives (u_i', p_i') of (u_i, p_i). First, we give some preliminary tangential and shape calculus.

Lemma 5.1 *Let E be any smooth vector field defined on Γ. Then,*

$$\nabla_\Gamma(V.E) = {}^*D_\Gamma V.E + {}^*D_\Gamma E.V$$

Proof *We denote by $\overline{E}$ any extension of E such that $\overline{E}_{|\Gamma} = E$. We know that $\nabla(V.\overline{E}) = {}^*DV.\overline{E} + {}^*D\overline{E}.V$ and ${}^*D\overline{E} = {}^*D_\Gamma E + n.\,{}^*(D\overline{E}.n)$. Then,*

$$\nabla_\Gamma(V.E) + \frac{\partial}{\partial n}(V.\overline{E})\vec{n} = ({}^*D_\Gamma V + n.\,{}^*(DV.n)).E + ({}^*D_\Gamma E + n.\,{}^*(D\overline{E}.n)).V$$
$$= {}^*D_\Gamma V.E + {}^*D_\Gamma E.V + ({}^*(DV.n)).E)n + ({}^*(D\overline{E}.n)).V)n$$

If we consider the tangential component of each vector, on the both sides of the equality, we obtain the desired result.

Lemma 5.2 *The shape derivative of $u.n = 0$ on Γ gives*

$$u'(\Omega, V)_{|\Gamma}.n = div_\Gamma((V.n)u) \quad \text{on } \Gamma.$$

Proof $\int_{\Gamma_t} u_t.n_t\varphi d\Gamma_t = 0$ *for all* $\varphi \in C^1(\Gamma_t)$.

$$\int_\Gamma u_t \circ T_t.n_t \circ T_t\psi\omega(t)dt = 0, \quad \forall\psi \in C^1(\Gamma)$$

Since $\omega(o) = 1$, we obtain $\dot{u}.n + u.\dot{n} = 0$.
Recall $u' = \dot{u} - Du.V$ the shape derivative of u and $n_\Gamma' = \dot{n} - D_\Gamma n.V_\Gamma = -\,{}^*D_\Gamma V.n - D_\Gamma n.V_\Gamma$ the boundary shape derivative of n (cf. [4]). Then,

$$\begin{aligned} u'.n &= -< Du.V, n > -u.(n_\Gamma' + D_\Gamma n.V_\Gamma) \\ u'.n &= -< Du.V, n > + < u, \nabla_\Gamma(V.n) > - < u, D_\Gamma n.V_\Gamma > \end{aligned}$$

$$\begin{aligned} < {}^*Du.n, V > &= < ({}^*D_\Gamma u + n.\,{}^*(Du.n))n, V >=< {}^*D_\Gamma u.n, V > + < Du.n, n > V.n \\ u'.n &= < u, \nabla_\Gamma(V.n) > - < D_\Gamma n.u, V_\Gamma > - < {}^*D_\Gamma u.n, V > - < Du.n, n > V.n \end{aligned}$$

Using the fact that $\nabla_\Gamma V.E = {}^*D_\Gamma V.E + {}^*D_\Gamma E.V$ and ${}^*D_\Gamma n = D_\Gamma n$ ([6]), we get

$$\begin{aligned} u'.n &= < u, \nabla_\Gamma(V.n) > - < \nabla_\Gamma(u.n), V_\Gamma > - < Du.n, n > V.n \\ &= < u, \nabla_\Gamma(V.n) > - < Du.n, n > V.n =< u, \nabla_\Gamma(V.n) > + div_\Gamma u(V.n) \\ &= div_\Gamma((V.n)u) \end{aligned}$$

(since $divu = 0$ we can replace $- < Du.n, n >$ by $div_\Gamma u$).

Lemma 5.3

$$\int_\Gamma \varepsilon(u)..\varepsilon(v)V.nd\Gamma = -\int_\Gamma div_\Gamma((V.n)\varepsilon(u)).vd\Gamma + \int_\Gamma < Dv.n + Hv, \varepsilon(u).n > V.nd\Gamma$$

Proof

$$\int_\Gamma \varepsilon(u)..\varepsilon(v)V.nd\Gamma = \int V.n\varepsilon(u)..Dvd\Gamma$$

Indeed, $2\varepsilon(u)..\varepsilon(v) = \varepsilon(u)_{ij}\partial_j v_i + \varepsilon(u)_{ij}\partial_i v_j = 2\varepsilon(u)..Dv$

$$\begin{aligned}\int_\Gamma V.n\varepsilon(u)..Dvd\Gamma &= \int_\Gamma V.n(\varepsilon(u)_i.\nabla v_i)d\Gamma \\ &= -\int_\Gamma v_i div_\Gamma(V.n\varepsilon(u)_i)d\Gamma + \int_\Gamma (\frac{\partial}{\partial n}v_i + Hv_i)(\varepsilon(u)_i.n)V.nd\Gamma \\ &= -\int_\Gamma div_\Gamma((V.n)\varepsilon(u)).vd\Gamma + \int_\Gamma < Dv.n + Hv, \varepsilon(u).n > V.nd\Gamma\end{aligned}$$

Remark 5.2 *In particular if* $divv = 0$ *and* $v.n = 0$, *then*

$$\int_\Gamma \varepsilon(u)..\varepsilon(v)V.nd\Gamma = -\int_\Gamma div_\Gamma((V.n)\varepsilon(u)).vd\Gamma + \int_\Gamma div_\Gamma v\, div_\Gamma u V.nd\Gamma$$

Lemma 5.4 *For any sufficiently smooth* v *such that* $v.n = 0$ *on* Γ, $divv = 0$ *in* Ω_i, $i = 1, 2$, *we have*

$$\int_\Gamma p_i < DV.v - Dv.V, n > d\Gamma = \int_\Gamma p_i div_\Gamma((V.n)v)d\Gamma = -\int_\Gamma \nabla_\Gamma p_i.v(V.n)d\Gamma$$

Proof *recall* ${}^*Dv = {}^*D_\Gamma v + n.{}^*(Dv.n)$, $D_\Gamma n = {}^*D_\Gamma n$

$$\begin{aligned}< DV.v - Dv.V, n > &= < D_\Gamma V.v, n > - < Dv.V, n > \\ &= < D_\Gamma V.v, n > - < V, {}^*D_\Gamma v.n > - < V, (n.{}^*(Dv.n))n > \\ &= < D_\Gamma V.v, n > + < V, D_\Gamma n.v > - < Dv.n, n > V.n \\ &= < v, {}^*D_\Gamma V.n + D_\Gamma n.V > + div_\Gamma v(V.n) \\ &= < v, \nabla_\Gamma(V.n) > + div_\Gamma v(V.n) = div_\Gamma((V.n)v)\end{aligned}$$

Lemma 5.5 *With the hypothesis of the previous lemma, we have*

$$\int_\Gamma < \varepsilon(u).n, DV.v - Dv.V > d\Gamma == \int_\Gamma \nabla_\Gamma(div_\Gamma u).v(V.n)d\Gamma$$

Proof *Since* $(\varepsilon(u).n)_\Gamma = 0$,

$$< \varepsilon(u).n, DV.v - Dv.V >=< \varepsilon(u).n, n >< n, DV.v - Dv.V > .$$

Thus using the expression of $< n, DV.v - Dv.V >$ *given in the previous proof, we obtain the desired result.*

So, denoting by $\nu_i = 2k_i$, we get

Lemma 5.6 *For all* $v \in H^1(\Omega_i, I\!R^3)$, $divv = 0$ *in* Ω_i, $v.n = 0$ *on* Γ *we have*

$$\int_{\Omega_i} \nu_i\varepsilon(u')..\varepsilon(v)dx + \int_{\Omega_i} < Du'.u + Du.u', v > dx =$$

$$\int_\Gamma (-div\sigma_{\Omega_i} + \nabla_\Gamma(\sigma_{\Omega_i}.n.n)).v(V.n)d\Gamma - \int_\Gamma \nu_i\varepsilon(u)..\varepsilon(v)(V.n)d\Gamma$$

Proof $\forall v \in H^1(\Omega_i, I\!R^3),\ divv = 0,\ v.n = 0\ on\ \Gamma,$

$$\nu_i \int_{\Omega_i} \varepsilon(u)..\varepsilon(v)dx + \int_{\Omega_i} < Du.u, v > dx - \int_{\Omega_i} fvdx = 0$$

$v_t = (DT_tv) \circ T_t^{-1},\ v_t \in H^1(\Omega_{i,t}, I\!R^3),\ divv_t = 0\ \ in\ \Omega_{i,t},\ \ v_t.n_t = 0\ \ on\ \Gamma_t,$

$$\nu_i \int_{\Omega_{i,t}} \varepsilon(u_t)..\varepsilon(v_t)dx + \int_{\Omega_{i,t}} < Du_t.u_t, v_t > dx - \int_{\Omega_{i,t}} fv_tdx = 0$$

Deriving with respect to t, we obtain

$$v' = DV.v - Dv.V$$

$$\int_{\Omega_i} \nu_i\varepsilon(u')..\varepsilon(v)dx + \int_{\Omega_i} < Du'.u + Du.u', v > dx =$$
$$\int_{\Omega_i} fv'dx - \int_{\Omega_i} \nu_i\varepsilon(u)..\varepsilon(v')dx - \int_{\Omega_i} < Du.u, v' > dx - \nu_i \int_\Gamma \varepsilon(u)..\varepsilon(v)(V.n)d\Gamma$$
$$- \int_\Gamma (V.n) < Du.u, v > d\Gamma + \int_\Gamma fv(V.n)d\Gamma$$

We have, in addition, $divv' = 0$ *and*

$$\int_{\Omega_i} \nu_i\varepsilon(u)..\varepsilon(v') + < Du.u, v' > dx - \nu_i \int_\Gamma < \varepsilon(u)n, n > v'.nd\Gamma + \int_\Gamma p_iv'.nd\Gamma = \int_{\Omega_i} fv'dx$$

$$\int_{\Omega_i} \nu_i\varepsilon(u)..\varepsilon(v') + < Du.u, v' > dx - \int_\Gamma (\sigma_{\Omega_i}.n.n)(v'.n)d\Gamma = \int_{\Omega_i} fv'dx$$

then,

$$\nu_i \int_{\Omega_i} \varepsilon(u')..\varepsilon(v)dx + \int_{\Omega_i} < Du'.u + Du.u', v > dx =$$
$$- \int_\Gamma (\sigma_{\Omega_i}.n.n) div_\Gamma((V.n)v)d\Gamma - \int_\Gamma \nu_i\varepsilon(u)..\varepsilon(v)(V.n)d\Gamma - \int_\Gamma (< Du.u, v > - fv)V.nd\Gamma$$

We know that
$< DV.v - Dv.V, n >= v.\nabla_\Gamma(V.n) + div_\Gamma v(V.n) = div_\Gamma\ ((V.n)v)$ *and*
$Du.u - f = \nu_i div\varepsilon(u) - \nabla p = div\sigma_{\Omega_i}$ *in* Ω_i *then,*

$$\int_\Gamma (< Du.u, v > - fv)V.nd\Gamma = \int_\Gamma < div\sigma_{\Omega_i}, v > V.nd\Gamma$$

Remark 5.3 *We deduce that*

$$\nu_i \int_\Gamma < (\varepsilon(u')n)_\Gamma, v > d\Gamma = \int_\Gamma (-div\sigma_{\Omega_i} + \nabla_\Gamma(\sigma_{\Omega_i}.n.n)).v(V.n)d\Gamma - \int_\Gamma \nu_i\varepsilon(u)..\varepsilon(v)(V.n)d\Gamma.$$

Proposition 5.2 *Let* Ω *be a sufficiently smooth optimal domain for problem (12). Then,*

$$\left[\frac{1}{2}\nu|\varepsilon(u)|^2 + < -\nu\nabla_\Gamma(div_\Gamma u) + \nabla p, U\!\!\!U > - \nu\varepsilon(u)..\varepsilon(U\!\!\!U) - \nabla_\Gamma P.u + \rho g x_3\right]_\Gamma + \theta H = cst$$

Proof *The shape derivative of*

$$e_1(\Omega) = \sum_{i=1}^{2} \int_{\Omega_i} \frac{\nu_i}{2} |\varepsilon(u)|^2 + \frac{1}{2}|u|^2 dx$$

is given by

$$de_1(\Omega, V) = \sum_{i=1}^{2} \int_{\Omega_i} \nu_i \varepsilon(u)..\varepsilon(u') + u.u'dx + \int_\Gamma \left[\frac{\nu}{2}|\varepsilon(u)|^2\right]_\Gamma V.nd\Gamma.$$

The adjoint state $\mathcal{W} \in H_0^1(div, D)$ *satisfies* $\mathcal{W}.n = 0$ *on* Γ *and*

$$\sum_{i=1}^{2} \int_{\Omega_i} \nu_i \varepsilon(\mathcal{W})..\varepsilon(v)dx + \int_{\Omega_i} < Dv.u + Du.v, \mathcal{W} > dx + \int_{\Omega_i} \nabla P_i v dx = \int_{\Omega_i} \nu_i \varepsilon(u)..\varepsilon(v) + uvdx.$$

Then, for $v = u'$, *we obtain*

$$\sum_{i=1}^{2} \int_{\Omega_i} \nu_i \varepsilon(u)..\varepsilon(u') + uu'dx = \int_{\Omega_i} \nu_i \varepsilon(\mathcal{W})..\varepsilon(u')dx + \int_{\Omega_i} < Du'.u + Du.u', \mathcal{W} > dx + \int_{\Omega_i} \nabla P_i u'dx.$$

Using the equation satisfied by u', *we get*

$$\sum_{i=1}^{2} \int_{\Omega_i} \nu_i \varepsilon(u')..\varepsilon(\mathcal{W})dx + \int_{\Omega_i} < Du'.u + Du.u', \mathcal{W} > dx = \int_\Gamma < [\nu\varepsilon(u')n]_\Gamma, \mathcal{W} > d\Gamma$$

$$\sum_{i=1}^{2} \int_{\Omega_i} \nu_i \varepsilon(u)..\varepsilon(u') + uu'dx = \int_\Gamma [P_i(u'.n)]_\Gamma d\Gamma + \int_\Gamma < [\nu\varepsilon(u')n]_\Gamma, \mathcal{W} > d\Gamma.$$

But

$$\int_\Gamma [P_i(u'.n)]_\Gamma d\Gamma = \int_\Gamma [P_i div_\Gamma((V.n)u)]_\Gamma d\Gamma = -\int_\Gamma [\nabla_\Gamma P_i]_\Gamma.u(V.n)d\Gamma$$

and

$$\int_\Gamma < [\nu\varepsilon(u')n)]_\Gamma, \mathcal{W} > d\Gamma = \int_\Gamma [-div\sigma_{\Omega_i} + \nabla_\Gamma(\sigma_{\Omega_i}.n.n)]_\Gamma.\mathcal{W}(V.n)d\Gamma - \int_\Gamma \nu\varepsilon(u)..\varepsilon(\mathcal{W})V.nd\Gamma$$

Thus,

$$\sum_{i=1}^{2} \int_{\Omega_i} \nu_i \varepsilon(u)..\varepsilon(u') + uu'dx = -\int_\Gamma [\nabla_\Gamma P_i]_\Gamma.u(V.n)d\Gamma + \int_\Gamma [-div\sigma + \nabla_\Gamma(\sigma.n.n)]_\Gamma.\mathcal{W}(V.n)d\Gamma$$
$$-\int_\Gamma \nu\varepsilon(u)..\varepsilon(\mathcal{W})V.nd\Gamma$$

and

$$de_1(\Omega; V) = \int_\Gamma \left[\frac{1}{2}\nu|\varepsilon(u)|^2 - \nabla_\Gamma P.u + < -div\sigma + \nabla_\Gamma(\sigma.n.n),\ \mathcal{W} > -\nu\varepsilon(u)..\varepsilon(\mathcal{W})\right]_\Gamma V.nd\Gamma$$

Hence the necessary optimality condition can be derived.

6 References

[1] M.C. Delfour and J.P. Zolésio, Shape optimization, Comett Matari Programme, Mathematical Toolkit for Artificial Intelligence and Regulation of Macro-systems, INRIA Sophia-Antipolis, 1993.

[2] H. Federer, Geometric Measure theory, Berlin Heidberg New-York. Springer, 1969.

[3] V. Girault and P.A. Raviart, Finite element methods for Navier-Stokes equations, theory and algorithms, SCM, 5, Springer-Verlag, 1986.

[4]J. Sokolowski and J.P.Zolésio, Introduction to shape optimization, SCM,vol16, Springer-Verlag,1992.

[5] J.P. Zolésio, Identification de domaines par déformations, doctorat d'état, univesity of Nice, 1979.

[6]————, Numerical algorithms and existence results for Bernoulli like steady free boundary problems, Large scale systems, theory and applications,North-Holland,Amsterdam, 1984.

[7]————, Weak shape formulation of free boundary problems, Annali della Scuola Normale di Pisa, IV. vol XXI. Fasc 1, pp. 11-44, 1994.

Dynamic Programming Techniques in the Approximation of Optimal Stopping Time Problems in Hilbert Spaces

Roberto Ferretti

Dipartimento di Matematica, Seconda Universita' di Roma "Tor Vergata"
Via Fontanile di Carcaricola, 00133 Roma, Italy

1. Introduction.

In this paper we consider numerical approximations of optimal stopping time problems for evolution equations in the general form

$$\begin{cases} y'(t) = A(y(t)) \\ y(0) = x \end{cases} \tag{1.1}$$

posed in some Hilbert space H. The problem is to choose (for each initial state x) a stopping time for equation (1.1) so as to minimize the cost

$$J(x,t) = \int_0^t e^{-\lambda s} g(y(s))ds + \Phi(y(t)). \tag{1.2}$$

over all $t \geq 0$. Here, the two terms play the roles respectively of a running and of a stopping cost. A numerical discretization of (1.1), (1.2) requires a finite dimensional (semi–discrete) approximation of (1.1), that is

$$\begin{cases} y_n'(t) = A_n(y_n(t)) \\ y_n(0) = P_n x \end{cases} \tag{1.1_n}$$

and a corresponding discretization for (1.2), that is

$$J_n(P_n x, t) = \int_0^t e^{-\lambda s} g(y_n(s))ds + \Phi(y_n(t)). \tag{1.2_n}$$

We have shown in [Fe] that, under suitable assumptions, semi–discrete approximations of a large class of non–quadratic control problems provide minimizing sequences of approximate optimal solutions. However, although the convergence theory is quite satisfactory, the application of Dynamic Programming techniques for the numerical computation of optimal solutions poses severe complexity problems. Our aim is precisely to use (1.1), (1.2) as a simple test problem to examine complexity and reliability of numerical schemes for its approximation.

The outline of the paper is the following. In section 2 we state the problem and prove convergence of approximate value functions adapting the arguments in [Fe]. Section 3 is devoted to the convergence of optimal solutions, whereas in section 4 we show the fully discrete algorithm and discuss the results of a numerical test.

2. Basic assumptions and convergence of value functions.

We will assume a state equation in the form (1.1), where x and $y(t) = y(x,t)$ belong to a separable real Hilbert space H, and A is a (possibly nonlinear) operator on H. We assume moreover that, for any initial state x, (1.1) has a unique continuous global solution, that is

$$y(x, \cdot) \in C([0, +\infty[; H). \tag{$h1$}$$

We consider a semi-discrete approximation of (1.1) in the form (1.1_n) (see [GO], [LT], [RT]). Let $H_n \subset H$ be a sequence of vector spaces, and $P_n : H \to H_n$ be a sequence of projections, and assume:

$$\dim H_n = k_n \tag{$h2a$}$$

$$\lim_{n\to\infty} \|x - P_n x\| = 0 \text{ for any } x \in H \tag{$h2b$}$$

with $k_n \to \infty$.

Approximation (1.1_n) is assumed to be convergent, in the sense that, for all $t > 0$ and $x \in H$:

$$\lim_{n\to\infty} \|y(x,t) - y_n(x,t)\| = 0. \tag{$h3$}$$

We recall our optimal control problem: given the evolution equation (1.1) and the initial state x, find a stopping time $t^* \in [0, +\infty]$ minimizing the cost (1.2), assuming $(h1)$-$(h3)$, and:

$$\lambda \geq 0 \tag{$h4a$}$$

$$g, \Phi : H \to R$$

$$|g(y)| \leq M_g \text{ for any } y \in H \tag{$h4b$}$$

$$|g(y_1) - g(y_2)| \leq L_g \|y_1 - y_2\| \text{ for any } y_1, y_2 \in H \tag{$h4c$}$$

$$|\Phi(y)| \leq M_\Phi \text{ for any } y \in H \tag{$h4d$}$$

$$|\Phi(y_1) - \Phi(y_2)| \leq L_\Phi \|y_1 - y_2\| \text{ for any } y_1, y_2 \in H. \tag{$h4e$}$$

The approximate version of this problem is, given the evolution equation (1.1_n) with initial state $x_n = P_n x$, to find a stopping time $t_n^* \in [0, +\infty]$ minimizing the cost (1.2_n). Even if both optimal stopping times t^*, t_n^* depend on the initial state, we will drop this dependence in the sequel to simplify notation.

In order to apply the dynamic programming approach, we define the value functions for both the original and the approximate problem:

$$v(x) := \inf_{t \geq 0} J(x,t) \tag{2.1}$$

$$v_n(x_n) := \inf_{t\geq 0} J_n(x_n, t). \tag{2.2}$$

Note that, since $v(x) \leq \Phi(x)$, the boundedness of Φ implies that v is bounded from above on H. We will assume moreover that $v(x)$ is also bounded from below. This happens, for instance, if $g(x) \geq 0$ or $\lambda > 0$.

By Dynamic Programming arguments, it is possible to show that, once defined the sets

$$S = \{x \in H : v(x) = \Phi(x)\} \tag{2.3}$$

$$S_n = \{x_n \in H_n : v_n(x_n) = \Phi(x_n)\} \tag{2.4}$$

the optimal stopping times might be characterized as:

$$t^* = \inf\{t \geq 0 : y(x,t) \in S\} \tag{2.5}$$

$$t_n^* = \inf\{t \geq 0 : y_n(P_n x, t) \in S_n\} \tag{2.6}$$

thus providing the optimal solution in some "feedback form" (the boundaries of S and S_n are stopping surfaces for the state equations).

Our plan is the following. We will prove in theorem 2.1 the convergence of $v_n(P_n x)$ to $v(x)$. In the next section, we will first prove that the approximate optimal stopping times t_n^* are minimizers for the cost (1.2); then, we will discuss the convergence of S_n to the exact stopping set S.

Theorem 2.1. *Assume (h1)-(h4). Assume moreover that either $\lambda > 0$ or*

$$\|y(x,t) - y_n(P_n x, t)\| \leq C(x,n)(1+t)^{-\alpha} \tag{2.7}$$

with $\lim_{n\to\infty} C(x,n) = 0$, and $\alpha > 1$. Then, for any $x \in H$, $|v_n(P_n x) - v(x)| \to 0$ as $n \to \infty$.

Proof. By the definition of $v(x)$, $v_n(P_n x)$, for $x \in H$ and any given $\varepsilon > 0$, it is possible to find two finite stopping times t^ε, t_n^ε (depending on ε) such that:

$$v(x) \leq J(x, t^\varepsilon) \leq v(x) + \varepsilon \tag{2.8}$$

$$v_n(P_n x) \leq J_n(P_n x, t_n^\varepsilon) \leq v_n(P_n x) + \varepsilon. \tag{2.9}$$

Again by the definition of v_n one has:

$$v_n(P_n x) \leq \int_0^{t^\varepsilon} e^{-\lambda s} g(y_n(s))ds + e^{-\lambda t^\varepsilon}\Phi(y_n(t^\varepsilon)). \tag{2.10}$$

Adding the terms $\pm J(x, t^\varepsilon)$, and using (2.8), we obtain:

$$v_n(P_n x) \leq \int_0^{t^\varepsilon} e^{-\lambda s}[g(y_n(s)) - g(y(s))]ds + e^{-\lambda t^\varepsilon}[\Phi(y_n(t^\varepsilon)) - \Phi(y(t^\varepsilon))] + v(x) + \varepsilon \leq$$

$$\leq \int_0^{t^\varepsilon} \min[L_g\|y(s)-y_n(s)\|, 2M_g]ds + L_\Phi\|y(t^\varepsilon)-y_n(t^\varepsilon)\| + v(x) + \varepsilon \tag{2.11}$$

whence we have, using the convergence of the scheme and the dominated convergence theorem:

$$\limsup_{n\to\infty} v_n(P_n x) \leq v(x).$$

On the other hand, using the definition of v, we have:

$$v(x) \leq \int_0^{t_n^\varepsilon} e^{-\lambda s} g(y(s))ds + e^{-\lambda t_n^\varepsilon}\Phi(y(t_n^\varepsilon)). \tag{2.12}$$

Operating as before, one has:

$$v(x) \leq \int_0^{t_n^\varepsilon} e^{-\lambda s} \min[L_g\|y(s)-y_n(s)\|, 2M_g]ds +$$

$$+e^{-\lambda t_n^\varepsilon} \min[L_\Phi\|y(t_n^\varepsilon)-y_n(t_n^\varepsilon)\|, 2M_\Phi] + v_n(P_n x) + \varepsilon. \tag{2.13}$$

We cannot conclude as in (2.11) since the sequence t_n^ε may not be bounded as $n \to \infty$. However, if $\lambda > 0$, we obviously have

$$\int_0^{t_n^\varepsilon} e^{-\lambda s} \min[L_g\|y(s)-y_n(s)\|, 2M_g]ds \leq \frac{2M_g}{\lambda}, \tag{2.14}$$

and if (2.7) holds, we have

$$\int_0^{t_n^\varepsilon} e^{-\lambda s} \min[L_g\|y(s)-y_n(s)\|, 2M_g]ds \leq \frac{L_g C(x,n)}{\alpha - 1} \tag{2.15}$$

so that it is possible again to apply the dominated convergence theorem to the first term in the right–hand side of (2.13). If $\lambda > 0$, the second term satisfies the bounds:

$$e^{-\lambda t_n^\varepsilon} \min[L_\Phi\|y(t_n^\varepsilon)-y_n(t_n^\varepsilon)\|, 2M_\Phi] \leq e^{-\lambda t_n^\varepsilon} L_\Phi\|y(t_n^\varepsilon)-y_n(t_n^\varepsilon)\| \tag{2.16}$$

$$e^{-\lambda t_n^\varepsilon} \min[L_\Phi\|y(t_n^\varepsilon)-y_n(t_n^\varepsilon)\|, 2M_\Phi] \leq e^{-\lambda t_n^\varepsilon} 2M_\Phi. \tag{2.17}$$

Now, (2.16) may be used to prove convergence to zero of its left–hand side on bounded subsequences of t_n^ε, whereas (2.17) accounts for the convergence on subsequences tending to $+\infty$. Lastly, if (2.7) holds, then we have

$$e^{-\lambda t_n^\varepsilon} \min[L_\Phi\|y(t_n^\varepsilon)-y_n(t_n^\varepsilon)\|, 2M_\Phi] \leq L_\Phi C(x,n) \tag{2.18}$$

which again ensures the convergence of the second term in (2.13). Therefore, we may conclude that

$$\liminf_{n\to\infty} v_n(P_n x) \geq v(x)$$

which completes the proof.

∎

Remark. Under proper assumptions (see [L]), the value function $v_n(x)$ may be shown to be the unique (viscosity) solution on R^{k_n} of the first order obstacle problem

$$\max\,[\Phi(x), \lambda v_n(x) - < A_n(x), \nabla v_n(x) > -g(x)] = 0.$$

Example. Consider the heat equation on R^1:

$$\begin{cases} y_t(z,t) = y_{zz}(z,t) \\ y(z,0) = x \\ y(0,t) = y(\pi,t) = 0 \end{cases} \tag{2.19}$$

Here, $z \in]0,\pi[$ denotes the space variable, and for simplicity we set $H = H^1([0,\pi])$. Operating a Fourier expansion on the solution $y(z,t)$, we obtain

$$Y_i(t) = X_i e^{-i^2 t} \tag{2.20}$$

where Y_i represents the i–th component of the Fourier expansion, and

$$X_i = \frac{2}{\pi}\int_0^\pi x(z)\sin izdz.$$

Assume now that $\lambda = 0$, $g(x) \equiv 1$ and $\Phi(x) = \|x\|^2_{L^2}$ (note that assumptions ($h4$) are satisfied on any bounded set of H). Then, the cost may be rewritten as:

$$J(x,t) = t + \frac{\pi}{2}\sum_{i=1}^{\infty} X_i^2 e^{-2i^2 t}. \tag{2.21}$$

This form allows us to explicitly compute the optimal solution. Indeed, differentiating (2.21) with respect to t, we obtain

$$\frac{dJ}{dt} = 1 - \frac{\pi}{2}\sum_{i=1}^{\infty} 2i^2 X_i^2 e^{-2i^2 t} = 1 - \pi\sum_{i=1}^{\infty} i^2 Y_i(t)^2 \tag{2.22}$$

and therefore, for any initial state x, the unique minimum of (2.21) satisfies the condition

$$\pi\sum_{i=1}^{\infty} i^2 Y_i(t^*)^2 = 1 \tag{2.23}$$

that is, the stopping surface is an ellipsoid given by (2.23).

Equation (2.19) may be discretized by truncating the Fourier expansion, so that

$$H_n = \operatorname{span}\{\sin z, \dots, \sin nz\},$$

$$y_{n,i}(t) = Y_i(t) = X_i e^{-i^2 t} \quad (i = 1, \dots, n), \tag{2.24}$$

and the semi-discrete solution is $y_n = (Y_1 \cdots Y_n)$. The corresponding approximate cost will then be given by

$$J_n(P_n x, t) = t + \frac{\pi}{2} \sum_{i=1}^{n} X_i^2 e^{-2i^2 t}. \tag{2.25}$$

Hence, imposing the condition $dJ_n/dt = 0$, one obtains the stopping surface

$$\pi \sum_{i=1}^{n} i^2 y_{n,i}(t_n^*)^2 = 1 \tag{2.26}$$

which is in fact the restriction to H_n of the exact stopping surface.

3. Approximate optimal stopping times.

In the previous section we have proved the convergence of the approximate value function $v_n(P_n x)$ to $v(x)$. We will show now that the approximate optimal stopping times are quasi–optimal for the exact state equation (1.1). In order to do this, let $t_n^* \in [0, +\infty]$ be defined by (2.6). The first result is the following

Theorem 3.1. *Under the assumptions of theorem 2.1, for any $x \in H$, $J(x, t_n^*) \to v(x)$ as $n \to \infty$.*

Proof. We first note that:

$$|v(x) - J(x, t_n^*)| \le |v(x) - v_n(P_n x)| + |v_n(P_n x) - J(x, t_n^*)|. \tag{3.1}$$

The first term in the right–hand side of (3.1) vanishes by theorem 2.1. Moreover, by the definition of v_n and t_n^* we have:

$$|v_n(P_n x) - J(x, t_n^*)| \le \int_0^{t_n^*} e^{-\lambda s} |g(y_n(s)) - g(y(s))| ds. \tag{3.2}$$

The right–hand side of (3.2) may be bounded as in (2.14), (2.15) (replacing t_n^ε with t_n^*) to show that it also vanishes as $n \to \infty$.

■

In general, we do not expect that the approximate stopping sets S_n converge in any sense to the exact stopping set S, as shown by the following

Example. Consider again the example of the previous section, but set now:

$$g(y) = \min\left[1, \pi \sum_{i=1}^{\infty} i^2 Y_i^2\right] \tag{3.3}$$

(note that g is locally Lipschitz continuous since $x \in H^1$). For the exact problem, dJ/dt has the form (2.22) when y is outside the stopping surface (2.23), and is identically zero inside. Therefore the exact stopping set is also the same as above, and if the discretization (2.24) is used, one obtains again the approximate stopping surface (2.26).

On the other hand, let us consider the discretization

$$y_{n,i}(t) = X_i e^{-(i+1/n)^2 t} \quad (i = 1, \ldots, n).$$

If $\pi \sum_i i^2 Y_i(t)^2 > 1$ (that is, if $g = 1$), then

$$\frac{dJ_n}{dt} = 1 - \pi \sum_{i=1}^{n} (i + 1/n)^2 y_{n,i}(t)^2 <$$

$$< \pi \sum_{i=1}^{n} [i^2 - (i + 1/n)^2] y_{n,i}(t)^2$$

and hence $dJ_n/dt < 0$. If $\pi \sum_i i^2 Y_i(t)^2 < 1$ (that is, if $g = \pi \sum_i i^2 Y_i^2$), then

$$\frac{dJ_n}{dt} = \pi \sum_{i=1}^{n} [i^2 - (i + 1/n)^2] y_{n,i}(t)^2$$

and again $dJ_n/dt < 0$. Therefore the cost is a decreasing function of t and the approximate stopping set is empty for any n. Note that this example does not contradict theorem 3.1 since the stopping time $t_n^* \equiv +\infty$ is a minimizer for the cost defined by (3.3).

4. Discretization and numerical tests.

The numerical computation of v_n has been carried out by means of a fully discrete scheme based on the Discrete Dynamic Programming Principle. We will only give a rough idea of its structure, and refer to [CM], [C], [CI], [F], [FF1–2] for a complete convergence theory.

Let $h > 0$ be a time step, $\{x_j\}_{j \in Q}$ be a given set of grid points in the state space R^{k_n}, and $\{\psi_j\}_{j \in Q}$ a set of base functions (defined on the same space), such that $\psi_j(x_j) = 1$, $\psi_j(x_k) = 0$ for any $k \neq j$. Then the numerical scheme reads:

$$v_j = \min\left[\Phi(x_j), hg(x_j) + e^{-\lambda h} \sum_{m \in Q} v_m \psi_m(x_j + hA_n(x_j))\right] \quad (j \in Q). \tag{4.1}$$

Here, v_j approximates $v_n(x_j)$, and $\sum_j v_j\psi_j(x)$ is the fully discrete approximation of $v_n(x)$; therefore $\sum_m v_m\psi_m(x_j+hA_n(x_j))$ approximates (via an Euler scheme) $v_n(y_j(h))$, $y_j(h)$ denoting the solution at time h of (1.1) with initial state x_j. Hence, (4.1) turns out to be a discretization of the Dynamic Programming Principle for problem (1.1), (1.2).

As a test problem, we consider again the heat equation (2.19), with the cost

$$J(x,t) = t + \|y(t)\|_{L^2}^2.$$

The semi-discretization has been carried out as in section 2 by a truncated Fourier expansion. Moreover, to simplify programming, the base $\{\psi_j\}$ is made of characteristic functions (that is, the fully discrete approximation of $v(x)$ is piecewise constant). Since the cost is convex with respect to time, the minimum is unique and we may also expect convergence of stopping times.

We show in figure 1 the section of v_n (with $n=5$) on the X_1X_2 plane as computed by the numerical scheme. Even with only 15 nodes on each side of the grid, the total number of grid points is $15^5 \approx 7.6 \cdot 10^5$.

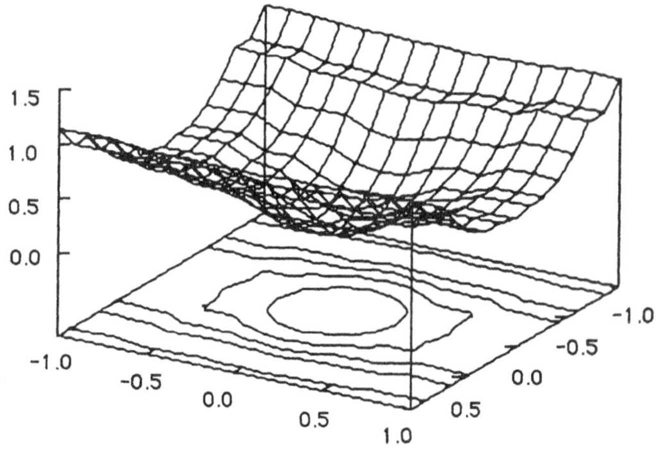

Figure 1. Section of the value function on the X_1X_2 plane.

Figure 2 compares exact and approximate stopping sets, the latter being characterized by the relationship

$$\sum_{j\in Q} v_j\psi_j(x) = \sum_{j\in Q} \Phi(x_j)\psi_j(x).$$

Note that even though the space step is not small enough to provide a good resolution, still the computation is quite accurate.

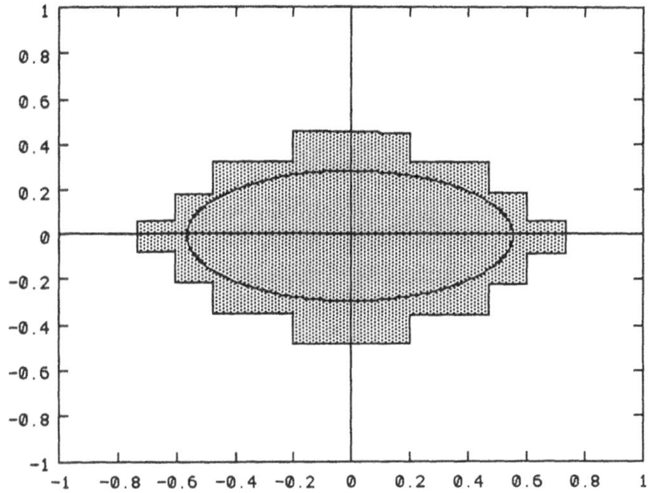

Figure 2. Exact and approximate stopping sets on the X_1X_2 plane.

As it is well known, the main drawback of Dynamic Programming is the high computational complexity. We list in the table below the number of grid points used by the numerical scheme and the CPU time on a VAX 6000–410 for fixed space step and increasing dimension of the state space.

n	nodes	CPU time
2	225	00:00.5
3	3375	00:01
4	50625	00:05
5	759375	02:43

As a concluding remark, we point out that, in spite of its complexity, the Dynamic Programming approach still has two great avantages. One is the fact that it does not require to compute a different solution for each initial state. The second, that it ensures that the solution is a global minimum. In this respect, it is well known that numerical schemes based on variational methods may converge to local minima depending on the initial guess. This suggests to use Dynamic Programming techniques (up to a feasible complexity) to provide a first guess of the global minimum, and then to improve accuracy by direct minimization.

BIBLIOGRAPHY

[C] I. Capuzzo Dolcetta, *On a discrete approximation of the Hamilton - Jacobi equation of dynamic programming*, Appl.Math.and Optim. **10** (1983), 367-377

[CI] I. Capuzzo Dolcetta and H. Ishii, *Approximate solutions of the Bellman equation of deterministic control theory*, Appl. Math.and Optim. **11** (1984), 161-181

[CM] I. Capuzzo Dolcetta and M. Matzeu, *On the dynamic programming inequalities associated with the deterministic optimal stopping time problem in discrete and continuous time*, Num. Funct. Anal. and Optim. **3** (1981),

[Fa] M. Falcone, *A numerical approach to the infinite horizon problem of deterministic control theory*, Applied Mathematics and Optimization **15**,1987, 1-13 and **23**, 1991, 213-214.

[Fe] R. Ferretti, *Internal approximation schemes for optimal control problems in Hilbert spaces*, to appear

[FF1] M. Falcone and R. Ferretti,*Discrete time high order schemes for Hamilton–Jacobi–Bellman equations*, to appear on Numerische Mathematik.

[FF2] M. Falcone and R. Ferretti, *Fully discrete high order schemes for Hamilton–Jacobi–Bellman equations*, preprint

[GO] D. Gottlieb and S. A. Orszag, *Numerical analysis of spectral methods*, SIAM, Philadelphia, 1977

[L] P. L. Lions, *Generalized solutions of Hamilton–Jacobi equations*, Pitman, London, 1982

[LT] I. Lasiecka and R. Triggiani, *Differential and algebraic Riccati equations with applications to boundary/point control problems: continuous theory and approximation theory*, Springer–Verlag, New York, 1991

[RT] P. A. Raviart and J. M. Thomas, *Introduction a l'analyse numérique des équations aux derivées partielles*, Masson, Paris, 1983

Strong Solutions for Kolmogorov Equation in Hilbert Spaces

Fausto GOZZI

Dipartimento di Matematica
Università di Pisa
Via F. Buonarroti n.2, 56127 Pisa, Italy

Key words: Kolmogorov equations, Bernstein's polynomials,

Abstract

Let X be a separable Hilbert space and $T > 0$. We consider uniformly continuous and bounded functions $\varphi : X \to \mathbb{R}$ and functions $\mathcal{F} : [0,T] \times X \to \mathbb{R}$, which are uniformly continuous in the second variable with an integrable singularity at time 0. We prove that these functions can be approximated by regular functions in a suitable convergence and we apply this result to show that mild solutions of nonhomogeneus Kolmogorov equations in X can be seen as the limit of classical solutions improving results of [3].

1 Introduction

The aim of this paper is to show an approximation result for solutions of the following nonhomogeneus Kolmogorov equation

$$\begin{cases} \dfrac{\partial u}{\partial t} = \dfrac{1}{2} \operatorname{Tr} [Q u_{xx}] + < Ax, u_x > + \mathcal{F}(t,x), \quad t \in]0,T],\ x \in D(A) \\ \\ u(0,x) = \varphi(x),\ x \in X, \end{cases} \tag{1}$$

in a separable Hilbert space X. Here A is the infinitesimal generator of a strongly continuous semigroup e^{tA} on X, Q is a nonnegative selfadjoint operator in X (possibly, but not necessarily nuclear), $\mathcal{F} : [0,T] \times X \to \mathbb{R}$ is a bounded continuous function and $\varphi : X \to \mathbb{R}$ is uniformly continuous and bounded. In [5] and [3] it is proved that (1) has a unique "mild solution" that can be written as

$$u(t,x) = [P_t\varphi](x) + \int_X [P_{t-s}\mathcal{F}(s,\cdot)](x)ds \tag{2}$$

where $\{P_t,\ t \geq 0\}$ is the transition semigroup generated by the realization, in the space of uniformly continuous and bounded functions, of the operator $\mathcal{A}_0$ defined as follows

$$\mathcal{A}_0 u = \operatorname{Tr} [Q u_{xx}] + < Ax, u_x > .$$

Moreover in [3] it is also proved that mild solutions of (1) are also $\mathcal{K}$-strong solutions, in the sense that they are the limit, uniformly on compact subsets, of equibounded classical solutions. In this paper we prove a generalization of this result by showing that

(i) If the data $\mathcal{F}$ has an integrable singularity at time $t = 0$, (e.g. $|\mathcal{F}(t,x)| \le \frac{C}{t^\alpha}$ for some $C > 0$ and $\alpha \in]0,1[$, then the result still holds true.

(ii) If the data φ and $\mathcal{F}$ are k-times differentiable with respect to the space variable x, then the convergence of classical solutions to the mild solutions holds also for the derivatives up to the order $k+1$.

(iii) The convergence can be uniform also on bounded subsets of X.

These results can be applied to the study of nonlinear perturbations of (1) since they help to obtain good estimates of the local solutions (see [7]) and they obtain approximation results for solutions in a more general setting than the one given in [5] and in [3].

For studies of equation (1) see e.g. [1], [9], [3], [5] and the references quoted therein.

The outline of the paper is the following. In §2 we give some preliminaries, including a lemma on Bernstein polynomials; in §3 we show an approximation result in spaces of uniformly continuous and bounded functions; §4 is devoted to the application of the approximation result to (1).

2 Preliminaries

Troughout this paper X will denote a real separable Hilbert space endowed with the scalar product $< \cdot, \cdot >$ and the norm $|\cdot|$. For $i \in \mathbb{N}$ we denote by $\mathcal{L}^i(X)$ the Banach space of the continuous linear operators $T : X \to X^i$ with norm $\|\cdot\|_{\mathcal{L}^i}$, (we will omit the subscript when no confusion is possible) and we agree that $\mathcal{L}^0(X) = X$. We set

$$\mathcal{L}^+(X) = \{T \in \mathcal{L}(X),\ T \text{ self-adjoint},\ \langle Tx, x\rangle \ge 0\ \forall x \in X\} \tag{3}$$

Moreover we denote by $\mathcal{L}_1(X)$ the set of all nuclear operators $T : X \to X$. It is well known that $\mathcal{L}_1(X)$ is a Banach space with the norm

$$\|T\|_{\mathcal{L}_1(X)} = \sum_{k=1}^{\infty} \langle |T| e_k, e_k\rangle \overset{\text{def}}{=} \operatorname{Tr} |T| \tag{4}$$

where $|T| = (T^*T)^{\frac{1}{2}}$ and $\{e_k\}_{k\in\mathbb{N}}$ is any orthonormal complete system in X.

Let X and Y be two Hilbert spaces and define $B(X,Y)$ and $UC_b(X,Y)$ to be the Banach space of all functions $\varphi : X \to Y$ which are, respectively, bounded and uniformly continuous and bounded on X with the usual norm

$$\|\varphi\|_0 = \sup_{x\in X} |\varphi(x)|_Y.$$

For $k \in \mathbb{N}$, we denote by $UC_b^k(X,Y)$ the set of all functions $\varphi : H \to \mathbb{R}$ which are uniformly continuous and bounded on X togheter with all their Fréchet derivatives up to the order k. We set

$$\|\varphi\|_k = \sum_{h=0}^{k} \sup_{x\in H} \left|D^h\varphi(x)\right|_Y.$$

If $Y = \mathbb{R}$ then we write $UC_b^k(X)$ instead of $UC_b^k(X, \mathbb{R})$.

A continuity modulus is a continuous subadditive function $\rho : [0, +\infty[\to [0, +\infty[$ with $\rho(0) = 0$. For any $\varphi \in UC_b(X)$ we denote by $\rho(\varphi; \cdot)$ the continuity modulus of φ i.e. the function

$$\rho(\varphi; \delta) \stackrel{def}{=} \sup_{|x-y|\leq\delta} |\varphi(x) - \varphi(y)|$$

Let $I \subset \mathbb{R}$ (or $\mathbb{R}^d$). We denote by $B(I \times X, Y)$, $C_b(I \times X, Y)$, $UC_b(I \times X, Y)$ the Banach spaces of all functions $\varphi : I \times X \to Y$ which are, respectively, bounded, continuous and bounded, uniformly continuous and bounded on $I \times X$ with the norm

$$\|\varphi\|_0 = \sup_{(t,x)\in I\times X} |\varphi(t,x)|_Y.$$

We define

$$\widetilde{U}C_b(I\times X; Y) = \left\{\varphi \in C_b(I\times X; Y) : \varphi(t, \cdot) \in UC_b(X), \quad \text{uniformly in } t \in I\right\} \tag{5}$$

If $Y = \mathbb{R}$ then we write $B(I \times X)$, $C_b(I \times X)$, $UC_b(I \times X)$, $\widetilde{U}C_b(I \times X)$ instead of $B(I \times X; \mathbb{R})$, $C_b(I \times X; \mathbb{R})$, $UC_b(I \times X; \mathbb{R})$, $\widetilde{U}C_b(I \times X; \mathbb{R})$

The space $\widetilde{U}C_b(I \times X; Y)$ will be very important in our setting, so we recall some simple property of it which will be useful in the following (see [3] and [6]).

Remark 2.1 If $I \subset \mathbb{R}$ is compact, and X, Y are Hilbert spaces then a function $u \in \widetilde{U}C_b(I \times X; Y)$ if and only if

(i) the map $u(t, \cdot)$ is uniformly continuous on X, uniformly with respect to $t \in I$; that is, for every $x, y \in X$ we have

$$\sup_{t\in I} |u(t, x) - u(t, y)|_Y \leq \rho(|x - y|) \tag{6}$$

where $\rho : \mathbb{R}^+ \to \mathbb{R}^+$ is a continuity modulus;

(ii) for every $x \in X$, $u(\cdot, x)$ is continuous on I. ■

Remark 2.2

1. Given $I \subset \mathbb{R}$ compact, by standard arguments it follows that if $u \in \widetilde{U}C_b(I \times X; Y)$, then for every compact set $K \subset X$, $u|_{I\times K} \in UC_b(I \times K; Y)$ where $u|_{I\times K}$ means the restriction of u to $I \times K$. In particular $u(\cdot, x)$ is uniformly continuous on I, uniformly with respect to $x \in K$, that is for every $t, s \in I$ we have
$$\sup_{x\in K} |u(t, x) - u(s, x)|_Y \leq \rho_K(|t - s|) \tag{7}$$
where ρ_K is a continuity modulus depending on the compact set K.

2. For every compact set $K \subset X$ we have
$$\begin{gathered} u \in \widetilde{U}C_b(I \times X; Y) \Rightarrow \\ u \in B(I; UC_b(X; Y)) \text{ and } u|_{I\times K} \in C(I; C(K; Y)) \end{gathered} \tag{8}$$

3. $\widetilde{UC}_b(I\times X;Y)$ is a closed subspace of $C_b(I\times X;Y)$ because uniform continuity is stable with respect to the convergence in the norm $\|\cdot\|_0$. We can say that the space $\widetilde{UC}_b(I\times X;Y)$ is "in the middle" between $C_b(I\times X;Y)$ and $UC_b(I\times X;Y)$. If we consider $\varphi\in UC_b(X)$ and $\{e^{tA},\ t\geq 0\}$ a strongly continuous semigroup on X, then
$$w(t,x)=\varphi(e^{tA}x)$$
is a natural example of a function belonging to $\widetilde{UC}_b(I\times X)$ but not to $UC_b(I\times X)$. ■

Finally we introduce the following Banach spaces of functions blowing up at 0 at a prescribed rate $\theta\in[0,1]$.

$$\begin{aligned}\Sigma^0_{T,\theta} &= \left\{v\in\widetilde{UC}_b([\tau,T]\times X)\ \forall\tau\in]0,T[,\ t^\theta v\in\widetilde{UC}_b([0,T]\times X)\right\}\\ \|v\|_{\Sigma^0_{T,\theta}} &= \|t^\theta v\|_0=\sup\{|t^\theta v(t,x)|:(t,x)\in]0,T]\times X\}\end{aligned}\tag{9}$$

$$\begin{aligned}\Sigma^1_{T,\theta} &= \Big\{v\in\widetilde{UC}_b([0,T]\times X):\\ &\qquad v_x\in\widetilde{UC}_b([\tau,T]\times X;X)\ \forall\tau\in]0,T[,\ t^\theta v_x\in\widetilde{UC}_b([0,T]\times X;X)\Big\}\\ \|v\|_{\Sigma^1_{T,\theta}} &= \|v\|_0+\|t^\theta v_x\|_0\end{aligned}\tag{10}$$

and

$$\begin{aligned}\Sigma^n_{T,\theta} &= \Big\{v\in\widetilde{UC}_b([0,T]\times X):\\ &\qquad D_xv,D^2_xv(\xi_1),..,D^{n-1}_xv(\xi_1,..,\xi_{n-2})\in\widetilde{UC}_b([0,T]\times X;X),\\ &\qquad D^n_xv(\xi_1,..,\xi_{n-1})\in\widetilde{UC}_b([\tau,T]\times X;X)\ \text{ and}\\ &\qquad t^\theta D^n_xv(\xi_1,..,\xi_{n-1})\in\widetilde{UC}_b([0,T]\times X;X),\ \ \forall\xi_1,..,\xi_{n-1}\in X,\ \forall\tau\in]0,T[,\\ &\qquad \text{and }\ t^\theta D^h_xv\in B([0,T]\times X;\mathcal{L}^{h-1}(X))\ \ h=2,..,n\Big\}\\ \|v\|_{\Sigma^n_{T,\theta}} &= \|v\|_0+\|Dv\|_0+..+\|D^{n-1}_xv\|_0+\|t^\theta D^n_xv\|_0\end{aligned}\tag{11}$$

Remark 2.3 The spaces $\Sigma^n_{T,\theta}$ are introduced for technical reason, due to the properties of the Ornstein-Uhlenbeck transition semigroup (see §4.1) that will be a very important tool to solve problem (1) (see [6] and [7]). ■

We now recall the definition of $\mathcal{K}$-convergence (see [3] and [6]) by extending it to our needs.

Definition 2.4 *A sequence* $(\varphi_n) \subset UC_b(X;Y)$ *is said to be* $\mathcal{K}$-convergent *to* $\varphi \in UC_b(X;Y)$ *if*

$$\begin{cases} \sup_{n\in\mathbb{N}} \|\varphi_n\|_0 < +\infty \\ \\ \lim_{n\to+\infty} \sup_{x\in K} |\varphi_n(x) - \varphi(x)|_Y = 0 \end{cases} \tag{12}$$

for every compact set $K \subset X$. *In this case we will write*

$$\varphi = \mathcal{K}\text{-}\lim_{n\to+\infty} \varphi_n \quad in \quad UC_b(X;Y).$$

In a similar way a sequence $(\mathcal{F}_n) \subset \widetilde{U}C_b(I\times X;Y)$ *is said to be* $\mathcal{K}$-convergent *to* $\mathcal{F} \in \widetilde{U}C_b(I\times X;Y)$ *if*

$$\begin{cases} \sup_{n\in\mathbb{N}} \|\mathcal{F}_n\|_0 < +\infty \\ \\ \lim_{n\to+\infty} \sup_{(t,x)\in I\times K} |\mathcal{F}_n(t,x) - \mathcal{F}(t,x)|_Y = 0 \end{cases} \tag{13}$$

for any compact set $K \subset H$. *We will write*

$$\mathcal{F} = \mathcal{K}\text{-}\lim_{n\to+\infty} \mathcal{F}_n \quad in \quad \widetilde{U}C_b(I\times X;Y).$$

Definition 2.5 *Let* $k = 0,1,2,\ldots$ *A sequence* $(\varphi_n) \subset UC_b^k(X)$ *is said to be* $\mathcal{K}$-convergent *to* $\varphi \in UC_b^k(X)$ *if*

$$\varphi = \mathcal{K}\text{-}\lim_{n\to+\infty} \varphi_n \quad in \quad UC_b(X).$$

and, for $j = 1,..,k$,

$$\mathcal{K}\text{-}\lim_{n\to+\infty} D^j\varphi_n = D^j\varphi \quad in \quad UC_b(X;\mathcal{L}^{j-1}(X)).$$

In this case we will write

$$\varphi = \mathcal{K}\text{-}\lim_{n\to+\infty} \varphi_n \quad in \quad UC_b^k(X).$$

A sequence $(\mathcal{F}_n) \subset \Sigma^k_{T,\alpha}$, *for* $\alpha \in]0,1[$ *is said to be* $\mathcal{K}$-convergent *to* $\mathcal{F} \in \Sigma^k_{T,\alpha}$ *if (for* $k \geq 1$*)*

$$\mathcal{K}\text{-}\lim_{n\to+\infty} \mathcal{F}_n = \mathcal{F} \quad in \quad \widetilde{U}C_b([0,T]\times X),$$

$$\mathcal{K}\text{-}\lim_{n\to+\infty} D_x^j\mathcal{F}_n = D_x^j\mathcal{F} \quad in \quad \widetilde{U}C_b([0,T]\times X;\mathcal{L}^{j-1}(X)), \quad for\ j = 1,..,k-1\ (if\ k\geq 2),$$

and finally

$$\begin{aligned} &\sup_{n\in\mathbb{N}} \sup_{t\in]0,T]} \|t^\alpha D_x^k\mathcal{F}_n(t,\cdot)\|_0 < +\infty \\ &\mathcal{K}\text{-}\lim_{n\to+\infty} D_x^k\mathcal{F}_n = D_x^k\mathcal{F} \quad in \quad \widetilde{U}C_b([\varepsilon,T]\times X;\mathcal{L}^{k-1}(X)). \end{aligned} \tag{14}$$

for every $\varepsilon > 0$. *We will write*

$$\mathcal{F} = \mathcal{K}\text{-}\lim_{n\to+\infty} \mathcal{F}_n \quad in \quad \Sigma^k_{T,\alpha}.$$

2.1 Bernstein polynomials

Before proving the Proposition we need a simple Lemma about Bernstein polynomials (see [10], p.9 and [3], Lemma 3.3).

Lemma 2.6 *Let $\mathcal{G} \in \Sigma^0_{T,\alpha}$, $\alpha \in]0,1[$ and set, for $\delta > 0$*

$$\overline{\rho}(\mathcal{G};\delta) \stackrel{def}{=} \sup_{t\in[0,T]} \rho(t^\alpha \mathcal{G}(t,\cdot);\delta) = \sup_{\substack{|x-y|\le\delta \\ t\in[0,T]}} |t^\alpha \mathcal{G}(t,x) - t^\alpha \mathcal{G}(t,y)|$$

Take the Bernstein-like polynomials

$$\mathcal{P}^{\mathcal{G}}_n(t,x) = T^{-n} \sum_{k=0}^{n} \binom{n}{k} t^k (T-t)^{n-k} \mathcal{G}\left(\frac{k+1}{n}T, x\right) \tag{15}$$

where we set $\mathcal{G}(\frac{n+1}{n}T,x) \stackrel{def}{=} \mathcal{G}(T,x)$. Then for every $n \in \mathbb{N}$, $\mathcal{P}^{\mathcal{G}}_n \in C([0,T];UC_b(X))$ and,

$$\mathcal{K} - \lim_{n\to\infty} \mathcal{P}^{\mathcal{G}}_n = \mathcal{G} \quad in \ \Sigma^0_{T,\alpha} \tag{16}$$

Moreover for every $\delta > 0$, $n \in \mathbb{N}$,

$$\overline{\rho}(\mathcal{P}^{\mathcal{G}}_n;\delta) \stackrel{def}{=} \sup_{\substack{|x-y|\le\delta \\ t\in[0,T]}} t^\alpha |\mathcal{P}^{\mathcal{G}}_n(t,x) - \mathcal{P}^{\mathcal{G}}_n(t,y)| \le C(T)\overline{\rho}(\mathcal{G};\delta) \tag{17}$$

Where $C(T)$ is a positive constant depending only on T.

Proof. First we remark that we are obliged to change the classical definition of Bernstein polynomials since the function $\mathcal{F}(t,x)$ blows up as $t \to 0$. It easy to check, by (15) that $\mathcal{P}^{\mathcal{G}}_n \in C([0,T];UC_b(X))$.

By [3], Lemma 3.3 it is easy to see that, for every $\varepsilon > 0$,

$$\mathcal{K} - \lim_{n\to\infty} \mathcal{P}^{\mathcal{G}}_n = \mathcal{G} \quad \text{in } \tilde{U}C_b([\varepsilon,T]\times X). \tag{18}$$

Then, by definition 2.5 it is enough to prove that

$$\sup_{t\in]0,T]} \|t^\alpha \mathcal{P}^{\mathcal{G}}_n(t,\cdot)\|_0 \le C(T) \tag{19}$$

for some positive constant $C(T)$ independent of n. To prove (19) we first set for simplicity $T=1$. Then, for $t \in]0,\frac{1}{n}]$ we have, by definition of $\mathcal{P}^{\mathcal{G}}_n$,

$$\left|t^\alpha \mathcal{P}^{\mathcal{G}}_n(t,x)\right| \le \frac{1}{n^\alpha} \sum_{k=0}^{n} \binom{n}{k} t^k (1-t)^{(n-k)} \left(\frac{n}{k+1}\right)^\alpha \|\mathcal{G}\|_{\Sigma^0_{T,\alpha}}$$

$$\le \sum_{k=0}^{n} \binom{n}{k} t^k (1-t)^{(n-k)} \|\mathcal{G}\|_{\Sigma^0_{T,\alpha}} = \|\mathcal{G}\|_{\Sigma^0_{T,\alpha}}$$

Now let $h \in \mathbb{N}$, $1 \le h \le n-1$. If $t \in \left]\frac{h}{n}, \frac{h+1}{n}\right]$, then for every given $\delta \in]0,1]$ we have

$$\begin{aligned}\left|t^\alpha \mathcal{P}_n^{\mathcal{G}}(t,x)\right| \le \Bigg|\left(\frac{h+1}{n}\right)^\alpha \Bigg[&\sum_{\substack{k=0 \\ \left|\frac{k+1}{n}-t\right|>\delta}}^{n} \binom{n}{k} t^k(1-t)^{(n-k)}\mathcal{G}\left(\frac{k+1}{n},x\right) \\ + &\sum_{\substack{k=0 \\ \left|\frac{k+1}{n}-t\right|\le\delta}}^{n} \binom{n}{k} t^k(1-t)^{(n-k)}\mathcal{G}\left(\frac{k+1}{n},x\right)\Bigg]\Bigg| = I + II\end{aligned} \tag{20}$$

The first term can be estimated as follows, by using well known identities on Bernstein polynomials (see [10] p.9)

$$I \le \left(\frac{h+1}{n}\right)^\alpha \sum_{k=0}^{n} \left(\frac{k+1-nt}{n\delta}\right)^2 \binom{n}{k} t^k(1-t)^{(n-k)}\mathcal{G}\left(\frac{k+1}{n},x\right)$$

$$\le \left(\frac{h+1}{n}\right)^\alpha \frac{1}{n^2\delta^2}[nt(1-t)+1]n^\alpha \|\mathcal{G}\|_{\Sigma^0_{T,\alpha}} \le \frac{(h+1)^\alpha}{n^2\delta^2}[h+2]\|\mathcal{G}\|_{\Sigma^0_{T,\alpha}},$$

so that, setting $\delta = \frac{h}{2n}$ we obtain

$$I \le \frac{4(h+1)^\alpha}{h^2}[h+2]\|\mathcal{G}\|_{\Sigma^0_{T,\alpha}} \le 24\|\mathcal{G}\|_{\Sigma^0_{T,\alpha}}.$$

For the second term we observe first that, setting $\delta = \frac{h}{2n}$, then

$$\left(t \in \left]\frac{h}{n}, \frac{h+1}{n}\right] \text{ and } \left|\frac{k+1}{n} - t\right| \le \delta\right) \implies k+1 \ge \frac{h}{2}$$

so that

$$II \le \left(\frac{h+1}{n}\right)^\alpha \sum_{\substack{k=0 \\ k+1\ge\frac{h}{2}}}^{n} \binom{n}{k} t^k(1-t)^{(n-k)}\mathcal{G}\left(\frac{k+1}{n},x\right)$$

$$\le \left(\frac{h+1}{n}\right)^\alpha \sum_{\substack{k=0 \\ k+1\ge\frac{h}{2}}}^{n} \binom{n}{k} t^k(1-t)^{(n-k)}\frac{n^\alpha}{(k+1)^\alpha}\|\mathcal{G}\|_{\Sigma^0_{T,\alpha}}$$

$$\le \frac{(h+1)^\alpha}{(\frac{h}{2})^\alpha}\|\mathcal{G}\|_{\Sigma^0_{T,\alpha}} \le 2^{2\alpha}\|\mathcal{G}\|_{\Sigma^0_{T,\alpha}}.$$

We have obtained that, for $t \in \left]\frac{h}{n}, \frac{h+1}{n}\right]$

$$\left|t^\alpha \mathcal{P}_n^{\mathcal{G}}(t,x)\right| \le [24 + 2^{2\alpha}]\|\mathcal{G}\|_{\Sigma^0_{T,\alpha}}.$$

which gives the claim, since the estimate does not depend on h and n. If $T \neq 1$ then we only have to multiply the constant in the latter inequality by T^α.

The statement about the continuity modulus can be proved exactly in the same way by getting the same constant, so we will omit it. ∎

Remark 2.7 This result is optimal in the sense that we cannot obtain that for every compact subset K of X

$$\sup_{t\in[0,T]}\sup_{x\in K}(\mathcal{P}_n^{\mathcal{F}}-\mathcal{F})\overset{n\to\infty}{\longrightarrow}0$$

It is enough to check the case $\mathcal{F}(t,x)=\frac{1}{t^\alpha}$.

3 Approximation results

In this section we prove an approximation result for functions in $UC_b^k(X)$ and in $\Sigma_{T,\alpha}^k$ ($k=0,1,..,\ \alpha\in]0,1[$) which generalize the one obtained in [3].

Proposition 3.1 *Let X be a separable Hilbert space and $T>0$. Let $\varphi\in UC_b^k(X)$, $\mathcal{F}\in\Sigma_{T,\alpha}^k$ and let $\{P_n\}$ be a sequence of finite dimensional projectors on X increasing to the identity. Then there exists two sequences φ_n and $\mathcal{F}_n$ such that*

$$\begin{gathered}\varphi_n\in UC_b^\infty(X),\quad \text{and}\quad \varphi_n(x)=\varphi_n(P_nx)\ \forall x\in X\\ \varphi_n(x)=0\ \text{if}\ |P_nx|\geq n+1\\[1ex] \mathcal{F}_n\in C([0,T];UC_b^\infty(X))\quad\text{and}\quad \mathcal{F}_n(t,x)=\mathcal{F}_n(t,P_nx)\ \forall x\in X\\ \mathcal{F}_n(t,x)=0\ \text{if}\ |P_nx|\ \text{is sufficiently big}\end{gathered}\tag{21}$$

and,

$$\begin{aligned}&\mathcal{K}-\lim_{n\to\infty}\varphi_n=\varphi &&\text{in}\ UC_b^k(X)\\ &\mathcal{K}-\lim_{n\to\infty}\mathcal{F}_n=\mathcal{F} &&\text{in}\ \Sigma_{T,\alpha}^k.\end{aligned}\tag{22}$$

Proof. We present the proof of the Proposition for the case $k=1$, since the proof for the case $k=0$ and for $k\geq 1$ can be obtained from this proof with minor modifications.

Part 1.

Let $\varphi\in UC_b(X)$, and assume that P_n is of the following kind

$$P_n:X\to X;\qquad P_nx=\sum_{i=1}^n x_ie_i.$$

where $\{e_n\}_{n\in\mathbb{N}}$ is any orthonormal complete system in X and $x_i=\langle x,e_i\rangle$, for $i\in\mathbb{N}$. Consider the following linear mapping.

$$\Pi_n:X\to\mathbb{R}^n;\qquad \Pi_nx=(x_1,..,x_n)$$

$$T_n:\mathbb{R}^n\to X;\qquad T_n(x_1,..,x_n)=\sum_{i=1}^n x_ie_i$$

where

$$T_n^* = \Pi_n, \qquad \Pi_n^* = T_n;$$
$$\Pi_n T_n = Id|_{\mathbb{R}^n} \qquad T_n \Pi_n = P_n$$

Define

$$\psi_n : \mathbb{R}^n \to \mathbb{R}^n; \qquad \psi_n(x_1, .., x_n) = \varphi(T_n(x_1, .., x_n))$$

(so that $\psi_n(\Pi_n x) = \varphi(P_n x)$), and take, for $n, k \in \mathbb{N}$

$$\overline{\psi_n^k} = \psi_n * \eta_k$$

where η_k is a regularizing sequence in $\mathbb{R}^n$ and $*$ means the convolution product. By properties of convolution it is clear that, for $n, k \in \mathbb{N}$, $\overline{\psi_n^k} \in UC_b^\infty(\mathbb{R}^n)$ and, for $n \in \mathbb{N}$,

$$\overline{\psi_n^k} \stackrel{k\to\infty}{\longrightarrow} \psi_n, \qquad \text{in } UC_b(\mathbb{R}^n)$$

and

$$\|\overline{\psi_n^k}\|_1 \le \|\psi_n\|_1 \le \|\varphi\|_1$$

Take $\theta_n \in UC_b^\infty(\mathbb{R})$ such that

$$\theta_n(r) = \begin{cases} 0 & r \ge n+1 \\ 1 & r \le n \end{cases}$$

and

$$\|D\theta_n\|_0 \le 2 \qquad \|D^2\theta_n\|_0 \le 8 \tag{23}$$

Setting

$$\psi_n^k(x) = \overline{\psi_n^k}(x)\theta_n(x)$$

we have, for $n, k \in \mathbb{N}$, $\psi_n^k \in UC_b^\infty(\mathbb{R}^n)$ and, for $n \in \mathbb{N}$,

$$\psi_n^k \stackrel{k\to\infty}{\longrightarrow} \psi_n$$

$$D\psi_n^k \stackrel{k\to\infty}{\longrightarrow} D\psi_n$$

uniformly on bounded subsets of $\mathbb{R}^n$. For $n \in \mathbb{N}$ let $k(n) \in \mathbb{N}$ such that

$$\sup_{x \in B(0,n)} \left\{ |\psi_n^{k(n)}(x) - \psi_n(x)| + |D\psi_n^{k(n)}(x) - D\psi_n(x)| \right\} \le \frac{1}{n}$$

and define, for $n \in \mathbb{N}$ and $x \in X$,

$$\varphi_n(x) \stackrel{def}{=} \psi_n^{k(n)}(\Pi_n x) \tag{24}$$

By construction it is clear that

$$\varphi_n \in UC_b^\infty(X), \quad \text{and} \quad \varphi_n(x) = \varphi_n(P_n x) \;\; \forall x \in X$$

$$\varphi_n(x) = 0 \text{ if } |P_n x| \ge n+1,$$

and that

$$\sup_{\substack{x,y\in B(0,n)\\|x-y|\le\delta}} |\varphi_n(x)-\varphi_n(y)| \le \rho(\varphi;\delta)$$

$$\sup_{\substack{x,y\in B(0,n)\\|x-y|\le\delta}} |D\varphi_n(x)-D\varphi_n(y)| \le \rho(D\varphi;\delta) \tag{25}$$

and finally that, by (23)

$$\|\varphi_n\|_1 \le 3\|\varphi\|_1.$$

Moreover, let $K \in X$, K, compact, and set $R \stackrel{def}{=} \sup_{x\in K}|x|$. Then, if $n > R$ we have

$$\begin{aligned}
&\sup_{x\in K}|\varphi_n(x)-\varphi(x)| \\
\le &\sup_{x\in B(0,n)}\left\{|\psi_n^{k(n)}(\Pi_n x)-\psi_n(\Pi_n x)|\right\} + \sup_{x\in K}|\psi_n(\Pi_n x)-\varphi(x)| \\
&\le \frac{1}{n} + \sup_{x\in K}|\varphi(P_n x)-\varphi(x)|
\end{aligned}$$

which goes to 0 as $n\to\infty$ thanks to the Ascoli-Arzelà Theorem. Similarly

$$\begin{aligned}
&\sup_{x\in K}|D\varphi_n(x)-D\varphi(x)| \\
\le &\sup_{x\in B(0,n)}\left\{|T_n D\psi_n^{k(n)}(\Pi_n x)-T_n D\psi_n(\Pi_n x)|\right\} + \sup_{x\in K}|T_n D\psi_n(\Pi_n x)-D\varphi(x)| \\
&\le \frac{1}{n} + \sup_{x\in K}|P_n D\varphi(P_n x)-\varphi(x)|
\end{aligned}$$

which, as before, goes to 0 as $n\to\infty$ thanks to the Ascoli-Arzelà Theorem. This concludes the proof for the approximation of φ.

Part 2.

Consider now $\mathcal{F}\in\Sigma^1_{T,\alpha}$ and, as in Lemma 2.6 let us take the Bernstein-like polynomials

$$\mathcal{P}_n^{\mathcal{F}}(t,x) = T^{-n}\sum_{k=0}^{n}\binom{n}{k}t^k(T-t)^{n-k}\mathcal{F}\left(\frac{k+1}{n}T,x\right) \tag{26}$$

where we set $\mathcal{F}(\frac{n+1}{n}T,x) \stackrel{def}{=} \mathcal{F}(T,x)$. Clearly $\mathcal{P}_n^{\mathcal{F}}$ is differentiable with respect to x and

$$D_x\mathcal{P}_n^{\mathcal{F}}(t,x) = \mathcal{P}_n^{D_x\mathcal{F}}(t,x) = T^{-n}\sum_{k=0}^{n}\binom{n}{k}t^k(T-t)^{n-k}D_x\mathcal{F}\left(\frac{k+1}{n}T,x\right)$$

Then by Lemma 2.6 we obtain $\mathcal{P}_n^{\mathcal{F}}\in C([0,T];UC_b^1(X))$ and

$$\mathcal{K}-\lim_{n\to\infty}P_n = \mathcal{F} \quad\text{in } \Sigma^1_{T,\alpha} \tag{27}$$

Let $\mathcal{S} = \{ x_p \, ; \, p \in \mathbb{N} \}$ be a dense sequence in X. Since $\mathcal{F} \in \Sigma^1_{T,\alpha}$, by the first part of the proof for every $n \in \mathbb{N}$, $k = 1, .., n$ we can find an element ψ^n_k satisfying (21) such that

$$|\mathcal{F}(\frac{k}{n}T, x_p) - \psi^n_k(x_p)| \leq \frac{1}{n}, \quad \forall p \leq n, \tag{28}$$

and,

$$|D_x\mathcal{F}(\frac{k}{n}T, x_p) - D\psi^n_k(x_p)| \leq \frac{1}{n}, \quad \forall p \leq n. \tag{29}$$

Moreover ψ^n_k can be taken such that

$$\|\psi^n_k\|_1 \leq 3 \left\| \mathcal{F}\left(\frac{k}{n}T, \cdot\right) \right\|_1 \tag{30}$$

and that, for $\delta > 0$, (see (25))

$$\begin{aligned} &\rho\left(\psi^n_k|_{B(0,R(n))}; \delta\right) \leq \rho\left(\mathcal{F}\left(\frac{k}{n}T, \cdot\right); \delta\right) \\ &\rho\left(D\psi^n_k|_{B(0,R(n))}; \delta\right) \leq \rho\left(D_x\mathcal{F}\left(\frac{k}{n}T, \cdot\right); \delta\right) \end{aligned} \tag{31}$$

where $R(n) \stackrel{def}{=} \sup\{|x_p| + 1; \;\; p \leq n\}$. Then we define

$$\mathcal{F}_n(t, x) = T^{-n} \sum_{k=0}^{n} \binom{n}{k} t^k (T-t)^{(n-k)} \psi^n_{k+1}(x). \tag{32}$$

where we set $\psi^n_{n+1} \stackrel{def}{=} \psi^n_n$. By the properties of ψ^n_k it is easy to check that $\mathcal{F}_n$ satisfies (21) for every $n \in \mathbb{N}$. We first remark that from (31) (32) and Lemma 2.6 it follows that for every $\delta \in [0, 1]$, $n \in \mathbb{N}$ and $t \in [0, T]$ we have

$$\begin{aligned} &\sup_{\substack{x,y \in B(0,R(n)) \\ |x-y|<\delta}} |\mathcal{F}_n(t,x) - \mathcal{F}_n(t,y)| \\ &\leq T^{-n} \sum_{k=0}^{n} \binom{n}{k} t^k (T-t)^{(n-k)} \sup_{\substack{x,y \in B(0,R(n)) \\ |x-y|<\delta}} |\psi^n_{k+1}(x) - \psi^n_{k+1}(y)| \\ &\leq T^{-n} \sum_{k=0}^{n} \binom{n}{k} t^k (T-t)^{(n-k)} \sup_{|x-y|<\delta} \left| \mathcal{F}\left(\frac{k+1}{n}, x\right) - \mathcal{F}\left(\frac{k+1}{n}, y\right) \right| \\ &\leq \sup_{\substack{|x-y|<\delta \\ t \in [0,T]}} |\mathcal{F}(t,x) - \mathcal{F}(t,y)| \leq \rho(\mathcal{F}; \delta). \end{aligned} \tag{33}$$

Moreover

$$\sup_{\substack{x,y\in B(0,R(n))\\|x-y|<\delta}} t^\alpha |D_x\mathcal{F}_n(t,x) - D_x\mathcal{F}_n(t,y)|$$

$$\leq t^\alpha T^{-n}\sum_{k=0}^{n}\binom{n}{k} t^k(T-t)^{(n-k)} \sup_{\substack{x,y\in B(0,R(n))\\|x-y|<\delta}} |D\psi^n_{k+1}(x) - D\psi^n_{k+1}(y)|$$

$$\leq t^\alpha T^{-n}\sum_{k=0}^{n}\binom{n}{k} t^k(T-t)^{(n-k)} \sup_{|x-y|<\delta}\left|D_x\mathcal{F}\left(\frac{k+1}{n},x\right) - D_x\mathcal{F}\left(\frac{k+1}{n},y\right)\right|$$

$$\leq C(T)\overline{\rho}(D_x\mathcal{F};\delta)$$

where we used the proof of (17) to get the last inequality. Then, given $\varepsilon > 0$, we obtain

$$\sup_{t\in[\varepsilon,T]}\sup_{\substack{x,y\in B(0,R(n))\\|x-y|<\delta}} |D_x\mathcal{F}_n(t,x) - D_x\mathcal{F}_n(t,y)| \leq \frac{C(T)}{\varepsilon^\alpha}\overline{\rho}(D_x\mathcal{F};\delta) \tag{34}$$

At this point, by reasoning exactly as in the proof of Lemma 3.3 of [3], we can prove, by using (33), that

$$\mathcal{K}-\lim_{n\to\infty}\mathcal{F}_n = \mathcal{F} \qquad \text{in } \widetilde{U}C_b([0,T]\times X) \tag{35}$$

It remains to prove the convergence of the derivative. First, for every $p \in \mathbb{N}$ and $\varepsilon > 0$,

$$\lim_{n\to+\infty}\sup_{t\in[\varepsilon,T]} |D_x\mathcal{F}_n(t,x_p) - D_x\mathcal{F}(t,x_p)| = 0. \tag{36}$$

Indeed, by (28), (26), for all $t \in]0,T]$,

$$|D_x\mathcal{F}_n(t,x_p) - D_x\mathcal{P}^{\mathcal{F}}_n(t,x_p)|$$

$$\leq T^{-n}\sum_{k=1}^{n}\binom{n}{k} t^k(T-t)^{(n-k)}|D_x\mathcal{F}\left(\frac{k+1}{n}T,x_p\right) - D\psi^n_{k+1}(x_p)| \leq \frac{1}{n}, \quad \forall n\geq p, \tag{37}$$

so that we have, by Lemma 2.6

$$\sup_{t\in[\varepsilon,T]} |D_x\mathcal{F}_n(t,x_p) - D_x\mathcal{F}(t,x_p)|$$

$$\leq \sup_{t\in[\varepsilon,T]}\left[|D_x\mathcal{F}_n(t,x_p) - D_x\mathcal{P}^{\mathcal{F}}_n(t,x_p)| + |D_x\mathcal{P}^{\mathcal{F}}_n(t,x_p) - D_x\mathcal{F}(t,x_p)|\right]$$

$$\leq \frac{1}{n} + \sup_{t\in[\varepsilon,T]} |D_x\mathcal{P}^{\mathcal{F}}_n(t,x_p) - D_x\mathcal{F}(t,x_p)| \to 0, \quad \text{as } n\to+\infty.$$

Now fix $\varepsilon > 0$ and let K be a compact set in X. For every $\sigma > 0$ let us choose $0 < \delta_\sigma$ such that

$$\frac{C(T)}{\varepsilon^\alpha}\overline{\rho}(D_x\mathcal{F};\delta) \leq \sigma \tag{38}$$

and choose $h_\sigma \in \mathbb{N}$ and $\{x_{p_i}\,;\, i = 1,\ldots,h_\sigma\,;\, p_1 < \cdots < p_{h_\sigma}\} \subset \mathcal{S}$, such that

$$K \subset \bigcup_{i=1}^{h_\sigma} B(x_{p_i},\delta_\sigma). \tag{39}$$

Consequently from (39) for every $x \in K$ we can take $j = j(x) \in \{1,\ldots,h_\sigma\}$ such that $|x - x_{p_j}| \leq \delta_\sigma$ which gives, by (34) and (38), that, for every $n \geq p_{h_\sigma}$ and $\varepsilon > 0$,

$$\sup_{t\in[\varepsilon,T]}\left\{|D_x\mathcal{F}_n(t,x) - D_x\mathcal{F}_n(t,x_{p_j})| + |D_x\mathcal{F}(t,x) - D_x\mathcal{F}(t,x_{p_j})|\right\} < 2\sigma.$$

Now $\forall\,(t,x) \in [\varepsilon,T] \times X$ we have

$$\begin{aligned}|D_x\mathcal{F}_n(t,x) - D_x\mathcal{F}(t,x)| &\leq |D_x\mathcal{F}_n(t,x) - D_x\mathcal{F}_n(t,x_{p_j})| \\ &+|D_x\mathcal{F}_n(t,x_{p_j}) - D_x\mathcal{F}(t,x_{p_j})| + |D_x\mathcal{F}(t,x_{p_j}) - D_x\mathcal{F}(t,x)| \\ &\leq 2\sigma + |D_x\mathcal{F}_n(t,x_{p_j}) - D_x\mathcal{F}(t,x_{p_j})|.\end{aligned}$$

Then if we choose $n > p_{h_\sigma}$, from (37) we have for every $(t,x) \in [\varepsilon,T] \times K$

$$|D_x\mathcal{F}_n(t,x) - D_x\mathcal{F}(t,x)| \leq 2\sigma + \frac{1}{n} + \sup_{i=1,..,h_\sigma}|D_x\mathcal{P}_n^{\mathcal{F}}(t,x_{p_i}) - D_x\mathcal{F}(t,x_{p_i})|$$

which by (27) and Lemma 2.6 implies

$$\limsup_{n\to+\infty}\sup_{t\in[\varepsilon,T]\times K}|D_x\mathcal{F}_n(t,x) - D_x\mathcal{F}(t,x)| \leq 2\sigma,$$

and the claim follows by the arbitrariness of σ and ε.

Finally by the definition of $\mathcal{F}_n$ and (19) it holds, for $(t,x) \in]0,T] \times X$, $n \in \mathbb{N}$,

$$|t^\alpha D_x\mathcal{F}_n(t,x)| \leq t^\alpha T^{-n}\sum_{k=0}^{n}\binom{n}{k}t^k(T-t)^{(n-k)}|D\psi_{k+1}^n(x)|$$

so that, by (30) and Lemma 2.6,

$$\begin{aligned}\|t^\alpha D_x\mathcal{F}_n(t,\cdot)\|_0 &\leq t^\alpha T^{-n}\sum_{k=0}^{n}\binom{n}{k}t^k(T-t)^{(n-k)}3\left\|D_x\mathcal{F}\left(\frac{k+1}{n}T,\cdot\right)\right\|_0 \\ &\leq C_0C(T)\|\mathcal{F}\|_{\Sigma^1_{T,\alpha}}.\ \blacksquare\end{aligned}$$

4 Kolmogorov equations

In this section we apply the results of the previous section to prove that the mild solutions of the linear problem

$$\begin{cases} \dfrac{\partial u}{\partial t} = \dfrac{1}{2} \operatorname{Tr}\,[Qu_{xx}] + < Ax, u_x > +\mathcal{F}(t,x) = 0, \ t \in]0,T], \ x \in D(A) \\ \\ u(0,x) = \varphi(x), \ x \in X, \end{cases} \tag{40}$$

can be approximated by classical solution in a suitable sense by generalizing results of [3]. We assume the following

Hypothesis 4.1

(i) X is a separable real Hilbert space.

(ii) The linear operator $A : D(A) \subset X \to X$ is the infinitesimal generator of a strongly continuous semigroup e^{tA} on X and

$$\|e^{tA}\|_{\mathcal{L}(X)} \leq Me^{\omega t} \qquad \forall t \geq 0$$

fo $M \geq 1$ and $\omega \in \mathbb{R}$

(iii) $Q \in \mathcal{L}^+(X)$ (not necessarily invertible).

(iv) For any $t > 0$ we have

$$\operatorname{Tr}\,[Q_t] < +\infty, \tag{41}$$

where $Q_t = \int_0^t e^{sA} Q e^{sA^*} ds$.

(v) $e^{tA}(X) \subset Q_t^{1/2}(X)$.

(vi) If we define the operator $\Gamma(t)$ as

$$\Gamma(t) = Q_t^{-1/2} e^{tA}(X).$$

then there exists $\alpha \in]0,1[$ and $\gamma > 0$ such that:

$$\|\Gamma(t)\|_{\mathcal{L}(X)} \leq \frac{\gamma}{t^\alpha}. \tag{42}$$

Remark 4.2 1. We recall, see [5] p.263, that Hypothesis 4.1–(v) is equivalent to the null controllability of the deterministic system

$$\xi'(t) = A\xi(t) + \sqrt{Q}z(t), \ \xi(0) = \xi_0.$$

The null controllability assumption is crucial to guarantee the regularity, with respect to x of the solution of (40). In the finite dimensional case it reduces to the Hörmander hypoellipticity condition

$$Q_t > 0 \quad \forall t > 0$$

(see [8] and [4]). Finally if Hypothesis 4.1-(v) holds, then the operator $\Gamma(t)$ given by

$$\Gamma(t) = Q_t^{-1/2} e^{tA}(X).$$

is well defined and bounded due to the Closed Graph Theorem. This makes sense also when Q is not invertible since we can take $Q_t^{-1/2}$ the pseudo–inverse of $Q_t^{1/2}$ (see [5] p.407).

2. Hypothesis 4.1-(iv) and (v) imply that the semigroup e^{tA}; $t \geq 0$ is Hilbert Schmidt and, in particular, compact, since

$$e^{tA} = Q_t^{1/2}\Gamma(t)$$

3. Hypothesis 4.1-(vi) yields local integrability of the map $t \to \|\Gamma(t)\|_{\mathcal{L}(X)}$ and will be used to build an integral solution of the equation (40). ∎

To clarify the notion of solution for equation (40) we recall some knowledge about the Ornstein-Uhlenbeck semigroup that can be found e.g. in [5].

4.1 Ornstein-Uhlenbeck semigroup

Let $(\Omega, \mathcal{F}, \mathbf{P})$ be a probability space and W be a cylindrical Brownian Motion with values in X. Consider the following stochastic differential equation

$$dZ(t) = AZ(t)dt + \sqrt{Q}dW(t), \quad Z(0) = x \in X \tag{43}$$

where we assume that Hypothesis 4.1 holds true. As showed in [6], under Hypothesis 4.1 the stochastic convolution

$$\int_0^t e^{(t-s)A} dW(s)$$

is a Gaussian process with mean 0 and covariance Q_t. Moreover the problem (43) has a unique weak solution given by

$$Z(t,x) = e^{tA}x + \int_0^t e^{(t-s)A}\sqrt{Q}dW(s). \tag{44}$$

(see [5] p.118 for the concept of weak solution of (43)). We shall denote by P_t, $t > 0$ the corresponding transition semigroup

$$\begin{aligned} P_t\, \varphi(x) &= \mathbb{E}[\eta(Z(t,x))] = \mathbb{E}[\eta(e^{tA}x + \int_0^t e^{(t-s)A}\sqrt{Q}dW(s))] \\ &= \int_X \varphi(y + e^{tA}x)\mu_t(dy) \quad x \in X, t > 0;\ \varphi \in UC_b(X). \end{aligned} \tag{45}$$

where μ_t is the Gaussian measure in X with mean 0 and covariance operator Q_t.

Remark 4.3

1. The semigroup P_t is not strongly continuous on $UC_b(X)$ in general (see [2] for the counterexample). In fact it is only weakly continuous, as defined in [2].
2. Hypothesis 4.1–(v) guarantees the smoothing property of P_t, namely that, for $\varphi \in UC_b(X)$, $P_t\, \varphi \in UC_b^\infty(X)$ (see e.g [5] p.262). ■

The following Proposition collects the regularity properties of P_t that we will use throughout this section. The proof can be found in [6].

Proposition 4.4 *Assume Hypothesis* 4.1. *Let* $\varphi \in UC_b(X)$. *Then the map*

$$u : [0, +\infty[\times X \to \mathbb{R}, \quad u(t,x) = P_t\varphi(x)$$

satisfies the following:

(i) For every $t > 0$

$$u(t,\cdot) \in UC_b^\infty(X) \tag{46}$$

and

$$\begin{aligned} \|D_x u(t,\cdot)\|_0 &\leq \|\Gamma(t)\| \|\varphi\|_0 \\ \|D_{xx} u(t,\cdot)\|_0 &\leq 2\|\Gamma(t)\|^2 \|\varphi\|_0 \end{aligned}$$

Moreover, if $\varphi \in UC_b^{n-1}(X)$, $(n \geq 1)$ *then*

$$\|D_x^n u(t,\cdot)\|_0 \leq \|\Gamma(t)\| \|e^{tA}\|^{n-1} \|D_x^{n-1}\varphi\|_0. \tag{47}$$

(ii)

$$u \in \widetilde{U}C_b([0,T] \times X) \tag{48}$$

and, if $\varphi \in UC_b^{n-1}(X)$ $(n \geq 1)$, *for every* $(h_1, .., h_n) \in X^n$ *it holds*

$$D_x^n u(h_1, .., h_n) \in \Sigma^0_{T,\alpha}. \tag{49}$$

Remark 4.5

(i) We cannot ask uniform continuity of $P_t\varphi$ on the joint variables (t,x) since it would imply strong continuity of the semigroup P_t (see [3])

(ii) . Moreover we cannot ask the continuity of $D^n u, n \geq 2$ from $[0,T] \times X$ to $\mathcal{L}^{n-1}(X)$ because of the strong continuity of the semigroup e^{tA}. (see [6]) ■

If Hypothesis 4.1 (i)–(iv) holds, Q is nuclear and $\varphi \in UC_b^2(X)$, then the function $u(t,x) = P_t\varphi(x)$ is a classical solution of the linear problem

$$\begin{cases} \dfrac{\partial u}{\partial t} = \dfrac{1}{2}\, \mathrm{Tr}\, [Q u_{xx}] + < Ax, u_x >, \; t > 0, \; x \in D(A) \\ \\ u(0,x) = \varphi(x), \; x \in X, \end{cases} \tag{50}$$

where "classical" means that

$$\begin{array}{ll} (i) & u \in \tilde{U}C_b([0,T] \times X) \\ (ii) & u(t,\cdot) \in UC_b^2(X),\ \forall t \geq 0 \\ (iii) & u(\cdot,x) \in C^1([0,T])\ \forall x \in D(A) \\ (iv) & u \text{ satisfy } (40)\ \forall x \in D(A), t \geq 0 \end{array} \tag{51}$$

(see [5] p. 261). We pass now to the general case.

4.2 Strong solutions

We first recall some definitions given in [3]. Let $\mathcal{A}$ be the infinitesimal generator of the weakly continuous semigroup $\{P_t;\ t \geq 0\}$ on $UC_b(X)$ (see [2] and [3]), let $\varphi \in UC_b(X)$ and $\mathcal{F} \in \Sigma^0_{T,\alpha}$. Consider the Cauchy problem

$$\begin{cases} u'(t) = \mathcal{A}u(t) + \mathcal{F}(t,\cdot) \ \ t \in]0,T] \\ u(0) = \varphi \in UC_b(X). \end{cases} \tag{52}$$

Define the mild solution of (52) as

$$u(t,x) = [P_t\varphi](x) + \int_0^t \left[P_{t-s}\mathcal{F}(s,\cdot)\right](x)ds \tag{53}$$

and the operator $\mathcal{A}_0$ as follows:

$$\begin{cases} D(\mathcal{A}_0) = \Big\{\varphi \in UC_b^2(X)\ :\varphi_{xx} \in UC_b(X,\mathcal{L}_1(X)); A^*\varphi_x \in UC_b(X); \\ \qquad\qquad\qquad\qquad\qquad \text{and } x \to \langle x, A^*\varphi_x\rangle\ \in UC_b(X)\Big\} \\ \mathcal{A}_0\eta = \frac{1}{2}\mathrm{Tr}[Q\varphi_{xx}] + \langle x, A^*\varphi_x\rangle \end{cases} \tag{54}$$

It easy to see that $D(\mathcal{A}_0)$, endowed with the norm

$$\|\eta\|_* \stackrel{def}{=} \|\eta\|_0 + \|A^*\eta_x\|_0 + \| < x, A^*\eta_x > \|_0 + \sup_{x\in X} \|\eta_{xx}(x)\|_{\mathcal{L}_1(X)} \tag{55}$$

is a Banach space. Moreover $\mathcal{A}_0 \subset \mathcal{A}$ and $D(\mathcal{A}_0)$ is $\mathcal{K}$-dense in $UC_b(X)$ i.e. for every $\varphi \in UC_b(X)$ there exists a sequence $\varphi_n \subset D(\mathcal{A}_0)$ such that

$$\varphi_n \xrightarrow{\mathcal{K}} \varphi \qquad \text{in } UC_b(X).$$

We recall also the following definitions of solution for the equation (40).

Definition 4.6 *A function $u : [0,T] \times X \to \mathbb{R}$ is a strict solution of equation (40) if u has the following regularity properties*

$$\begin{cases} u(\cdot, x) \in C^1([0,T]), \quad \forall x \in X \\ u \in B([0,T]; D(\mathcal{A}_0)) \\ u_t, \quad \mathcal{A}_0 u \in \widetilde{U}C_b([0,T] \times X) \end{cases} \tag{56}$$

and satisfies (40) in the classical sense.

Definition 4.7 *A function $u \in \Sigma^1_{T,\alpha}$ is a $\mathcal{K}$-strong solution of equation (40) if there exist three sequences $\{u_n\}$, $\{\mathcal{F}_n\} \subset \widetilde{U}C_b([0,T] \times X)$ and $\{\varphi n\} \subset D(\mathcal{A}_0)$ such that for every $n \in \mathbb{N}$, u_n is a strict solution of the Cauchy problem:*

$$\begin{cases} w_t = \mathcal{A}_0 w + \mathcal{F}_n \\ w(0) = \varphi_n \end{cases} \tag{57}$$

and moreover, for $n \to +\infty$

$$\begin{array}{ll} \varphi_n \xrightarrow{\mathcal{K}} \varphi & \text{in } UC_b(X) \\ \mathcal{F}_n \xrightarrow{\mathcal{K}} \mathcal{F} & \text{in } \Sigma^0_{T,\alpha} \\ u_n \xrightarrow{\mathcal{K}} u & \text{in } \Sigma^1_{T,\alpha} \end{array} \tag{58}$$

■

In [3] it is proved that mild solutions and $\mathcal{K}$-strong solutions coincide. Here we prove this fact under more general assumptions and we show that the approximating sequence can be chosen in a more regular way. We start by the following lemma.

Lemma 4.8 *Assume that Hypotheses 4.1 hold and let $\varphi \in UC_b(X)$, $\mathcal{F} \in \Sigma^0_{t,\alpha}$. Then there exists a sequence of finite dimensional projectors (P_n) such that $P_n X \subset D(A^*)$ and so there exists two sequences φ_n and $\mathcal{F}_n$ such that (21), (22) hold and*

$$\varphi_n \in D(\mathcal{A}_0) \quad \text{and} \quad \mathcal{F}_n \in C([0,T]; D(\mathcal{A}_0)). \tag{59}$$

Moreover, for every $k \geq 1$ and $(h_1, .., h_k) \in X^k$,

$$< D^k \varphi_n(h_1, .., h_k) \in D(\mathcal{A}_0) \quad \text{and} \quad t \to D^k_x \mathcal{F}_n(t, \cdot)(h_1, .., h_k) \in C([0,T]; D(\mathcal{A}_0)) \tag{60}$$

Proof. Let $\{e_n\}_{n \in \mathbb{N}}$ be an orthonormal complete system contained in $D(A^*)$. Such a system exists due to the density of $D(A^*)$ in X and can be found by applying an orthonormalization procedure to a sequence $\{v_n\}_{n \in \mathbb{N}} \subset D(A^*)$ which is dense in X. For every $i \in \mathbb{N}$ and $x \in X$ set $x_i =< x, e_i >$. For every $n \in \mathbb{N}$, the following linear mapping

$$P_n : X \to X; \qquad P_n x = \sum_{i=1}^{n} x_i e_i.$$

satisfies $P_nX \in D(A^*)$. Finally, taking φ_n and $\mathcal{F}_n$ associated to P_n as in Proposition 3.1, we can verify, by straightforward calculations that they satisfy (59) and (60). ∎

Remark 4.9 By Lemma 4.8 it follows, in particular, that $D(\mathcal{A}_0)$ is $\mathcal{K}$-dense in X. ∎

An important consequence of Proposition 3.1 is the following proposition which generalizes a result proven in [3].

Proposition 4.10 *Assume that Hypothesis 4.1 holds. Let* $k = 0, 1, 2, .., \varphi \in UC_b^k(X)$, $\mathcal{F} \in \Sigma_{T,\alpha}^k$ *and let*

$$u(t,x) = [P_t\varphi](x) + \int_0^t [P_{t-s}\mathcal{F}(s,\cdot)](x)ds$$

be the mild solution of (40) Let φ_n *and* $\mathcal{F}_n$ *be as in Lemma lm:s5 and*

$$u_n(t,x) = [P_t\varphi_n](x) + \int_0^t [P_{t-s}\mathcal{F}_n(s,\cdot)](x)ds$$

Then for every $n \in \mathbb{N}$ *and* $j \in \mathbb{N}$ u_n *and* $D_x^j u_n$ *satisfy the second and the third of (56) and*

$$\mathcal{K} - \lim_{n\to\infty} u_n = u \qquad in\ \Sigma_{T,\alpha}^{k+1} \tag{61}$$

To start the proof of Proposition 4.10 we give a useful Lemma. We omit the proof because it is similar to the one of Lemma 4.7 in [6].

Lemma 4.11 *Assume that Hypothesis 4.1 holds. Let* Y *be another Hilbert space and let* $\psi \in \widetilde{U}C_b([0,T]^2 \times X; Y)$. *Then the function*

$$(s,t,x) \to [P_t\psi(s,t,\cdot)](x)$$

belongs to $\widetilde{U}C_b([0,T]^2 \times X; Y)$ *and*

$$(t,x) \to \int_0^t [P_{t-s}\psi(s,t,\cdot)](x)ds$$

belongs to $\widetilde{U}C_b([0,T] \times X; Y)$.

Proof of Proposition 4.10. First we prove the regularity of u_n for every $n \in \mathbb{N}$. We have to prove that, for every $n \in \mathbb{N}$ and $j \in \mathbb{N}$

$$u_n(t,\cdot) \in UC_b^\infty(X) \ \forall t \in [0,T]$$

$$D_x^{j+2}u_n(t,\cdot)(h_1,..,h_j,\cdot,\cdot) \subset \widetilde{U}C_b([0,T] \times X, \mathcal{L}_1(X)) \qquad \forall (h_1,..,h_j) \in X^j$$

$$D_x^{j+1}u_n(t,x)(h_1,..,h_j) \subset D(A^*), \quad \forall (t,x) \in [0,T] \times X, \ \forall (h_1,..,h_j) \in X^j$$

$$A^*D_x^{j+1}u_n(h_1,..,h_j) \in \widetilde{U}C_b([0,T] \times X; X) \qquad \forall (h_1,..,h_j) \in X^j$$

$$(t,x) \to \langle x, A^*D_x^{j+1}u_n(t,x)(h_1,..,h_j)\rangle \in \widetilde{U}C_b([0,T] \times X) \qquad \forall (h_1,..,h_j) \in X^j \tag{62}$$

By the property (46) of the semigroup P_t it easy to see that the first of (62) holds. Moreover let us write, by the definition of P_t,

$$u_n(t,x) = \int_X \varphi_n(z + e^{tA}x)d\mu_t + \int_0^t \int_X \mathcal{F}_n(s,(z + e^{(t-s)A}x))d\mu_{t-s}ds \tag{63}$$

and, by differentiating $j+2$ times ($j \geq 0$) under the integral sign

$$\begin{aligned} D_x^{j+2}u_n(t,x) &= \int_X \varphi_n(z + e^{tA}x)(e^{tA})^{\otimes(j+2)}d\mu_t(z) \\ &+ \int_0^t \int_X \mathcal{F}_n(s,(z + e^{(t-s)A}x))(e^{(t-s)A})^{\otimes(j+2)}d\mu_{t-s}(z)ds \end{aligned} \tag{64}$$

Let $j \geq 0$, and let $h_1,..,h_j \in X$ (if $j = 0$ we don't need them) and consider the function

$$(t,x) \in \mathrm{Tr}\ D_x^{j+2}u_n(t,x)(h_1,..,h_j).$$

First we observe that this function is finite and bounded on $[0,T] \times X$. In fact if we take $\{e_i\}_{i\in\mathbb{N}}$ as the orthonormal basis of X that determines P_n in the Proposition 3.1, we have, by (64)

$$\begin{aligned} \mathrm{Tr}\ D_x^{j+2}u_n(t,x)(h_1,..,h_j) &= \sum_{i=1}^{+\infty} D_x^{j+2}u_n(t,x)(h_1,..,h_j,e_i,e_i) \\ &= \int_X \sum_{i=1}^{+\infty} D^{j+2}\varphi_n(z + e^{tA}x)(e^{tA}h_1,..e^{tA}h_j,e^{tA}e_i,e^{tA}e_i)d\mu_t(z) \\ &+ \int_0^t \int_X \sum_{i=1}^{+\infty} D_x^{j+2}\mathcal{F}_n(s,(z + e^{(t-s)A}x)) \\ &\quad (e^{(t-s)A}h_1,..,e^{(t-s)A}h_j,e^{(t-s)A}e_i,e^{(t-s)A}e_i)d\mu_{t-s}(z)ds \end{aligned} \tag{65}$$

so that

$$\begin{aligned} &\left|\mathrm{Tr}\ D_x^{j+2}u_n(t,x)(h_1,..,h_j)\right| \\ &\leq \|D^{j+2}\varphi_n\|_0\|e^{tA}\|^j|h_1|\cdot\cdot|h_j|\sum_{i=1}^{+\infty}|P_ne^{tA}e_i|^2 \\ &+ \int_0^t \|D_x^{j+2}\mathcal{F}_n(s,\cdot)\|_0\|e^{(t-s)A}\|^j|h_1|\cdot\cdot|h_j|\sum_{i=1}^{+\infty}|P_ne^{(t-s)A}e_i|^2 \end{aligned} \tag{66}$$

which is bounded since, for $t \in [0,T]$,

$$\begin{aligned} \sum_{i=1}^{+\infty}|P_ne^{tA}e_i|^2 &= \sum_{i=1}^{+\infty}\sum_{r=1}^{n} < e^{tA}e_i, e_r >^2 \\ &= \sum_{r=1}^{n}\sum_{i=1}^{+\infty} < e_i, e^{tA^*}e_r >^2 = \sum_{r=1}^{n}|e^{tA^*}e_r|^2 \leq n\|e^{tA^*}\|^2 \end{aligned} \tag{67}$$

To prove regularity we use Lemma 4.8 by observing that (65) can be written as

$$\mathrm{Tr}\, D_x^{j+2}u(t,x)(h_1,..,h_j) = P_t\left(\sum_{i=1}^{+\infty} D^{j+2}\varphi_n(e^{tA}h_1,..e^{tA}h_j, e^{tA}e_i, e^{tA}e_i)\right)$$

$$+\int_0^t P_{t-s}\left(\sum_{i=1}^{+\infty} D_x^{j+2}\mathcal{F}_n(s,\cdot)(e^{(t-s)A}h_1,..e^{(t-s)A}h_j, e^{(t-s)A}e_i, e^{(t-s)A}e_i)\right) ds.$$

This concludes the proof of the second of (62).

In the following we fix $(h_1,..,h_j) \in X^j$, and consider the following map

$$\begin{array}{ccc} (t,x) & \longrightarrow & D_x^{j+1}u_n(t,x)(h_1,..,h_j) \\ [0,T]\times X & \longrightarrow & X \end{array} \tag{68}$$

First by (64) it is clear that, for every $(t,x) \in [0,T]\times X$, $h \in D(A)$ we have

$$< D_x^{j+1}u_n(t,x)(h_1,..,h_j), Ah >$$

$$= \int_X < D^{j+1}\varphi_n(z+e^{tA}x)(e^{tA}h_1,..,e^{tA}h_j), e^{tA}Ah > d\mu_t(z)$$

$$+\int_0^t\int_X < D_x^{j+1}\mathcal{F}_n(s,(z+e^{(t-s)A}x))(e^{(t-s)A}h_1,..,e^{(t-s)A}h_j), e^{(t-s)A}Ah > d\mu_{t-s}(z)ds \tag{69}$$

so that, by the regularity of φ_n and $\mathcal{F}_n$,

$$| < D_x^{j+1}u_n(t,x)(h_1,..,h_j), Ah > |$$

$$\leq \left[\|A^*D^{j+1}\varphi_n\|_0 + T\|A^*D_x^{j+1}\mathcal{F}_n\|_0\right] M^{j+1}e^{(j+1)\omega T}|h_1|\cdot\cdot|h_j||h|$$

which implies that

$$D_x^{j+1}u_n(t,x)(h_1,..,h_j) \subset D(A^*), \quad \forall (t,x) \in [0,T]\times X,$$

and that the function $[0,T]\times X \to X$, $A^*D_x^{j+1}u_n(h_1,..,h_j)$, is bounded on $[0,T]\times X$. So the third of (62) is proved. Moreover we can write, by (69)

$$< A^*D_x^{j+1}u_n(t,x)(h_1,..,h_j), h >$$

$$= \int_X < A^*D^{j+1}\varphi_n(z+e^{tA}x)(e^{tA}h_1,..,e^{tA}h_j), e^{tA}h > d\mu_t(z)$$

$$+\int_0^t\int_X < A^*D_x^{j+1}\mathcal{F}_n(s,(z+e^{(t-s)A}x))(e^{(t-s)A}h_1,..,e^{(t-s)A}h_j), e^{(t-s)A}h > d\mu_{t-s}(z)ds$$

$$= P_t\left(< A^*D^{j+1}\varphi_n(e^{tA}h_1,..e^{tA}h_j), e^{tA}h >\right)(x)$$

$$+\int_0^t P_{t-s}\left(< A^*D_x^{j+1}\mathcal{F}_n(s,\cdot)(e^{(t-s)A}h_1,..e^{(t-s)A}h_j), e^{(t-s)A}h >\right)(x)ds. \tag{70}$$

which implies the fourth claim of (62) by Lemma 4.8. Finally take the function

$$(t,x) \to < x, A^* D_x^{j+1} u_n(t,x)(h_1,..,h_j) >$$

and, as before, observe that

$$< A^* D_x^{j+1} u_n(t,x)(h_1,..,h_j), x >$$

$$= \int_X < A^* D^k \varphi_n(z + e^{tA}x)(e^{tA}h_1,..,e^{tA}h_j), e^{tA}x > d\mu_t(z)$$

$$+ \int_0^t \int_X < A^* D_x^{j+1} \mathcal{F}_n(s, (z + e^{(t-s)A}x))(e^{(t-s)A}h_1,..,e^{(t-s)A}h_j), e^{(t-s)A}x > d\mu_{t-s}(z)ds \tag{71}$$

Now we prove first the continuity with respect to t observing that by (71) we can write

$$< A^* D_x^{j+1} u_n(t,x)(h_1,..,h_j), x >$$

$$=< P_t \left(< A^* e^{tA^*} D^{j+1} \varphi_n(e^{tA}h_1,..,e^{tA}h_j)\right)(x), x > \tag{72}$$

$$+ \int_0^t < P_{t-s} \left(< A^* e^{(t-s)A^*} D_x^k \mathcal{F}_n(s,\cdot)(e^{(t-s)A}h_1,..,e^{(t-s)A}h_{k-1})\right)(x), x >$$

which is continuous with respect to t thanks to Lemma 4.11. Moreover, being $e^{tA}x = z + e^{tA}x - z$, we can rewrite (71) as follows

$$< A^* D_x^{j+1} u_n(t,x)(h_1,..,h_j), x >= P_t \left(< A^* D^{j+1} \varphi_n(e^{tA}h_1,..,e^{tA}h_j), \cdot >\right)(x)$$

$$+ \int_0^t P_{t-s} \left(< A^* D_x^{j+1} \mathcal{F}_n(s,\cdot)(e^{(t-s)A}h_1,..,e^{(t-s)A}h_j), \cdot >)\right)(x)ds$$

$$+ \int_X < A^* D^{j+1} \varphi_n(z + e^{tA}x)(e^{tA}h_1,..,e^{tA}h_j), z > d\mu_t(z)$$

$$+ \int_0^t \int_X < A^* D_x^{j+1} \mathcal{F}_n(s, (z + e^{(t-s)A}x))(e^{(t-s)A}h_1,..,e^{(t-s)A}h_j), z > d\mu_{t-s}(z)ds. \tag{73}$$

The first part belongs to $\widetilde{U}C_b([0,T] \times X)$ thanks to Lemma 4.8. For the second one we observe that, by the Hölder inequality

$$\left| \int_X < A^* D^{j+1} \varphi_n(z + e^{tA}x)(e^{tA}h_1,..,e^{tA}h_j), z > d\mu_t(z) \right|$$

$$\leq \|A^* D^{j+1} \varphi_n\|_0 M^j e^{j\omega T} |h_1| \cdot \cdot |h_j| \left(\int_X |z|^2 d\mu_t(z) \right)^{\frac{1}{2}}$$

$$= \|A^* D^{j+1} \varphi_n\|_0 M^j e^{j\omega T} |h_1| \cdot \cdot |h_j| (\mathrm{Tr}\ Q_t)^{\frac{1}{2}}$$

and similarly for the term containing $\mathcal{F}_n$. For the continuity with respect to x of the second part we have, for $x, y \in X$, arguing as before

$$\left| \int_X < A^* D^{j+1} \varphi_n(z + e^{tA}x)(e^{tA}h_1, .., e^{tA}h_j), z > d\mu_t(z) \right.$$

$$\left. - \int_X < A^* D^{j+1} \varphi_n(z + e^{tA}y)(e^{tA}h_1, .., e^{tA}h_j), z > d\mu_t(z) \right|$$

$$\leq \rho\left(A^* D^{j+1}\varphi_n; Me^{\omega T}|x-y|\right) M^j e^{j\omega T}|h_1| \cdot \cdot |h_j| \, (\text{Tr } Q_t)^{\frac{1}{2}}$$

and similarly for the term containing $\mathcal{F}_n$. This ends the proof of (62).

We now prove (61). By [5], p. 264, we can write

$$u_n(t,x) = \int_X \varphi_n(z + e^{tA}x)\mu_t(dz) + \int_0^t ds \int_X \mathcal{F}_n(s, z + e^{(t-s)A}x)\mu_{t-s}(dz) \tag{74}$$

and, for $j = 1, .., k$, and $(h_1, .., h_j) \in X^j$,

$$\langle D_x^j u_n(h_1, .., h_j)\rangle = \int_X D^j \varphi_n(z + e^{tA}x), (e^{tA}h_1, .., e^{tA}h_j)\mu_t(dz)$$

$$+ \int_0^t ds \int_X D_x^j \mathcal{F}_n(s, z + e^{(t-s)A}x)(e^{(t-s)A}h_1, .., e^{(t-s)A}h_j)\mu_{t-s}(dz). \tag{75}$$

Finally

$$D_x^{k+1} u(h_1, .., h_{k+1}) = \int_X D^k \varphi_n(z + e^{tA}x)(e^{tA}h_1, .., e^{tA}h_k)\langle\Gamma(t)h_{k+1}, Q_t^{-\frac{1}{2}}z\rangle\mu_t(dz) \tag{76}$$

$$+ \int_0^t ds \int_X D_x^k \mathcal{F}_n(s, z + e^{(t-s)A}x)(e^{(t-s)A}h_1, .., e^{(t-s)A}h_k)\langle\Gamma(t-s)h_{k+1}, Q_{t-s}^{-\frac{1}{2}}z\rangle\mu_{t-s}(dz)$$

The last three equations give the claim by simply applying the Dominated Convergence Theorem. ■

Remark 4.12 By using that the semigroup $\{e^{tA};\ t \geq 0\}$ is compact (see Remark 4.2-2) it can be shown that the convergence in (61) is uniform on bounded subsets on $[0,T] \times X$. In fact, if $B \subset X$ is bounded, then the set

$$B_t \stackrel{def}{=} \left\{e^{tA}x;\ x \in B\right\}$$

is compact for all $t > 0$, so that, if $\varphi_n \xrightarrow{\mathcal{K}} \varphi$, then

$$\sup_{x\in B} |[P_t\varphi_n](x) - [P_t\varphi](x)| \leq \int_X \sup_{x \in B} |\varphi_n(z + e^{tA}x) - \varphi(z + e^{tA}x)| d\mu_t(z)$$

$$= \int_X \sup_{y \in B_t} |\varphi_n(z+y) - \varphi(z+y)| d\mu_t(z)$$

which goes to 0 thanks to the Dominated Convergence Theorem. The proof of the convergence for the other terms is completely similar. ∎

A straightforward consequence of the last Proposition is the following

Theorem 4.13 *Assume that Hypothesis 4.1 holds. Let $\varphi \in UC_b^k(X)$ and $\mathcal{F} \in \Sigma_{T,\alpha}^k$ Then a function $u \in \Sigma_{T,\alpha}^{k+1}$ is the mild solution of (1) if and only if it is a $\mathcal{K}$-strong solution. Moreover the approximating sequence can be chosen so that (21), (61), (59) and (60) hold. If $\varphi \in D(\mathcal{A}_0)$ and $\mathcal{F} \in C([0,T]; D(\mathcal{A}_0))$ then the mild solution is strict.*

Proof. The fact is that the mild solution is also a $\mathcal{K}$-strong solution and the properties of the approximating sequence follow by the previous Proposition. Moreover if u is a $\mathcal{K}$-strong solution, then the approximating sequence u_n satisfies

$$u_n(t,x) = [P_t\varphi_n](x) + \int_0^t [P_{t-s}\mathcal{F}_n(s,\cdot)](x)ds$$

for suitable φ_n and $\mathcal{F}_n$ as in Definition 4.7. The claim follows by the Dominated Convergence Theorem.

The final statement follows by the fact that $P_t(D(\mathcal{A}_0)) \subset D(\mathcal{A}_0)$, proved in [3].

Remark 4.14 The approximation result of Proposition 4.10 can be used to get estimates of the local solution of the nonlinear equation

$$\begin{cases} \dfrac{\partial u}{\partial t} = \dfrac{1}{2}\,\mathrm{Tr}\,[Qu_{xx}] + \langle Ax, u_x \rangle + H(x,u_x) + g(t,x) = 0, \ t \in]0,T], \ x \in D(A) \\ \\ u(0,x) = \varphi(x), \ x \in X, \end{cases} \tag{77}$$

when the nonlinearity H is Lipschitz continuous on bounded subsets of X (see [7]). ∎

References

[1] P. CANNARSA & G.DA PRATO *On a Functional Analysis approach to Parabolic Equations in infinite dimensions.* J. Func. Anal., 118, n.1, (1993) pp.22-42.

[2] S. CERRAI *A Hille-Yosida Theorem for weakly continuous semigroups,* Preprints di Matematica n.17, Scuola Normale Superiore, Pisa, (1993). Semigroup Forum n49, (1994) pp.

[3] S. CERRAI & F. GOZZI *Strong solutions of Cauchy problems associated to weakly continuous semigroups,* Preprint di Matematica n.35, Scuola Normale Superiore, Pisa, (1993). To appear in Differential and Integral Equations.

[4] G. DA PRATO & A. LUNARDI *On the Ornstein Uhlenbeck operator in spaces of continuous functions,* To appear in J. Funct. Anal.

[5] G. DA PRATO & J. ZABCZYK STOCHASTIC EQUATIONS IN INFINITE DIMENSIONS, Encyclopedia of Mathematics and its Applications, Cambridge University Press, (1992).

[6] F. GOZZI *Regularity of solutions of second order Hamilton-Jacobi equations and application to a control problem,* Preprint di Matematica n.09, Scuola Normale Superiore, Pisa, (1994). To appear in Communications in Partial Differntial Equations.

[7] F. GOZZI *Global existence of solutions for a second order Hamilton-Jacobi equation in Hilbert spaces* Preprint n, 2.192.852, Marzo 1995, Dipartimento di Matematica, Università di Pisa. To appear in J. of Math. Anal. Appl..

[8] L. HÖRMANDER *Hypoelliptic second order differential equations* Acta Math. **119** (1967), pp.147-171.

[9] A. SWIECH VISCOSITY SOLUTIONS OF FULLY NONLINEARPARTIAL DIFFERENTIAL EQUATIONS WITH "UNBOUNDED" TERMS IN INFINITE DIMENSIONS, Ph.D. Thesis, University of California (Santa Barbara), (July 1993).

[10] K. YOSIDA FUNCTIONAL ANALYSIS, sixth edition Springer-Verlag, Berlin, New-York, (1980).

Sufficient Conditions for Dirichlet Boundary Control Problems of Parabolic Type

Fausto Gozzi & Maria Elisabetta Tessitore

Dipartimento di Matematica, Università di Pisa
Via F. Buonarroti 2, 56127 Pisa, Italy
&
Dipartimento di Matematica, Università di Roma "La Sapienza"
Piazzale A. Moro 2, 00185 Roma, Italy

Abstract

This paper is devoted to the search for sufficient optimality conditions for a finite horizon optimal control problem with Dirichlet boundary control and with a semilinear state equation. We prove sufficient conditions for optimality of trajectory–control pairs showing a uniqueness result for a system of differential equations which is the Hamiltonian formulation of the classical necessary condition known as the Pontryagin Maximum Principle. This yields a sufficient condition for optimality that extends the results derived in [19] since the state equation is assumed to be linear in order to prove sufficient optimality conditions.

Key words: Boundary control, parabolic equations, Dirichlet boundary conditions, optimal controls, sufficient optimality conditions.

AMS (MOS) subject classifications: 49A22, 49B22, 49A52, 47D05, 49C20.

1 Introduction

In this paper we consider the following finite horizon optimal control problem

$$\text{minimize } J(t_0, x_0; \gamma) = \int_{t_0}^{T} L(s, x(s; t_0, x_0, \gamma), \gamma(s))ds + \phi(x(T; t_0, x_0, \gamma)) \;, \tag{1}$$

overall trajectory–control pairs $\{x, \gamma\}$, which are mild solutions of the abstract equation

$$\begin{cases} x'(t) + Ax(t) + F(x(t)) = A^{\beta} B\gamma(t) \\ x(t_0) = x_0 \end{cases} \tag{2}$$

Here L and ϕ are real–valued smooth function, $A : D(A) \subset X \to X$ is a self–adjoint accretive operator, A^{β} is the β–fractional power of A, $\frac{3}{4} < \beta < 1$, B is a bounded linear operator. Moreover $F : X \to X$ is a Lipschiz continuous map and $\gamma : [t_0, T] \to U$ is a measurable function. U and X are two real Hilbert spaces respectively denoted control space and state space. If, in addition, we consider $B = D_\beta = A^{1-\beta}D$, where D is the Dirichlet map, then system (2) is the abstract version of a parabolic partial differential equation controlled by a Dirichlet datum at the boundary

$$\begin{cases} \dfrac{\partial x}{\partial t}(t,\xi) = \Delta_\xi x(t,\xi) + f(x(t,\xi)) & \text{in } (t_0,T)\times\Omega \\ x(t_0,\xi) = x_0(\xi) & \text{on } \Omega \\ x(t,\xi) = \gamma(t,\xi) & \text{on } (t_0,T)\times\partial\Omega \end{cases}$$

Associated to the optimal control problem (1)–(2) we consider the value function

$$v(t_0,x_0) = \inf_{\gamma(t)\in U}\left\{\int_{t_0}^T L(s,x(s;t_0,x_0,\gamma),\gamma(s))ds + \phi(x(T;t_0,x_0,\gamma))\right\}. \tag{3}$$

A control $\overline{\gamma}(\cdot)$ is said to be optimal if the infimum in the above equation is attained at $\overline{\gamma}(\cdot)$.

The goal of this paper is to state sufficient optimality conditions for boundary control problems of the type (1)–(2). The literature on boundary control problems is huge. It began with the study of the Linear Quadratic problem, see for instance [23], [5], [22]. Among the nonlinear boundary control problems one of the first cases to be treated was the convex problem, see [1], [20], where the state equation is linear and the running cost is convex. As for general nonlinear boundary control problems, most of the results that are available in the literature are concerned with necessary optimality conditions, see e.g. [14], [15] and [29]. The Dynamic Programming approach to nonlinear boundary control problems is more recent and uses viscosity solutions, see [9], [10] and [11].

As far as sufficient optimality conditions are concerned, this paper generalizes the results obtained in [19] for the Dirichlet boundary problem and the ones derived in [12] for the Neumann boundary problem. Consider the system of differential equations which represents the Hamiltonian formulation of the classical necessary condition known as the Pontryagin Maximum Principle

$$\begin{cases} x'(t) = -Ax(t) - F(x(t)) - A^\beta D_p H(t,x(t),A^\beta p(t)) \\ p'(t) = Ap(t) + [DF(x(t))]^* p(t) + D_x H(t,x(t),A^\beta p(t)) \end{cases}$$

with the initial–terminal condition

$$\begin{cases} x(t_0) = x_0 \\ p(T) = D\phi(x(T)). \end{cases} \tag{4}$$

In [12] and in [19] , adapting the results of [6] for distributed control, it is shown, under suitable regularity assumptions on the data, a weak sufficient condition for optimality proving that the Hamiltonian system above has a solution which is an optimal trajectory. Moreover they show that a stronger sufficient condition holds when $F = 0$.

In this case they consider p_0 as a cluster point of the gradient of the value function v with respect to x, evaluated at (t_0, x_0). Then substituting the initial–terminal data above with an initial–initial condition of the form

$$\begin{cases} x(t_0) = x_0 \\ p(t_0) = p_0, \end{cases}$$

they show that the solution of (4) is unique and satisfies also the terminal condition $p(T) = D\phi(x(T))$.

In this paper, instead, we do not require the state nonlinearity F to be identically equal to zero. Nevertheless we are able to prove the same sufficient condition.

We refer to [18] for second order sufficient conditions for boundary control problems.

We briefly outline the paper. In §2 we recall the main assumptions on the data and the basic material on boundary control problems. In §3 we show a uniqueness result for the Hamiltonian system (4) which is a sufficient condition for optimality, see Theorem 3.2.

2 Preliminaries

Let X, Y be two real Hilbert spaces. We denote by $L^2(a, b; X)$ the space of all square integrable functions $\gamma : [a, b] \to X$. Let Ω be a subset of Y, then $C(\Omega; X)$ will denote the set of all continuous functions $f : \Omega \to X$. For $p \in [1, +\infty]$, $L^p(\Omega, X)$ will be the set of all functions $f : \Omega \to X$ such that $\|f\|_X^p$ is integrable on Ω. If $X = I\!R$ we will simply write $C(\Omega)$ and $L^p(\Omega)$. Finally $\mathcal{L}(Y; X)$ will denote the space of all bounded linear operators $T : Y \to X$.

Let $x_0 \in X$, $T > 0$, $t_0 \in [0, T]$, $\gamma \in L^2(t_0, T; U)$ where X and U are two real Hilbert spaces. Consider the mild solution $x(\cdot; t_0, x_0, \gamma)$ of the abstract equation

$$\begin{cases} x'(t) + Ax(t) + F(x(t)) = A^\beta B\gamma(t) \\ x(t_0) = x_0 \end{cases} \tag{5}$$

that is the solution of the integral equation

$$x(t) = e^{-(t-t_0)A}x_0 - \int_{t_0}^t e^{-(t-s)A}F(x(s))ds + A^\beta \int_{t_0}^t e^{-(t-s)A}B\gamma(s)ds\,. \tag{6}$$

We assume

(*i*) $A : D(A) \subset X \to X$ is a closed linear operator such that $A = A^*$ and $< Ax, x > \geq \omega |x|^2$ for some $\omega > 0$ and all $x \in D(A)$;

(*ii*) the inclusion $D(A) \subset X$ is dense and compact ;

(*iii*) $F : X \to X$, $|F(x) - F(y)| \leq L_F |x - y|$, $\forall x, y \in X$;

(*iv*) $\beta \in \left(\frac{3}{4}, 1\right)$;

(*v*) $B \in \mathcal{L}(U; D(A^\rho))$.

(7)

for some positive constant L_F and for some $\rho > 0$.
In (5), A^β denotes the fractional power of the operator A, see [27]. We note that (i) and (ii) imply that $-A$ is the infinitesimal generator of an analytic semigroup satisfying $||e^{-tA}|| \leq e^{-\omega t}$ for some $\omega > 0$ and all $t \geq 0$. Now we recall two estimates which will be useful in the sequel.

For every $\theta \in [0,1]$ there exists a constant $M_\theta > 0$ such that

$$|A^\theta e^{-tA} x| \leq \frac{M_\theta}{t^\theta}|x|, \quad \forall t > 0, \forall x \in X. \tag{8}$$

Moreover let $\eta \in (0,1]$ and $\alpha \in (0,\eta)$. Then, a well known interpolation inequality, see e.g. [27], states that for every $\sigma > 0$ there exists $C_\sigma > 0$ such that

$$|A^\alpha x| \leq \sigma |A^\eta x| + C_\sigma |x|, \quad \forall x \in D(A^\eta). \tag{9}$$

The regularity properties of the solution of equation (6) are well known, see [5]. Under the above assumptions, we have

Proposition 2.1 *Assume that (7) holds. Fix $0 \leq t_0 \leq T$ and let $\gamma : [t_0, T] \to U$. Then for any $x_0 \in X$ there exists a unique solution $x \in L^2(t_0, T; D(A^{1-\beta+\rho}))$ such that*

$$A^{\frac{1}{2}-\beta} x \in C([t_0, T]; X) \text{ and } |A^{\frac{1}{2}-\beta} x(t)| \leq C_0[|x| + |F(0)| + ||\gamma||_{L^2(t_0,T;U)}],$$

for every $t \in [t_0, T]$ and for some $C_0 > 0$.
Moreover, if $\gamma(\cdot)$ is bounded then $x \in C([t_0, T]; X)$.
Finally, if $x_0 \in D(A^{1-\beta})$ and $\gamma(\cdot)$ is bounded, then $x \in C([t_0, T]; D(A^{1-\beta}))$.

We recall the definition of some generalized gradients which will be used in the sequel. Let O be an open subset of X. The superdifferential of a function $w : O \to I\!R$ at a point $x_0 \in R$ is the (possibly empty) set

$$D^+ w(x_0) = \left\{ p \in X \ : \ \limsup_{x \to x_0} \frac{w(x) - w(x_0) - < p, x - x_0 >}{|x - x_0|} \leq 0 \right\}.$$

We denote by $D^* w(x_0)$ the set of all vectors $p \in X$ for which there exists a sequence $\{x_n\}$ of points of O such that

$$\begin{cases} (i) & x_n \to x_0 \text{ as } n \to +\infty \\ (ii) & w \text{ is Fréchet differentiable at } x_n, \forall n \\ (iii) & Dw(x_n) \rightharpoonup p \text{ as } n \to +\infty \end{cases}$$

If the function w is Lipschitz continuous in a neighborhood R_0 of x_0, then w is Fréchet differentiable on a dense subset of R_0 (see [28]). Hence, $D^* w(x_0) \neq \emptyset$.

Now let us consider the problem of minimizing the functional

$$J(t_0, x_0; \gamma) = \int_{t_0}^T L(t, x(t; t_0, x_0, \gamma), \gamma(t))dt + \phi(x(T; t_0, x_0, \gamma)) \tag{10}$$

over all functions $\gamma \in L^2(t_0, T; U)$, where $x(t; t_0, x_0, \gamma)$ is the mild solution of (5).

The value function of problem (10)–(5) is defined as

$$v(t_0, x_0) = \inf_{\gamma(t) \in U} \left\{ \int_{t_0}^{T} L(t, x(t; t_0, x_0, \gamma), \gamma(t))dt + \phi(x(T; t_0, x_0, \gamma)) \right\}. \tag{11}$$

A control $\gamma^*(t) \in U$ at which the infimum in (11) is attained, is said to be optimal.

Here $L : [0, T] \times X \times U \to I\!R$ and $\phi : X \to I\!R$ are assumed to satisfy the following

(i) $L \in C([0, T] \times X \times U)$;

(ii) For some constant $C_L > 0$:
$|L(t, x, \gamma) - L(t, y, \gamma)| \leq C_L(1 + |x| + |y|)|x - y|, \forall t \in [0, T], \gamma \in U, |x|, |y| \in X$;

(iii) $L(t, x, \cdot)$ is strictly convex;
$\exists \lambda_0 > 0, \lambda_1 \in I\!R \ : \ L(t, x, \gamma) - L(t, x, 0) \geq \lambda_0 |\gamma|^2 + \lambda_1$,
and $L(t, x, \gamma) \geq \lambda_0 |\gamma|^2 + \lambda_1 \qquad \forall (t, x) \in [0, T] \times X, \ \gamma \in U$;

(iv) ϕ is bounded from below and $\forall R > 0$ and some constant $C_{\phi,R} > 0$:
$|\phi(x) - \phi(y)| \leq C_{\phi,R}|A^{\frac{1}{2}-\beta}(x - y)|, \forall x, y \in X$ such that $|A^{\frac{1}{2}-\beta}x|, |A^{\frac{1}{2}-\beta}y| \leq R$. (12)

Remark 2.2 *From assumptions* (7) *and* (12) *we derive, for every* $t_0 \in [0, T]$ *and* $x_0 \in X$, *the existence of optimal controls for problem* (5)-(11) *(see e.g. [2]). The two inequalities in hypothesis (12)-(iii) are necessary to guarantee the boundedness of optimal controls, see [19]. Moreover hypothesis (12)-(iv) is necessary in order to have a meaningful terminal cost.*

The value function v has the following regularity properties.

Proposition 2.3 *Assume* (7), (12). *Then, the value function* v *defined in* (11) *is continuous in* $[0, T] \times X$. *Moreover, for every* $R > \frac{1}{T}$ *and* $\theta \in [0, 1[$ *there exists a constant* $C_{\theta R} > 0$ *such that*

$$|v(t, x) - v(t, y)| \leq C_{\theta R}|A^{-\theta}(x - y)| \qquad \forall t \in [0, T - \frac{1}{R}], \ |x|, |y| \leq R$$

In particular v *is sequentially weakly continuous in* $[0, T] \times X$.

From this Proposition it follows that $D_x^* v(t, x)$ is not an empty set for every t and x. In order to state optimality conditions for the problem of minimizing $J(t_0, x_0; \gamma)$ overall controls $\gamma \in L^2(t_0, T; U)$ we assume

(i) F is continuously Fréchet differentiable;

(ii) L is continuously Fréchet differentiable with respect to x; (13)

(iii) ϕ is continuously Fréchet differentiable and $D\phi(A^{\beta-\frac{1}{2}})$ is continuous .

Let $\gamma \in L^2(t_0, T; U), x(\cdot) = x(\cdot; t_0, x_0, \gamma)$ and $p_T = D\phi(x(T))$. The co–state p associated to $\{\gamma, x, p_T\}$ is defined as the mild solution of the abstract equation

$$\begin{cases} p'(t) = Ap(t) + [DF(x(t))]^* p(t) - D_x L(t, x(t), \gamma(t)) \,, & t \in [t_0, T) \\ p(T) = p_T. \end{cases} \tag{14}$$

Remark that assumption (13) (iii) is necessary to have a meaningful terminal datum p_T.

Now we recall the classical formulation of the necessary optimality conditions via the Pontryagin Maximum Principle, see [19].

Theorem 2.4 *Assume* (7), (12), (13). *Let* $\{\overline{\gamma}, \overline{x}\}$ *be an optimal pair for problem* (11) – (5), *with starting point* $(t_0, x_0) \in [0, T] \times X$. *Moreover set* $p_T = D\phi(\overline{x}(T))$ *and let* $\overline{p}$ *be the corresponding co–state. Then it satisfies the co–state inclusion*

$$\overline{p}(t) \in D_x^+ v(t, \overline{x}(t))$$

for every $t \in [t_0, T]$ *and the Maximum Principle*

$$- < B\overline{\gamma}(t), A^\beta \overline{p}(t) > - L(t, \overline{x}(t), \overline{\gamma}(t)) = H(t, \overline{x}(t), A^\beta \overline{p}(t))$$

for a.e. $t \in [t_0, T]$ *Lebesgue point of* $\overline{\gamma}$, *where*

$$H(t, x, p) = \sup_{\gamma \in U} [- < B\gamma, p > - L(t, x, \gamma)].$$

For a proof of this result see [19] (see also [3] and [16] for similar results).

Next, by standard procedure, the Pontryagin Maximum Principle can be reformulated in terms of an Hamiltonian system (see e.g.[6]).

Theorem 2.5 *Assume* (7), (12), (13). *Let* $\{\overline{\gamma}, \overline{x}\}$ *be an optimal pair for problem* (11) – (5), *with starting point* $(t_0, x_0) \in [0, T] \times X$. *Moreover set* $p_T = D\phi(\overline{x}(T))$ *and let* $\overline{p}$ *be the corresponding co–state.*

If H *is Gâteaux differentiable with respect to* (x, p) *at* $(\overline{x}(t), \overline{p}(t))$, *for a. e.* $t \in [t_0, T]$, *then the pair* $(\overline{x}(t), \overline{p}(t))$ *is a mild solution of the Hamiltonian system*

$$\begin{cases} \overline{x}'(t) = -A\overline{x}(t) - F(\overline{x}(t)) - A^\beta D_p H(t, \overline{x}(t), A^\beta p(t)) & \overline{x}(t_0) = x_0 \\ \\ p'(t) = Ap(t) + [DF(\overline{x}(t)]^* p(t) + D_x H(t, \overline{x}(t), A^\beta p(t)) & p(T) = D\phi(x(T)). \end{cases}$$

3 Sufficient Conditions

As in [19], the next result may be directly derived following the reasonings contained in [6], Theorem 5.9.

Theorem 3.1 *Assume* (7), (12), (13) *and let* $H(t, \cdot, \cdot)$ *be Fréchet differentiable on* $X \times X$. *Moreover suppose that for all* $R > 0$,

$$|D_x H(t, x, p) - D_x H(t, y, q)| + |D_p H(t, x, p) - D_p H(t, y, q)| \leq C_R[|x - y| + |p - q|]$$

for some constant $C_R > 0$ *and all* $x, y, p, q \in X$ *satisfying* $|x|, |y| \leq R$. *Let* $(t_0, x_0) \in [0,T] \times X$ *and* $p_0 \in D_x^* v(t_0, x_0)$. *Then, the system*

$$\begin{cases} x'(t) = -Ax(t) - F(x(t)) - A^\beta D_p H(t, x(t), A^\beta p(t)) & x(t_0) = x_0 \\ p'(t) = Ap(t) + [DF(x(t))]^* p(t) + D_x H(t, x(t), A^\beta p(t)) & p(T) = D\phi(x(T)) \end{cases} \tag{15}$$

has a solution $(\overline{x}, \overline{p})$ *such that* $\overline{x}$ *is an optimal trajectory for problem* (11)–(5) *corresponding to some control* $\overline{\gamma}$. *Moreover,* $\overline{p}$ *is the co-state associated to* $\overline{\gamma}$ *and satisfies* $\overline{p}(t_0) = p_0$.

The above Theorem gives, in some sense, a sufficient condition for optimality. This condition would be more useful if one could guarantee uniqueness of solutions for (15). In the next Theorem we adapt the reasoning of [12] and [19] to the present case to show an existence and uniqueness result for the solution of an Hamiltonian system such as in (15). As in [12] and [19], we replace the terminal co-state datum with an initial one. Set $y(t) = A^{-\beta} x(t)$. Then the Hamiltonian system (15) becomes

$$\begin{cases} y'(t) = -Ay(t) - A^{-\beta} DF(A^\beta y(t)) - D_p H(t, A^\beta y(t), A^\beta p(t)), & y(0) = y_0 = A^{-\beta} x_0 \\ p'(t) = Ap(t) + [DF(A^\beta y(t))]^* p(t) + D_x H(t, A^\beta y(t), A^\beta p(t)), & p(0) = p_0 \end{cases} \tag{16}$$

Theorem 3.2 *Assume* (7), (12), (13) *and let* $H(t, \cdot, \cdot)$ *be Fréchet differentiable on* $X \times X$. *Moreover suppose that*

$$|D_x H(t,x,p) - D_x H(t,y,q)| + |D_p H(t,x,p) - D_p H(t,y,q)| \leq L_H[|x-y| + |p-q|] \tag{17}$$

and

$$|DF(x) - DF(y)| \leq L_{DF}|x-y| \tag{18}$$

for some constants $L_H, L_{DF} > 0$ *and all* $x, y, p, q \in X$. *Let* $(t_0, y_0) \in [0,T] \times D(A^{\frac{1}{2}})$ *and* $p_0 \in D_x^* v(t_0, y_0)$. *Then the Hamiltonian system* (16) *has a unique solution* $(\overline{y}, p)$ *such that* $\overline{y}, p \in C([t_0,T]; D(A^{\frac{1}{2}})) \cap L^2(t_0,T; D(A)) \cap W^{1,2}(t_0,T;X)$.

Proof–The existence part is a straightforward consequence of the previous Theorem. Without loss of generality we set $t_0 = 0$. Let $(y_1(t), p_1(t))$ and $(y_2(t), p_2(t))$ be two distinct solutions to system (16). Let $\theta \in C^1(I\!R)$ be a function such that

$$\theta(t) = \begin{cases} 1 & 0 \leq t \leq \frac{T}{2} \\ 0 & t = T \end{cases} \quad \text{and} \quad |\theta'(t)| \leq \frac{4}{T}$$

Now we set, for $k > 0$,

$$z(t) = e^{\frac{k(t-T)^2}{2}} \theta(t)(y_1(t) - y_2(t)) \text{ and } q(t) = e^{\frac{k(t-T)^2}{2}} \theta(t)(p_1(t) - p_2(t)) \tag{19}$$

Then, by easy computations (see e.g. [19]) $z(t)$ and $q(t)$ satisfy the system

$$\left\{\begin{array}{l} z'(t) = -Az(t) + k(t-T)z(t) - A^{-\beta}\tilde{F}(A^{\beta}z(t)) - D_q\tilde{H}(t, A^{\beta}z(t), A^{\beta}q(t)) + f_z(t)\ , \\ z(0) = z(T) = 0 \\ \\ q'(t) = Aq(t) + k(t-T)q(t) + [D\tilde{F}(A^{\beta}z(t))]^*q(t) + D_z\tilde{H}(t, A^{\beta}z(t), A^{\beta}q(t)) + f_q(t)\ , \\ q(0) = q(T) = 0 \end{array}\right. \tag{20}$$

where

$$\tilde{F}(A^{\beta}z(t)) = e^{\frac{k(t-T)^2}{2}}\theta(t)\left[F(A^{\beta}y_1(t)) - F(A^{\beta}y_2(t))\right], \tag{21}$$

$$[D\tilde{F}(A^{\beta}z(t))]^*q(t) = e^{\frac{k(t-T)^2}{2}}\theta(t)\left\{[DF(A^{\beta}y_1(t))]^*p_1(t) - [DF(A^{\beta}y_2(t))]^*p_2(t)\right\}. \tag{22}$$

Moreover

$$\begin{aligned} &D_q\tilde{H}(t, A^{\beta}z(t), A^{\beta}q(t)) \\ &= e^{\frac{k(t-T)^2}{2}}\theta(t)\left[D_pH(t, A^{\beta}y_1(t), A^{\beta}p_1(t)) - D_pH(t, A^{\beta}y_2(t), A^{\beta}p_2(t))\right], \end{aligned} \tag{23}$$

and analogously

$$\begin{aligned} &D_z\tilde{H}(t, A^{\beta}z(t), A^{\beta}q(t)) \\ &= e^{\frac{k(t-T)^2}{2}}\theta(t)\left[D_xH(t, A^{\beta}y_1(t), A^{\beta}p_1(t)) - D_xH(t, A^{\beta}y_2(t), A^{\beta}p_2(t))\right], \end{aligned} \tag{24}$$

finally

$$f_z(t) = e^{\frac{k(t-T)^2}{2}}\theta'(t)(y_1(t) - y_2(t)) \text{ and } f_q(t) = e^{\frac{k(t-T)^2}{2}}\theta'(t)(p_1(t) - p_2(t)), \tag{25}$$

We now divide the proof in several steps.

Step 1. *For $k > 0$ the following inequality holds*

$$\begin{aligned} k\int_0^T (|z(t)|^2 + |q(t)|^2)dt &\le \int_0^T (|f_z(t)|^2 + |f_q(t)|^2)dt \\ &+2\int_0^T |A^{-\beta}\tilde{F}(A^{\beta}z(t))|^2 + |[D\tilde{F}(A^{\beta}z(t))]^*q(t)|^2 dt \\ &+2\int_0^T (|D_q\tilde{H}(t, A^{\beta}z(t), A^{\beta}q(t))|^2 + |D_z\tilde{H}(t, A^{\beta}z(t), A^{\beta}q(t))|^2)dt\ . \end{aligned} \tag{26}$$

Indeed, multiplying the first equation of system (20) by $z'(t)$ and the second equation by $q'(t)$ we get

$$\begin{aligned} |z'(t)|^2 = &- < Az(t), z'(t) > + < k(t-T)z(t), z'(t) > - < A^{-\beta}\tilde{F}(A^{\beta}z(t)), z'(t) > \\ &- < D_q\tilde{H}(t, A^{\beta}z(t), A^{\beta}q(t)), z'(t) > + < f_z(t), z'(t) > \end{aligned}$$

and

$$|q'(t)|^2 =< Aq(t), q'(t) > + < k(t-T)q(t), q'(t) > + < [D\tilde{F}(A^\beta z(t))]^* q(t), q'(t) >$$

$$+ < D_z\tilde{H}(t, A^\beta z(t), A^\beta q(t)), q'(t) > + < f_q(t), q'(t) > .$$

The above equalities can be rewritten as

$$|z'(t)|^2 = \frac{1}{2}\frac{d}{dt}\{- < Az(t), z(t) > + k(t-T)|z(t)|^2\} - \frac{k}{2}|z(t)|^2$$

$$- < A^{-\beta}\tilde{F}(A^\beta z(t)), z'(t) > - < D_q\tilde{H}(t, A^\beta z(t), A^\beta q(t)), z'(t) > + < f_z(t), z'(t) >$$

and

$$|q'(t)|^2 = \frac{1}{2}\frac{d}{dt}\{< Aq(t), q(t) > + k(t-T)|q(t)|^2\} - \frac{k}{2}|q(t)|^2$$

$$+ < [D\tilde{F}(A^\beta z(t))]^* q(t), q'(t) > + < D_z\tilde{H}(t, A^\beta z(t), A^\beta q(t)), q'(t) > + < f_q(t), q'(t) > .$$

Integrating on $[0, T]$, recalling that z and q vanish at initial and terminal points, we get

$$\int_0^T (|z'(t)|^2 + \frac{k}{2}|z(t)|^2)dt$$

$$\leq \int_0^T |z'(t)|^2 dt + \frac{1}{2}\int_0^T (2|D_q H(t, A^\beta z(t), A^\beta q(t))|^2 + 2|A^{-\beta}\tilde{F}(A^\beta z(t))|^2 + |f_z(t)|^2)dt$$

and

$$\int_0^T (|q'(t)|^2 + \frac{k}{2}|q(t)|^2)dt$$

$$\leq \int_0^T |q'(t)|^2 dt + \frac{1}{2}\int_0^T (|D_z H(t, A^\beta z(t), A^\beta q(t))|^2 + |[D\tilde{F}(A^\beta z(t))]^* q(t)|^2 + |f_q(t)|^2)dt$$

from which we derive the claim.

Step 2. *Setting* $K_p = \sup_{t\in[0,T]} |p_1(t)|$ *for* $k > 0$ *we show that*

$$\begin{aligned} k\int_0^T (|z(t)|^2 + |q(t)|^2)dt &\leq \int_0^T (|f_z(t)|^2 + |f_q(t)|^2)dt + 4L_F^2 \int_0^T |q(t)|^2 dt \\ &+ 4L_H^2 \int_0^T |A^\beta q(t)|^2 dt + 2\left[2L_H^2 + 2K_p^2 L_{DF}^2 + L_F^2|A^{-\beta}|^2\right]\int_0^T |A^\beta z(t)|^2 dt. \end{aligned} \tag{27}$$

From the regularity assumption (18) on F and from (21) we derive

$$|A^{-\beta}\tilde{F}(A^\beta z(t))| \leq e^{\frac{k(t-T)^2}{2}}\theta(t)L_F|A^{-\beta}||A^\beta(y_1(t) - y_2(t))| \leq L_F|A^{-\beta}|\,|A^\beta z(t)|,$$

moreover from (22) we get

$$|[D\tilde{F}(A^\beta z(t))]^* q(t)|$$

$$\leq e^{\frac{k(t-T)^2}{2}}\theta(t)\left\{|[DF(A^\beta y_1(t)) - DF(A^\beta y_2(t))]^* p_1(t)| + |[DF(A^\beta y_2(t))]^*|\ |p_1(t)p_2(t)|\right\}$$

$$\leq K_p L_{DF}|A^\beta z(t)| + L_F|q(t)|$$

so that we derive the two inequalities

$$\begin{aligned} &|A^{-\beta}\tilde{F}(A^\beta z(t))|^2 \leq L_F^2|A^{-\beta}|^2\ |A^\beta z(t)|^2 \\ &|[D\tilde{F}(A^\beta z(t))]^* q(t)| \leq 2K_p^2 L_{DF}^2|A^\beta z(t)|^2 + 2L_F^2|q(t)|^2 \end{aligned} \tag{28}$$

Moreover, from the regularity assumption (17) on H and from (23), (24),

$$\begin{aligned} &|D_q\tilde{H}(t, A^\beta z(t), A^\beta q(t))|^2 + |D_z\tilde{H}(t, A^\beta z(t), A^\beta q(t))|^2 \\ &\leq 2L_H^2[|A^\beta z(t)|^2 + |A^\beta q(t)|^2]. \end{aligned} \tag{29}$$

The claim easily follows substituting estimates (28) and (29) in (26).

Step 3. *For $\theta \in]2\beta - 1, 1]$ and $k > 0$ the following inequality holds:*

$$\begin{aligned} &\int_0^T (|A^{\frac{\theta+1}{2}} z(t)|^2 + |A^{\frac{\theta+1}{2}} q(t)|^2)dt \leq kT\int_0^T |A^{\frac{\theta}{2}} q(t)|^2 dt + \int_0^T (|A^\theta z(t)|^2 + |A^\theta q(t)|^2)dt \\ &+[4L_H^2 + L_F^2|A^{-\beta}|^2 + 2K_p^2 L_{DF}^2]\int_0^T |A^\beta z(t)|^2 dt + 2L_F^2\int_0^T |q(t)|^2 dt \\ &+4L_H^2\int_0^T |A^\beta q(t)|^2]dt + \frac{1}{2}\int_0^T (|f_z(t)|^2 + |f_q(t)|^2)dt. \end{aligned} \tag{30}$$

Multiplying the first equation of (20) by $A^\theta z(t)$ we obtain

$$\begin{aligned} &< z'(t), A^\theta z(t) >= - < Az(t), A^\theta z(t) > +k(t-T) < z(t), A^\theta z(t) > \\ &- < A^{-\beta}\tilde{F}(A^\beta z(t)), A^\theta z(t) > - < D_q\tilde{H}(t, A^\beta z(t), A^\beta q(t)), A^\theta z(t) > \\ &+ < f_z(t), A^\theta z(t) > \end{aligned}$$

Integrating on $[0,T]$ and recalling that

$$< z'(t), A^\theta z(t) >= \frac{1}{2}\frac{d}{dt} < A^\theta z(t), z(t) >$$

and that

$$k(t-T) < z(t), A^\theta z(t) >= k(t-T)|A^{\frac{\theta}{2}} z(t)|^2 \leq 0$$

we have

$$\int_0^T |A^{\frac{\theta+1}{2}} z(t)|^2 dt \leq \int_0^T |A^\theta z(t)|^2 dt + \int_0^T [|A^{-\beta}\tilde{F}(A^\beta z(t))|^2$$

$$+|D_q\tilde{H}(t, A^\beta z(t), A^\beta q(t))|^2]dt + \frac{1}{2}\int_0^T |f_z(t)|^2 dt.$$

By the same reasoning, multiplying by $-A^\theta q(t)$ the second equation of (20) we obtain the similar estimate for $q(t)$.

$$\int_0^T |A^{\frac{\theta+1}{2}} q(t)|^2 dt \leq kT\int_0^T |A^{\frac{\theta}{2}} q(t)|^2 dt + \int_0^T |A^\theta q(t)|^2 dt$$

$$+\int_0^T [|[D\tilde{F}(A^\beta z(t))]^* q(t)|^2 + |D_z\tilde{H}(t, A^\beta z(t), A^\beta q(t))|^2]dt + \frac{1}{2}\int_0^T |f_q(t)|^2 dt$$

The claim follows by adding the last two inequalities and applying the estimates (28) and (29).

Step 4. *Now we show that if T is sufficiently small, then for every $k > 0$ there exist $C(k)$ such that*

$$\int_0^T |z(t)|^2 + |q(t)|^2 dt \leq C(k)\int_0^T \left(|f_z(t)|^2 + |f_q(t)|^2\right) dt. \tag{31}$$

Moreover, if k is big enough, $C(k) > 0$ and $C(k) \to 0$ as $k \to +\infty$.

Fix $\alpha \in]0,1[$ and $\gamma \in]\alpha, 1]$. First we observe that, by the interpolation inequality (9) there exists a constant C_γ such that, for every $\sigma > 0$ we have

$$|A^\alpha z(t)|^2 \leq \frac{C_1}{\sigma}|z(t)|^2 + \sigma|A^\gamma z(t)|^2$$

and

$$|A^\alpha q(t)|^2 \leq \frac{C_1}{\sigma}|q(t)|^2 + \sigma|A^\gamma q(t)|^2.$$

Exploiting these inequalities for (α, γ) equal respectively to $\left(\frac{\theta}{2}, \frac{\theta+1}{2}\right)$, $\left(\theta, \frac{\theta+1}{2}\right)$ and $\left(\beta, \frac{\theta+1}{2}\right)$ we transform (30) into the following inequality

$$\int_0^T (|A^\gamma z(t)|^2 + |A^\gamma q(t)|^2)dt \le \sigma kT \int_0^T |A^\gamma q(t)|^2 dt + \frac{C_1}{\sigma} kT \int_0^T |q(t)|^2 dt$$

$$+\sigma \int_0^T (|A^\gamma z(t)|^2 + |A^\gamma q(t)|^2)dt + \frac{C_1}{\sigma} \int_0^T (|z(t)|^2 + |q(t)|^2)dt$$

$$+[2L_H^2 + L_F^2|A^{-\beta}|^2 + 2K_p^2 L_{DF}^2][\sigma \int_0^T |A^\gamma z(t)|^2 dt + \frac{C_1}{\sigma} \int_0^T |z(t)|^2 dt]$$

$$+2L_F^2 \int_0^T |q(t)|^2 dt + \sigma 2L_H^2 \int_0^T |A^\gamma q(t)|^2] dt + \frac{C_1}{\sigma} 2L_H^2 \int_0^T |q(t)|^2] dt$$

$$+\frac{1}{2} \int_0^T (|f_z(t)|^2 + |f_q(t)|^2)dt$$

that is

$$\int_0^T (|A^\gamma z(t)|^2 + |A^\gamma q(t)|^2)dt \le +\sigma \left[1 + 2L_H^2 + L_F^2|A^{-\beta}|^2 + 2K_p^2 L_{DF}^2\right] \int_0^T |A^\gamma z(t)|^2 dt$$

$$+\sigma \left[kT + 1 + 2L_H^2\right] \int_0^T |A^\gamma q(t)|^2 dt + \frac{C_1}{\sigma} \left[kT + 1 + 2L_F^2 + 2L_H^2\right] \int_0^T |q(t)|^2 dt$$

$$+\frac{C_1}{\sigma} \left[1 + 2L_H^2 + L_F^2|A^{-\beta}|^2 + 2K_p^2 L_{DF}^2\right] \int_0^T |z(t)|^2 dt + \frac{1}{2} \int_0^T (|f_z(t)|^2 + |f_q(t)|^2)dt$$

Setting $B_1 = 1 + 2L_H^2 + L_F^2[2 + |A^{-\beta}|^2] + 2K_p^2 L_{DF}^2$, we can write

$$\int_0^T (|A^\gamma z(t)|^2 + |A^\gamma q(t)|^2)dt \le \sigma[kT + B_1] \int_0^T (|A^\gamma z(t)|^2 + |A^\gamma q(t)|^2)dt$$

$$+\frac{C_1}{\sigma} [kT + B_1] \int_0^T (|z(t)|^2 + |q(t)|^2)dt + \frac{1}{2} \int_0^T (|f_z(t)|^2 + |f_q(t)|^2)dt$$

so that, taking $\sigma = \frac{1}{2[kT + B_1]}$ we have

$$\int_0^T (|A^\gamma z(t)|^2 + |A^\gamma q(t)|^2)dt$$
$$\le 4C_1[kT + B_1]^2 \int_0^T (|z(t)|^2 + |q(t)|^2)dt + \int_0^T (|f_z(t)|^2 + |f_q(t)|^2)dt. \tag{32}$$

Now for every $\rho > 0$, we use (9) with $\alpha = \beta$ and $\gamma \in]\beta, 1]$ in (27) to derive,

$$k\int_0^T (|z(t)|^2 + |q(t)|^2)dt \le \int_0^T (|f_z(t)|^2 + |f_q(t)|^2)dt + 4L_F^2 \int_0^T |q(t)|^2 dt$$

$$+\frac{C_1}{\rho}4L_H^2 \int_0^T |q(t)|^2 dt + 2\rho \left[2L_H^2 + 2K_p^2 L_{DF}^2 + L_F^2 |A^{-\beta}|^2\right] \int_0^T |A^\gamma z(t)|^2 dt$$

$$+\rho 4L_H^2 \int_0^T |A^\gamma q(t)|^2 dt + 2\frac{C_1}{\rho}\left[2L_H^2 + 2K_p^2 L_{DF}^2 + L_F^2 |A^{-\beta}|^2\right] \int_0^T |z(t)|^2 dt$$

so that, setting $B_2 = 4L_H^2 + 4K_p^2 L_{DF}^2 + 2L_F^2 |A^{-\beta}|^2$, we have

$$\begin{aligned} k\int_0^T (|z(t)|^2 + |q(t)|^2)dt &\le \int_0^T (|f_z(t)|^2 + |f_q(t)|^2)dt \\ &+ \left[4L_F^2 + \frac{C_1}{\rho}B_2\right] \int_0^T (|z(t)|^2 + |q(t)|^2)dt + \rho B_2 \int_0^T \left(|A^\gamma z(t)|^2 + |A^\gamma q(t)|^2\right) dt \end{aligned} \tag{33}$$

Putting inequality (32) in (33) we obtain

$$k\int_0^T (|z(t)|^2 + |q(t)|^2)dt \le (1 + \rho B_2) \int_0^T (|f_z(t)|^2 + |f_q(t)|^2)dt$$

$$+ \left[4L_F^2 + \frac{C_1}{\rho}B_2 + 4\rho B_2 C_1 [kT + B_1]^2\right] \int_0^T (|z(t)|^2 + |q(t)|^2)dt$$

Now we take $\rho = \dfrac{1}{kT + B_1}$ so that the latter yields

$$\int_0^T (|z(t)|^2 + |q(t)|^2)dt \le C(k) \int_0^T (|f_z(t)|^2 + |f_q(t)|^2)dt$$

where

$$C(k) = \frac{1 + \rho B_2}{k[1 - C_1 T(1 + 4B_2)] - [4L_F^2 + C_1 B_1 + 4C_1 B_1 B_2)]}$$

which gives the claim for $T \le \dfrac{1}{C_1(1 + 4B_2)}$.

Step 5. *Conclusion.*

Let $T \le \dfrac{1}{C_1(1 + 4B_2)}$ and set $\tilde{y}(t) = y_1(t) - y_2(t)$ and $\tilde{p}(t) = p_1(t) - p_2(t)$. Then from (31) and from (25), (19) directly follows

$$\begin{aligned} &\int_0^T e^{k(t-T)^2} \theta^2(t)(|\tilde{y}(t)|^2 + |\tilde{p}(t)|^2)dt \\ &\le C(k) \int_0^T e^{k(t-T)^2} |\theta'(t)|^2 (|\tilde{y}(t)|^2 + |\tilde{p}(t)|^2)dt\,. \end{aligned} \tag{34}$$

On the other hand

$$\int_0^T e^{k(t-T)^2}\theta^2(t)(|\tilde{y}(t)|^2+|\tilde{p}(t)|^2)dt \geq \int_0^{\frac{T}{2}} e^{k(t-T)^2}(|\tilde{y}(t)|^2+|\tilde{p}(t)|^2)dt$$

$$\geq e^{k\frac{T^2}{4}} \int_0^{\frac{T}{2}} (|\tilde{y}(t)|^2+|\tilde{p}(t)|^2)dt \tag{35}$$

and the following holds

$$\int_0^T e^{k(t-T)^2}|\theta'(t)|^2(|\tilde{y}(t)|^2+|\tilde{p}(t)|^2)dt$$

$$\leq \left(\frac{4}{T}\right)^2 \int_{\frac{T}{2}}^T e^{k(t-T)^2}(|\tilde{y}(t)|^2+|\tilde{p}(t)|^2)dt \leq \left(\frac{4}{T}\right)^2 e^{k\frac{T^2}{4}} \int_{\frac{T}{2}}^T (|\tilde{y}(t)|^2+|\tilde{p}(t)|^2)dt\,. \tag{36}$$

In conclusion, from (34), (35) and (36) we get

$$e^{k\frac{T^2}{4}} \int_0^{\frac{T}{2}} (|\tilde{y}(t)|^2+|\tilde{p}(t)|^2)dt \leq C(k)\left(\frac{4}{T}\right)^2 e^{k\frac{T^2}{4}} \int_{\frac{T}{2}}^T (|\tilde{y}(t)|^2+|\tilde{p}(t)|^2)dt\,.$$

From the above inequality we obtain

$$\int_0^{\frac{T}{2}} (|\tilde{y}(t)|^2+|\tilde{p}(t)|^2)dt \to 0 \quad \text{as} \quad k \to \infty$$

and we conclude that $|\tilde{y}(t)| = |\tilde{p}(t)| = 0$ on $[0, \frac{T}{2}]$. Iterating this procedure we obtain the result on [0,T]. If $T > \frac{1}{C_1(1+4B_2)}$ we can easily obtain the claim by a finite number of steps. ∎

References

[1] V. Barbu, Boundary control problems with convex cost criterion, SIAM J. Control and Optim., 18, 227–254, 1980.

[2] V. Barbu and G. Da Prato, Hamilton-Jacobi equations in Hilbert spaces, Pitman, Boston, 1982.

[3] V. Barbu, E. N. Barron and R. Jensen, The necessary conditions for optimal control in Hilbert spaces, J. Math. Anal. Appl., 133 (1988), 151–162.

[4] E. N. Barron and R. Jensen, The Pontryagin maximum principle from the dynamic programming and viscosity solutions to first order partial differential equations, Trans. Amer. Math. Soc., 298 (1986), 635- 641.

[5] A. Bensoussan, G. Da Prato, M. C. Delfour and S. K. Mitter, Representation and control of infinite dimensional systems, Birkhäuser, Boston, 1992.

[6] P. Cannarsa, H. Frankowska, Value function and optimality conditions for semilinear control problems, Applied Math. Optim., 26, 139–169, 1992.

[7] P. Cannarsa, H. Frankowska, Value function and optimality conditions for semilinear control problems, II: parabolic case, Applied Math. Optim. (to appear).

[8] P. Cannarsa and F. Gozzi, On the smoothness of the value function along optimal trajectories, IFIP Workshop on Boundary Control and Boundary Variation (to appear in Lecture Notes in Control and Information Sciences, Springer–Verlag).

[9] P. Cannarsa, F. Gozzi and H. M. Soner, A dynamic programming approach to nonlinear boundary control problems of parabolic type, J. Funct. Anal. (to appear).

[10] P. Cannarsa and M. E. Tessitore, Cauchy problem for the dynamic programming equation of boundary control, Proceedings IFIP Workshop on Boundary Control and Boundary Variation (to appear).

[11] P. Cannarsa and M. E. Tessitore, Infinite dimensional Hamilton–Jacobi equations and Dirichlet boundary control problems of parabolic type, (pre–print).

[12] P. Cannarsa and M. E. Tessitore, Optimality conditions for boundary control problems of parabolic type, Proceedings International Conference on Distributed Control System: Nonlinear Problems" , Springer–Verlag (to appear).

[13] F. Clarke and R. B. Vinter, The relationship between the maximum principle and dynamic programming, SIAM J. Control Optim., 25, 1291–1311, 1989.

[14] H. O. Fattorini, Boundary control system, SIAM J. Control, 6, 349–385, 1968.

[15] H. O. Fattorini, A unified theory of necessary conditions for nonlinear nonconvex control systems, Appl. Math. Opt., 15, 141– 185, 1987.

[16] H. O. Fattorini and S. S. Sritharan, Optimal control theory for viscous flow problems, pre–print.

[17] A. Favini and A. Venni, On a two–point problem for a system of abstract differential equations, Numer. Funct. Anal. and Optimiz., 2(4), 301–322, 1980.

[18] H. Goldberg and F. Tröltzsch, Second–order sufficient optimality conditions for a class of nonlinear parabolic boundary control problems, SIAM J. Control and Optim., vol. 31, 4, 1007–1025, 1993.

[19] F. Gozzi and M.E.Tessitore, Optimality conditions for Dirichlet Boundary Control Problems of parabolic type, preprint Dipartimento di Matematica (Università di Pisa) n. 2.171.809, Agosto 1994. Submitted.

[20] V. Iftode, Hamilton–Jacobi equations and boundary convex control problems, Rev. Roumaine Math. Pures Appl., 34, (1989), 2, 117–127.

[21] G. E. Ladas and V. Lakshmikantham, Differential equations in abstract spaces, Academic Press, New York and London, 1972.

[22] I. Lasiecka and R. Triggiani, Differential and algebraic equations with application to boundary control problems:continuous theory and approximation theory, Lecture Notes in Control and Information Sciences, Springer–Verlag, 164, 1991.

[23] J. L. Lions, Optimal control of systems described by partial differential equations, Springer–Verlag, Wiesbaden, 1972.

[24] J. L. LIONS AND E. MAGENES, *Problèmes aux limites non homogènes et applications* II, Dunod, Paris, 1968.

[25] J. L. Lions and B. Malgrange, Sur l'unicité rétrograde dans les problèmes mixtes paraboliques, Math. Scand. 8, 277–286, 1960

[26] A. Lunardi, Analytic Semigroup and Optimal Regularity in Parabolic Problems, BirKhaüser, Basel, 1994 (to appear).

[27] A. Pazy, Semigroups of linear operators and applications to partial differential equations, Springer–Verlag, New York–Heidelberg- Berlin, 1983.

[28] D. Preiss, Differentiability of Lipschitz functions on Banach spaces, J. Funct. Anal., 91, 312–345, 1990.

[29] F. Tröltzsch, On the semigroup approach for the optimal control of semilinear parabolic equations including distributed and boundary control, Zeitschrift für Analysis und ihre Anwendungen Bd., 8, 431–443, 1989.

Shape Hessian for a Nondifferentiable Variational Free Boundary Problem

Y. GUIDO and J.P ZOLESIO
C.M.A, E.N.S.M.P, B.P 207 06904 Sophia Antipolis, France

Abstract

We consider a shape formulation of free boundary problem arising from plasmas physics. That problem has a nondifferentiable variational formulation in the variable y $\in H^1_0(\Gamma_D)$. By a "change of variable" we rewrite this variational formulation as a shape optimization problem for some shape cost functional $J(\Gamma)$ in the following form Min { $J_{in}(\Gamma,u)$ / u $\in H^1(\Gamma)$} + Min { $J_{out}(\Gamma^C,v)$ / v $\in H^1_0(\Gamma^C)$}. The "change of variable" is of the form $y = u\chi_\Gamma + (1-\chi_\Gamma)v$ and J turns to be smooth with respect to Γ and we compute the two first shape derivatives. In fact Γ is a subset of a smooth surface Γ_D and we shall do shape calculus using an intrinsic geometrical approach with generalized Laplace Beltrami operators.

1 A free boundary problem arising from plasmas physics

Γ_D being a smooth compact and oriented manifold, we assume that there exists a bounded open set D included in $\mathbb{R}^N$ such that $\Gamma_D = \partial D$. b_D is the oriented distance function to D defined in $\mathbb{R}^N$. Assuming Γ_D to have the regularity C^2, there exists a neighborhood $\mathcal{U}$ of Γ_D in $\mathbb{R}^N$ such the projection mapping p is defined onto Γ_D, with $p(x) = I - b_D \nabla b_D$. The tangential operators $\nabla_\Gamma\varphi, div_\Gamma\vec{\varphi}, \Delta_\Gamma\varphi$ are defined as being the restriction to Γ_D of $\nabla\Phi, div\vec{\Phi}, \Delta\Phi$ where they are defined on $\mathcal{U}$.
The Sobolev space $H^1(\Gamma_D) = \{\varphi \in L^2(\Gamma_D), \nabla_\Gamma\varphi \in L^2(\Gamma_D, \mathbb{R}^N)\}$, Γ_D being equipped with the surface measure $|.|_{\mathcal{H}^{N-1}}$.
We denote by γ the boundary of Γ_D in $\mathbb{R}^N$, e.g $\gamma = \partial_{\partial D}\Gamma_D$ and we recall the form of the Stokes theorem in Γ_D

$$\int_{\Gamma_D} < \nabla_\Gamma\varphi, \vec{V} > d\Gamma = \int_{\Gamma_D} - < \varphi, div_\Gamma\vec{V} > d\Gamma + \int_{\Gamma_D} < H\varphi, \vec{V}.\vec{n} > d\Gamma + + \int_\gamma < \varphi\vec{V}, \vec{\nu} > dl$$

where H is the mean curvature of Γ_D, $H = \Delta_\Gamma b_D$ on Γ_D, $\vec{n}$ is the unitary normal field vector on Γ_D out going of D ($\vec{n} = \nabla_\Gamma b_D$ and $H = div_\Gamma\vec{n}$ on Γ_D) and $\vec{\nu}$ is the unitary normal field vector contained in γ.
We have $\vec{\nu}(x).\vec{n}(x) = 0 \; \forall \, x \; \in \gamma, \vec{\nu}(x)$ orthogonal to γ at x (as we assume γ to have the C^1-regularity, γ being a N-2 dimensional submanifold of Γ_D, γ having no boundary in Γ_D.
$H^1_0(\Gamma_D)$ is the closure of $\mathcal{D}$ in $H^1(\Gamma_D)$. It is equipped with the equivalent norm

$\int_{\Gamma_D} |(\varphi|^2 + |\nabla_\Gamma \varphi|^2) d\Gamma$.
We consider the following variations calculus problem arising from plasmas physics (confinement of plasma under an electromagnetic field in a Tokomak in order to create thermonuclear fusion):

$$Min\{E(\varphi)/\varphi \in H_0^1(\Gamma_D) \bigcap H^2(\Gamma_D)\} \tag{1}$$

where E(φ) is equal to
$E(\varphi) = \int_{\Gamma_D} \frac{1}{2}|\nabla_\Gamma \varphi|^2 - |\Gamma_D|_{\mathcal{H}^{N-1}}(\varphi - 1)^+ d\Gamma - \int\int_{\Gamma_D \times \Gamma_D}[(\varphi(x) - 1)^+ - (\varphi(y) - 1)^+] d\Gamma_x d\Gamma_y$
We recall that $(\varphi(x) - 1)^+ = \begin{cases} 0 \; \forall \; \varphi(x) \leq 1 \\ \varphi(x) - 1 \; \forall \; \varphi(x) \geq 1 \end{cases}$

So the φ solution of (1) is the solution of the following problem

$$\begin{cases} -\Delta_\Gamma \varphi = \beta(x)(y)\chi_\Gamma \\ \varphi_{|\Gamma} = 1 \\ \varphi_{|\partial\Gamma} = 0 \end{cases} \tag{2}$$

χ_Γ is the characteristic function of Γ, we have $\chi_\Gamma = \begin{cases} 0 \; \forall x \notin \Gamma \\ 1 \; \forall x \in \Gamma \end{cases}$
and we have $\beta(x)(y) = |\{z \in \Gamma_D / y(z) < y(x)\}|_{\mathcal{H}^{N-1}}$.

Γ_D represents a section of the torus, Γ the plasma and Γ^C the vacuum.
This free boundary problem was modeled by H. Grad (adiabatic evolution of plasma equilibrium) and a simplified model was studied by R. Temam and J.P. Zolesio who did the variational analysis of (2)(see ref. [13] and [14], chap. 7).

As the new one contains the same difficulties of the previous one, in order to simplify the presentation the nondifferentiable energy functional E is given in the following form :
Given f and g in $L^\infty(I\!R^N) \times H^2(I\!R^N)$, for $\alpha > 0$,

$$for \; \varphi \in H_0^1(\Gamma_D) \bigcap H^2(\Gamma_D), \; E(\varphi) = \int_{\Gamma_D} \frac{1}{2}|\nabla_\Gamma \varphi|^2 - \alpha f(\varphi - g)^+ d\Gamma \tag{3}$$

2 Solutions existence

We assume that the previous functions f and g are strictly positive, Δ_Γ g is positive almost everywhere in Γ_D, the dimension N is strictly inferior to 4, and α is large enough.

Proposition 1 *There exists (at least one) $y \in H_0^1(\Gamma_D)$ minimizing E over $H_0^1(\Gamma_D)$.*
$y \in H_0^1(\Gamma_D) \cap H^2(\Gamma_D)$ solves the problem $-\Delta_\Gamma \; y = \alpha\chi_\Gamma f$ in Γ_D where
$\Gamma = \{x \in \Gamma_D / y(x) > g(x)\}$.$\Gamma \subset \Gamma_D$, with $|\partial\Gamma|_{\mathcal{H}^{N-1}} = 0$.

Proof.

Lemma 1 *Let y be a local minimum of E. If $dE(y, \Phi)$ the Hadamard semi derivative exists, then we have :*

$$dE(y, \Phi) \geq 0, \forall \; \Phi \in H_0^1(\Gamma_D). \tag{4}$$

Proof. E is Hadamard semidifferentiable: from Lebesgue convergence we have

$$\begin{aligned} d_h E(\varphi, \Phi) &= lim_{\substack{t>0 \\ \zeta \to \psi}} \frac{E(\varphi + \zeta t) - E(\varphi)}{t} \\ &= \int_{\Gamma_D} \nabla_\Gamma \varphi \nabla_\Gamma \zeta - \alpha \chi_{\varphi(x)>g(x)} f \zeta - \alpha f \zeta^+ \chi_{\varphi(x)=g(x)} d\Gamma \end{aligned}$$

At a minimum φ=y we get the following inequality E(y + ζ t) $\geq$ E(y) so d_hE(y,Φ) ≥ 0 so we obtain the finally variational inequality

$$\int_{\Gamma_D} \nabla_\Gamma y \nabla_\Gamma \zeta - \alpha \chi_{y(x)>g(x)} f \zeta d\Gamma \geq \int_{\Gamma_D} \alpha f \zeta^+ \chi_{y(x)=g(x)} d\Gamma \geq 0 \tag{5}$$

Obviously the two previous terms are equalities. With $\zeta \geq \chi_{y(x)>g(x)}$ we get
$|\{x \in \Gamma_D / y(x) = g(x)\}|_{\mathcal{H}^{N-1}} = 0$
>From (5) we obtain $\int_{\Gamma_D} -\Delta_\Gamma y \, \zeta d\Gamma = \int_{\Gamma_D} \alpha \chi_{y(x)>g(x)} f \zeta d\Gamma$.
If we denote $\Gamma = \{x \in \Gamma_D / y(x) > g(x)\}$ we finally get $-\Delta_\Gamma$y= $\alpha f \chi_\Gamma$

Lemma 2 *We have* $0 \leq y \leq g$ *in* $\Gamma^C = \Gamma_D \setminus \overline{\Gamma}$ *and* $y \geq g$ *in* Γ.

Proof. Assume $\tilde{H}_0^1(\Gamma) = \{u \in H_0^1(\Gamma_D) \ / \ u = 0$ q.e in $\Gamma^C\}$ and $H(\Gamma^C) = \{v = \tilde{v}|_{\Gamma^C} / \tilde{v} \in H_0^1(\Gamma_D), \tilde{v} = g$ q.e in $\Gamma\}$.
We have $(y-g)|_\Gamma \in \tilde{H}_0^1(\Gamma)$, using the weak maximum principle $(y-g)^+|_\Gamma \in \tilde{H}_0^1(\Gamma)$ and $\int_\Gamma |\nabla_\Gamma (y-g)^+|^2 - \alpha f (y-g)^+ d\Gamma \leq \int_\Gamma |\nabla_\Gamma (y-g)|^2 - \alpha f (y-g) d\Gamma$
As the problem gives u $\in \tilde{H}_0^1(\Gamma) \bigcap H^2(\Gamma_D)$ such that $-\Delta_\Gamma u = \alpha f$ in Γ has a unique solution, we get $(y-g)^+ = y-g$ in Γ. Similarly we have $(y-g)^- = y-g$ in Γ^C.

Lemma 3 *We have* $y < g$ *in* Γ^C *and so* $y > g$ *in* Γ.

Proof. Let z a point of Γ where y(z)=Min {y(x), x $\in \Gamma$}.
In a neighborhood U of z, y is a constant, effectively in a such point we have $\Delta_\Gamma y \leq 0$. Then y is subharmonic and cannot reache its minimum if it isn't constant. Then we would get $f = 0$ in U so this is a contradiction. So $y > g$ in Γ. Similarly we have $y < g$ in Γ^C.

3 Equivalent problem

$\varphi \mapsto$ E(φ) is not differentiable ($d_h E(\varphi, \Phi)$ is not linear with Φ).We study E(φ) without nonsmooth analysis but by using Γ as an independent variable in order to obtain differentiable functional.
So φ is split in two functions u and v such that u and v $\in \tilde{H}_0^1(\Gamma) \times H(\Gamma^C)$ and

$$\varphi = (u+g)\chi_\Gamma + v\chi_{\Gamma^C} \tag{6}$$

We define the set $\mathcal{O}$ that is equal to
$\{\Gamma$ submanifold in Γ_D, Γ open with $\overline{\Gamma} \subset \Gamma_D, |\Gamma_D|_{\mathcal{H}^{N-1}} > 0, |\partial\Gamma|_{\mathcal{H}^{N-1}} = 0\}$
We consider the following convex functionals:

$$\begin{cases} J_{in}(\Gamma, u) = \int_\Gamma \frac{1}{2} |\nabla_\Gamma (u+g)|^2 - \alpha f u d\Gamma \\ J_{out}(\Gamma^C, v) = \int_{\Gamma^C} \frac{1}{2} |\nabla_\Gamma v|^2 d\Gamma \end{cases}$$

Proposition 2 *For any set Γ in $\mathcal{O}$, there exists a unique element (u_Γ, v_Γ) such that (u_Γ, v_Γ) is in $\tilde{H}_0^1(\Gamma)\times H(\Gamma^C)$ and such that we have*

$$\forall u \in \tilde{H}_0^1(\Gamma), J_{in}(\Gamma, u_\Gamma) \leq J_{in}(\Gamma, u)$$

$$\forall v \in H(\Gamma^C), J_{out}(\Gamma^C, v_\Gamma) \leq J_{out}(\Gamma^C, v)$$

We introduce the following functional

$$\begin{aligned} J(\Gamma) &= Min\{J_{in}(\Gamma, u) + J_{out}(\Gamma^C, v)/(u, v) \in \tilde{H}_0^1(\Gamma) \times H(\Gamma)\} \\ &= Min\{J_{in}(\Gamma, u)/u \in H^1(\Gamma)\} + Min\{J_{out}(\Gamma^C, v)/v \in H_0^1(\Gamma^C)\}. \end{aligned}$$

and we consider this shape optimization problem :

$$Min\{J(\Gamma)/\Gamma \in \mathcal{O}\} \tag{7}$$

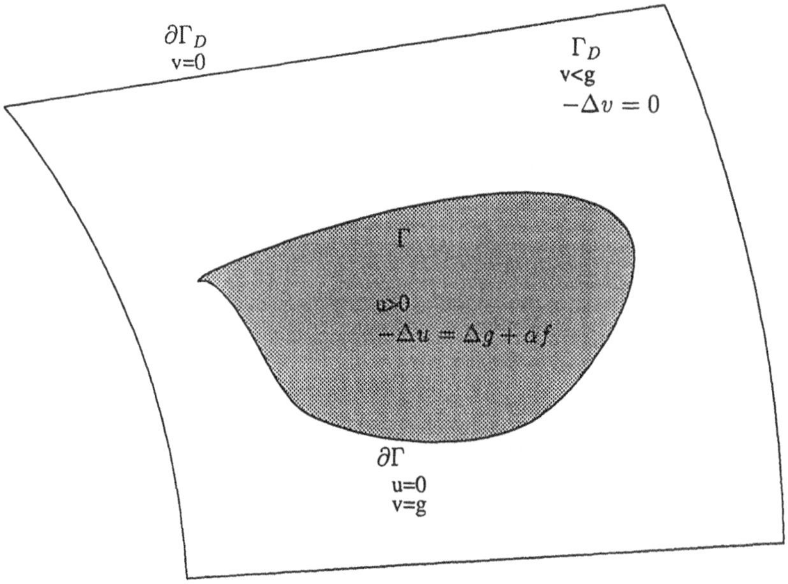

Figure 1: Equivalent problem

This problem turns to be equivalent to the problem (1).

Theorem 1 *Let $\mathcal{O}=\{\Gamma$ submanifold in Γ_D, Γ open with $\overline{\Gamma} \subset \Gamma_D, |\Gamma|_{\mathcal{H}^{N-1}} > 0, |\partial\Gamma|_{\mathcal{H}^{N-1}} = 0\}$.*
There exists a set Γ_y in $\mathcal{O}$ such that for any Γ in $\mathcal{O}$, $J(\Gamma_y) \leq J(\Gamma)$

Proof. The main idea is to return to the functional E.
E reach its minimum in some y in $H_0^1(\Gamma_D) \bigcap H^2(\Gamma_D)$. We know that $y \in c^1(\Gamma_{\overline{D}})$ and as $g \in H^2(\Gamma_D)$ g is $c^1(\overline{\Gamma}_D)$. The function $y - g$ is continuous in Γ_D so Γ_y in Γ_D. >From previously we know that $\int_D \chi_{\partial\Gamma}\, d\Gamma = 0$ and as α is large enough $|\Gamma|_{\mathcal{H}^{N-1}} > 0$. As $y = g$ on $\partial\Gamma_y$ and $g > 0$ in Γ_D we obtain $\overline{\Gamma}_y \subset \Gamma_D$, finally we have Γ_y in $\mathcal{O}$.

Lemma 4 *For $\Gamma = \Gamma_y$ we have $J(\Gamma) = E(y)$*

Proof. With the Stampacchia lemma as we have $y - g\,|_\Gamma \in \tilde{H}_0^1(\Gamma)$, $(y - g)^+|_{\Gamma_y} \in \tilde{H}_0^1(\Gamma)$. By definition of Γ_y as $y < g$ a.e in Γ_{Γ_y} then $(y - g)^+ = 0$ in Γ_{Γ_y}. So we have $u_{\Gamma_y} \in \tilde{H}_0^1(\Gamma)$. Assume that u_{Γ_y} doesn't solve (2) then it exists u in $\tilde{H}_0^1(\Gamma)$ such that $J_{in}(\Gamma,u) < J_{in}(\Gamma_y,u_{\Gamma_y})$.

Let Y in $H_0^1(\Gamma_D)$ and equal to $\begin{cases} u + g \text{ in } \Gamma \\ y \text{ in } \Gamma^C \end{cases}$

As $(Y - g)^+ = (y - g)^+ = 0$ in Γ^C and we have $J_{in}(\Gamma, u) < J_{in}(\Gamma_y, u_{\Gamma_y})$ we obtain $E(Y) < E(y)$, that is in contradiction with y minimizing E so u_{Γ_y} solves the proposition (2). The proof for v_{Γ_y}) is similar so we get the results of lemma.
Now prove that Γ_y solves (7). Assume that Γ_y doesn't solve then there would exist $\Gamma \neq \Gamma_y$ in $\mathcal{O}$ such that $J(\Gamma) < J(\Gamma_y)$. Let Y in $H_0^1(\Gamma_D)$ and equal to $(u + g)\chi_\Gamma + v\,\chi_{\Gamma^C}$. Let y in $H_0^1(\Gamma_D)$ and equal to $(u_{\Gamma_y} + g)\chi_{\Gamma_y} + v_{\Gamma_y}\,\chi_{\Gamma_y^C}$. From the previous lemma we get $J(\Gamma_y) = E(y)$ and $J(\Gamma) = E(Y)$ so we have $E(Y) < E(y)$, this is a contradiction. So Γ_y solves (7).
In the other way, the idea of the proof is similar. We have a set Γ_y in $\mathcal{O}$ that solves (7). We want to obtain that y that is equal to $(u_{\Gamma_y} + g)\chi_{\Gamma_y} + v_{\Gamma_y}\,\chi_{\Gamma_y^C}$ minimizes $E(\varphi)$. We have $J(\Gamma_y) = J_{in}(\Gamma_y,u_{\Gamma_y}) + J_{out}(\Gamma_y,v_{\Gamma_y}) = \int_{\Gamma_D} |\nabla_\Gamma y|^2 - \alpha\chi_{\Gamma_y} f\,(y - g)\, d\Gamma$.

Lemma 5 *We have $u_{\Gamma_y} > 0$ in Γ_y and $0 < v_{\Gamma_y} < g$ in Γ_y^C*

Proof. >From Stampacchia lemma as $u_{\Gamma_y} \in \tilde{H}_0^1(\Gamma_y)$ $u_{\Gamma_y}^+ \in \tilde{H}_0^1(\Gamma_y)$. We know that $J_{in}(\Gamma_y,u_{\Gamma_y}^+) < J_{in}(\Gamma_y,u_{\Gamma_y})$, by unicity of minimum $u_{\Gamma_y} = u_{\Gamma_y}^+ \geq 0$. >From strong maximum principle we obtain $u_{\Gamma_y} > 0$. For v_{Γ_y} this is similar and we have $0 < v_{\Gamma_y} < g$.

So $J(\Gamma_y) = E(y)$. If y does not minimize E then it would exist $Y \neq y \in H_0^1(\Gamma_D)$ such that $Y - g < 0$ in Γ_y^C, $Y - g > 0$ in Γ_y and $E(Y) < E(y)$ so $J(\Gamma_y) < J(\Gamma_y)$ that is not possible. So y minimizes E.
Finally the two problems (1) and (7) are equivalent.

4 Shape gradient

Assume $T_t(V(t,x))$ a smooth map, the flow of the velocity field V.
Let $\tau > 0$, $V \in V^{ad} = C^0([0,\tau[, C^2(D, I\!R^N))$.
We have $V(t,x) = \frac{\partial}{\partial t}(T_t(t,x)) \circ T_t^{-1}(t,x))$.
We denote by V(0) V(0,x).

Definition 1 *Given a speed field V, defined previously, $J(\Gamma)$ is said to have an Eulerian semiderivative at Γ in the direction V if*

$$\begin{aligned} DJ(\Gamma, V) &= lim_{t\searrow 0}\left[\frac{J(\Gamma_t(V)) - J(\Gamma)}{t}\right] \\ &= \frac{dJ(\Gamma_t)}{dt}\Big|_{t=0} \end{aligned}$$

exists and is finite. It is the shape gradient of $J(\Gamma)$ at Γ in the direction V.

Let $\Gamma_t = T_t(V)(\Gamma)$ and $j_V(t)= \det DT_t \|D^*T_t^{-1}.\vec{n}\|$. We have $u_t = u \circ T_t^{-1}$ in $\tilde{H}_0^1(\Gamma_t)$, $v_t= v \circ T_t^{-1}$ in $H(\Gamma_t^C)$.
We denote $L(\Gamma_t) = \int_{\Gamma_t} \frac{1}{2}|\nabla_{\Gamma_t}(u_t+g_t)|^2 - \alpha f_t u_t d\Gamma_t + \int_{\Gamma_t^C} \frac{1}{2}|\nabla_{\Gamma_t} v_t|^2 d\Gamma_t$.

Proposition 3 *Let $y \in H_0^1(\Gamma_D)$ such that $y = (u + g)\chi_\Gamma + v\chi_{\Gamma\setminus\Gamma}$.*
For any $V \in V^{ad}$ we obtain for (Γ, y) optimal,
$\int_{\Gamma_D} \frac{1}{2} <A'_V(0)\nabla_\Gamma y, \nabla_\Gamma y> d\Gamma = \int_{\Gamma_D} \alpha\, (f\, div_\Gamma V + \nabla f. V)\, (y - g)^+\, d\Gamma$

Proof. In order to derive $J(\Gamma)$ we must do a function and a variable change to return to a fixed domain.

Corollary 1 $\nabla_\Gamma y \circ T_t = D^* T_t^{-1} [\nabla(y \circ T_t) - <B(t).\vec{n}, \nabla(y \circ T_t) > \vec{n}\,]$ *where* $D^* T_t^{-1} = {}^T (DT_t^{-1})$, DT_t^{-1} *is the jacobian matrix of* T_t^{-1} *and*
$B(t)= n_t \circ T_t . n_t \circ T_t = \|D^* T_t^{-1} .\vec{n}\|^{-2} DT_t^{-1} D^* T_t^{-1}$

So $L(\Gamma_t) = \int_\Gamma [\frac{1}{2}|\nabla_\Gamma(u_t+g_t)|^2 \circ T_t - \alpha\, f_t \circ T_t\, u_t) \circ T_t]\, j_V(t)\, d\Gamma + \int_{\Gamma^C} \frac{1}{2}|\nabla_\Gamma v_t|^2 \circ T_t\, j_V(t)\, d\Gamma$.
We define $\dot{n}$ as the material derivative of $\vec{n}$, $F'_t|_{t=0}$ the shape derivative of F and

$$\left\{\begin{array}{l} A_V(t) = \det(\, DT_t)\, DT_t^{-1}. D^*T_t^{-1} \|D^*T_t^{-1}.\vec{n}_t\| \\ \frac{\partial}{\partial t}(A_V(t))|_{t=0} = A'_V\,(0) = (div_\Gamma V(0) Id \text{ - } (D^*V(0) + DV(0)) \\ \frac{\partial}{\partial t}(\vec{n}_t \circ T_t)|_{t=0} = \dot{n} = -(D^*V(0).\vec{n} - <D^*V(0).\vec{n}, \vec{n} > \vec{n}) = -(D^*V(0).n)_\Gamma \end{array}\right. \tag{8}$$

Remark 1 $A'_V\,(0)\nabla_\Gamma y = <div_\Gamma V(0) Id \text{ - } (D^*_\Gamma V(0) + D_\Gamma V(0)), \nabla_\Gamma y>$

Corollary 2 *For $F_t \in H^1(\Gamma_t)$ and any $V \in V^{ad}$*

$$\begin{aligned} \frac{\partial}{\partial t}\left[\int_{\Gamma_t} F_t d\Gamma_t\right]|_{t=0} &= \int_\Gamma F'_t|_{t=0} + HFV.\vec{n} d\Gamma + \int_\gamma FV.\vec{\nu} d\gamma. &(9)\\ &= \int_\Gamma \dot{F}_t|_{t=0} d\Gamma + \int_\Gamma F\, div_\Gamma V(0)\, d\Gamma &(10) \end{aligned}$$

Using the characterization of the parameter derivative of a minimum (cf. ref. [15]) we get

$$\begin{aligned} DJ(\Gamma, V) &= Min\{\frac{\partial}{\partial t}(L(\Gamma_t))|_{t=0} / (u, v) \in \tilde{H}_0^1(\Gamma) \times H(\Gamma^C)\} \\ &= \frac{\partial}{\partial t}[Min\{L(\Gamma_t)/(u, v) \in \tilde{H}_0^1(\Gamma) \times H(\Gamma^C)\}]_{t=0} \end{aligned}$$

Let (Γ,u,v) be optimal; we obtain

$L(\Gamma_t) = \int_\Gamma \frac{1}{2} < A_V(t)\nabla_\Gamma(u+g), \nabla_\Gamma(u+g) > d\Gamma - \int_\Gamma \alpha\, f_t \circ T_t(V)\, u\, j_V(t) d\Gamma +$
$+ \int_{\Gamma^C} \frac{1}{2} < A_V(t)\nabla_\Gamma v, \nabla_\Gamma v > d\Gamma$.

So $DJ(\Gamma,V)=\frac{1}{2} \int_{\Gamma_D} <A'_V(0)\nabla_\Gamma y, \nabla_\Gamma y> d\Gamma - \int_\Gamma \alpha(f div_\Gamma V + \nabla f.V) u\, d\Gamma$.
Finally the proposition derives using the

Lemma 6 *For (Γ, u, v) optimal we obtain $DJ(\Gamma, V) = 0$, for every $V \in V^{ad}$.*

5 Shape Hessian

Definition 2 *Given two velocities fields V and W belong to* $(V^{ad})^2$, *assuming that* $DJ(\Gamma_t(W),V)$ *exists for all* $t \in [0,\tau)$ *and* $\Gamma_{\Gamma_t(W)} = T_t(W)(\Gamma)$, *J is said to have a second-order Eulerian semiderivative at* Γ *in the directions* (V,W) *if* $D^2J(\Gamma,V,W) = lim_{t\searrow 0}[\frac{DJ(\Gamma_t(W),V)-DJ(\Gamma,V)}{t}]$ *exists and is finite.*

We take (V,W) as two non-autonomuous vector fields V and W, so we we have $D^2J(\Gamma,V,W) = D^2J(\Gamma,V(0),W(0)) + DJ(\Gamma,\frac{\partial V}{\partial t}|_{t=0})$.
From lemma 6, we obtain $D^2J(\Gamma,V,W) = D^2J(\Gamma,V(0),W(0))$.

Theorem 2 *Let* $y \in H_0^1(\Gamma_D)$ *such that* $y = (u+g)\chi_\Gamma + v\chi_{\Gamma^C}$. *Let* $P \in H_0^1(\Gamma_D)$ *such that* $P|_\Gamma = p$ *and* $P|_{\Gamma^C} = q$, *adjoint states of (u,v). For any* $V \in V^{ad}$ *we obtain for* (Γ,y) *optimal,*

$$\int_{\Gamma_D}\frac{1}{2} < (A'_V(0))^2\nabla_\Gamma y,\nabla_\Gamma y > d\Gamma \geq \int_{\Gamma_D} < \nabla_\Gamma P,\nabla_\Gamma P > d\Gamma +$$
$$+\int_{\Gamma_D} < D\epsilon(V).V(0)\nabla_\Gamma y,\nabla_\Gamma y > d\Gamma + \int_{\Gamma_D}\alpha\, div_\Gamma((f div_\Gamma V(0)+\nabla f.V(0))V(0))(y-g)^+ d\Gamma +$$
$$-\int_{\Gamma_D} < \nabla(div_\Gamma V(0)).V(0)\nabla_\Gamma y,\nabla_\Gamma y > d\Gamma +$$
$$+\int_{\Gamma_D}\alpha(y-g)^+ < \nabla(f div_\Gamma V(0)+\nabla f.V(0)),n > n.V(0)d\Gamma$$

Proof. The method to compute the shape hessian of the Lagrangian functional is to transform the problem into a saddle point formulation.
We denote (p_t,q_t) the adjoint states of (u_t,v_t), $p_t \in H_p(\Gamma_t) = \{p \in \tilde{H}_0^1(\Gamma_t)/\frac{\partial p}{\partial \nu} = 0 \text{ on } \partial\Gamma_t\}$ and $q_t \in H_q(\Gamma_t) = \{q \in H_0^1(\Gamma_t^C)/q|_{\Gamma_t} = 0, \frac{\partial q}{\partial \nu} = 0 \text{ on } \partial\Gamma_t \cup \partial\Gamma_D\}$.

We have the following lagrangian formulation

$$L_2(\Gamma_t(W)) = MinMax\,[\,DJ(\Gamma_t(W),V) + \int_{\Gamma_t(W)}\nabla_{\Gamma_t}(u_t+g_t)\nabla_{\Gamma_t}p_t\,d\Gamma_t +$$
$$+\int_{\Gamma_t^C(W)}\nabla_{\Gamma_t}v_t\nabla_{\Gamma_t}q_t\,d\Gamma_t - \int_{\Gamma_t(W)}\alpha f_t p_t\,d\Gamma_t/(u_t,v_t) \in \tilde{H}_0^1(\Gamma_t)\times H(\Gamma_t^C),(p_t,q_t)\in H_p(\Gamma_t)\times H_q(\Gamma_t^C)\,].$$

Lemma 7 *Let* (u_t,v_t,p_t,q_t) *be a saddle point.* (p_t,q_t) *are respectively solutions of the following systems :*

$$\begin{cases} -\Delta_{\Gamma_t}p_t = div_{\Gamma_t}(A'_V(0)\nabla_{\Gamma_t}(u_t+g_t)) + f_t div_{\Gamma_t}V + \nabla f_t.V \; on\,\Gamma_t \\ p_t = 0 \quad on\ \partial\Gamma_t \end{cases}$$

and

$$\begin{cases} -\Delta_{\Gamma_t}q_t = div_{\Gamma_t}(A'_V(0)\nabla_{\Gamma_t}v_t) \; on\,\Gamma_t^C \\ q_t = 0 \quad on\ \partial\Gamma_t^C \end{cases}$$

Proof. To find the systems we compute $\frac{\partial L}{\partial S}(\Gamma_t,u_t+S\gamma,v_t,p_t,q_t)|_{S=0} = 0$ and $\frac{\partial L}{\partial S}(\Gamma_t,u_t,v_t+S\gamma',p_t,q_t)|_{S=0} = 0$ for $\gamma \in H_p$ and $\gamma' \in H_q$.

Using the lagrangian formulation for any (u,v,p,q) saddle point,
$D^2J(\Gamma,V,W) = \frac{\partial}{\partial t}(L_2(\Gamma_t))|_{t=0}$

We use ϵ the distortion rate, $\epsilon_{ij}(V)=\frac{1}{2}(\partial_i V_j+\partial_j V_i)$.
We denote $D\epsilon(V).W$ the third order tensor, $\frac{1}{2}\partial_k(\partial_i V_j+\partial_j V_i).W_k$.
So we have

$$\begin{aligned}
L_2(\Gamma_t(W)) &= \tfrac{1}{2}\int_\Gamma < D^*T_t^{-1}A'_V(0)DT_t^{-1}\nabla_\Gamma(u+g),\nabla_\Gamma(u+g) > j_W(t)d\Gamma + \\
&+\tfrac{1}{2}\int_{\Gamma^C} < D^*T_t^{-1}A'_V(0)DT_t^{-1}\nabla_\Gamma v,\nabla_\Gamma v > j_W(t)d\Gamma + \\
&+\int_\Gamma < A_W(t)\nabla_\Gamma(u+g),\nabla_\Gamma p > d\Gamma + \int_\Gamma < A_W(t)\nabla_\Gamma v,\nabla_\Gamma q > d\Gamma + \\
&-\int_\Gamma \alpha f\circ T_t(W)p\,j_W(t)\,d\Gamma - \int_\Gamma \alpha(f\,\mathrm{div}_\Gamma V+\nabla f.V)\circ T_t(W)u j_W(t)\,d\Gamma
\end{aligned}$$

So for any $(V,W)\in(V^{ad})^2$ we have

$$\begin{aligned}
D^2J(\Gamma,V,W) &= \tfrac{1}{2}\int_\Gamma < A'_V(0)A'_W(0)\nabla_\Gamma(u+g),\nabla_\Gamma(u+g) > d\Gamma + \\
&+\tfrac{1}{2}\int_{\Gamma^C} < A'_V(0)A'_W(0)\nabla_\Gamma v,\nabla_\Gamma v > d\Gamma + \\
&+\tfrac{1}{2}\int_\Gamma < \tfrac{\partial}{\partial t}(A'_V(0)\circ T_t(W))|_{t=0}\nabla_\Gamma(u+g),\nabla_\Gamma(u+g) > d\Gamma + \\
&+\tfrac{1}{2}\int_{\Gamma^C} < \tfrac{\partial}{\partial t}(A'_V(0)\circ T_t(W))|_{t=0}\nabla_\Gamma v,\nabla_\Gamma v > d\Gamma + \\
&+\int_\Gamma < A'_W(0)\nabla_\Gamma(u+g),\nabla_\Gamma p > d\Gamma + \int_\Gamma < A'_W(0)\nabla_\Gamma v,\nabla_\Gamma q > d\Gamma + \\
&-\int_\Gamma \alpha p(f div_\Gamma W+\nabla f.W)d\Gamma - \int_\Gamma \alpha(f div_\Gamma V+\nabla f.V)\,u\,div_\Gamma W d\Gamma + \\
&-\int_\Gamma \alpha\,u\,\nabla(f div_\Gamma V+\nabla f.V).W d\Gamma
\end{aligned}$$

And we know that

$$\begin{cases}
\frac{\partial}{\partial t}(div_\Gamma V\circ T_t(W))|_{t=0} = \nabla(div_\Gamma V).W \\
\frac{\partial}{\partial t}(\epsilon(V)\circ T_t(W))|_{t=0}\nabla_\Gamma y =< D\epsilon(V).W,\nabla_\Gamma y >
\end{cases}$$

So we have $\frac{\partial}{\partial t}(A'_V(0)\circ T_t(W))|_{t=0}\nabla_\Gamma y =< (\nabla(div_\Gamma V)-D\epsilon(V)).W,\nabla_\Gamma y >$ and finally we obtain

$$\begin{aligned}
D^2J(\Gamma,V,W) &= \tfrac{1}{2}\int_{\Gamma_D} <A'_V(0)A'_W(0)\nabla_\Gamma y,\nabla_\Gamma y > d\Gamma + \int_{\Gamma_D} <A'_W(0)\nabla_\Gamma y,\nabla_\Gamma P> d\Gamma + \\
&-\int_{\Gamma_D}(div_\Gamma W+\nabla f.W)P\chi_\Gamma d\Gamma - \int_{\Gamma_D} < D\epsilon(V).W\nabla_\Gamma y,\nabla_\Gamma y > d\Gamma + \\
&-\int_{\Gamma_D} div_\Gamma((f div_\Gamma V+\nabla f.V)W)\alpha(y-g)^+ d\Gamma + \int_{\Gamma_D} < \nabla(div_\Gamma V).W\ \nabla_\Gamma y,\nabla_\Gamma y > d\Gamma + \\
&-\int_{\Gamma_D}\alpha(y-g)^+ < (\nabla(f div_\Gamma V+\nabla f.V),n > n.W d\Gamma
\end{aligned}$$

From lemma 7, P linearly depends on the field V so D^2J turns to be quadratic in (V,W).
Finally theorem 2 derives using the

Lemma 8 *For (Γ,u,v,p,q) optimal the hessian $D^2(\Gamma,V,V)\geq 0$, for any $V=W\in V^{ad}$.*

References

[1] H. BREZIS. *Analyse fonctionnelle, theorie et applications*, volume 1. MASSON, 1987.

[2] D. BUCUR and J.P. ZOLESIO. Shape Analysis on Manifold. article in preparation, 1995.

[3] R. CORREA and A. SEEGER. Directional Derivatives of a Minimax function. *Nonlinear Anal.9*, 13 -22, 1985.

[4] M. CUER and J.P. ZOLESIO. *control of singular problem via differentiation of a min max.* Systems Control Letters 11. North-Holland Publishing Compagny, 1988.

[5] M. DELFOUR and J.P. ZOLESIO. Anatomy of the shape hessian. Ann.Mat.Pura Appl., 1989.

[6] M. DELFOUR and J.P. ZOLESIO. Structure of Shape Derivatives for Nonsmooth Domains. Journal of Functional Analysis, February 1992. Vol.104, No.,pp.1-33.

[7] M. DELFOUR and J.P. ZOLESIO. Shape Optimization: Oriented Distance Function. Comett Cours, Sophia Antipolis, 1993.

[8] J. DIEUDONNE. *Eléments d'analyse.* Gauthier-Villars, PARIS, 1970.

[9] I. EKELAND and R. TEMAM. *Analyse convexe et problèmes variationnels.* Dunod, Paris, 1974.

[10] M. DELFOUR and J.P. ZOLESIO. Velocity Method and Lagrangian Formulation for the Computation of the Shape Hessian. SIAM Control and Optimization, vol 29, No.6 pp. 1414-1442, November 1991.

[11] O. PIRONNEAU. *Optimal Design for Elliptic Systems.* Springer-Verlag, New York, 1984.

[12] J. SOKOLOWSKI and J.P. ZOLESIO. *Introduction to Shape Optimisation*, volume 16 of *Computational Mathematics.* Springer-Verlag, 1991.

[13] J.P. ZOLESIO. Boundary Control and Boundary Variation. C.R. Acad. Sc. Paris, mai 1979.

[14] J.P. ZOLESIO. *Identification de Domaines par Déformations.* PhD thesis, Nice, 1979.

[15] J.P. ZOLESIO. *Semi Derivatives of Repeated Eigenvalues.* In Optimization of Distributed Parameter Structures. Sijthoff and Noordhoff, Alphen aan den Rijn , The Netherlands, 1981.

[16] J.P. ZOLESIO. *The material derivative (or speed) method for shape optimization*, volume II, pp 1089-1151 of *In Optimization of Distributed Parameter Structures.* Sijthoff and Noordhoff, Alphen aan den Rijn , The Netherlands, 1981.

[17] J.P. ZOLESIO. *Boundary Control and Boundary Variation*, volume 178 of *Lecture Notes in Control and Information Sciences.* Springer-Verlag, 1990.

[18] J.P. ZOLESIO. Shape Formulation for Free Boundary Problems with Nonlinearised Bernoulli Condition. Springer-Verlag, 1991. p. 362-392.

[19] J.P. ZOLESIO. *Boundary Control and Boundary Variation*, volume 178 of *Lecture Notes in Control and Information Sciences.* SPRINGER-VERLAG, 1992.

[20] J.P. ZOLESIO. Weak Shape Formulation of Free Boundary Problems. Annali della Scuola Normale Superiore di Pisa, Scienze Fisiche e Matematiche- Serie IV.VolXXI. Fasc. 1 p.11-44, 1994.

Carleman Estimates and Exact Boundary Controllability for a System of Coupled, Nonconservative Second-Order Hyperbolic Equations

Irena Lasiecka and Roberto Triggiani
Applied Mathematics, Thornton Hall
University of Virginia
Charlottesville, Virginia 22903

We consider a system of 2 (or more) coupled second-order hyperbolic equations in the difficult situation where the equations have first-order, lower-order terms, as well as first-order coupling terms in all independent variables. By using general differential multipliers we give a "friendly" proof of Carleman estimates and, consequently, of exact controllability results for the coupled system, under various combinations of boundary controls: Dirichlet/Dirichlet; Dirichlet/Neumann; Neumann/Neumann. The controls are active only on a suitable portion of the boundary. These resu̧lts *cannot* be obtained by standard multipliers.

1 Introduction

Problem formulation. Consider the following *coupled* system of two second-order hyperbolic equations in the unknowns $w(t,x)$ and $z(t,x)$:

$$\begin{cases} w_{tt} = \Delta w + F_1(w) + P_1(z) \quad \text{in } (0,T] \times \Omega \equiv Q; & (1.1) \\ z_{tt} = \Delta z + F_2(z) + P_2(w) \quad \text{in } Q, & (1.2) \end{cases}$$

defined on a bounded domain $\Omega \subset R^n$ with smooth boundary Γ, where F_1, F_2, P_1, P_2 are (linear) differential operators of order one in all variables t, $x_1, \ldots, x_n$, with $L_\infty(Q)$-coefficients, thus satisfying the pointwise bounds

$$|F_1(w)|^2 + |P_2(w)|^2 \leq c_T[w_t^2 + |\nabla w|^2 + w^2] \quad \forall t, x \in Q; \tag{1.3}$$

$$|F_2(z)|^2 + |P_1(z)|^2 \leq c_T[z_t^2 + |\nabla z|^2 + z^2] \quad \forall t, x \in Q. \tag{1.4}$$

The results of this paper were obtained in November 1992 when the authors were visiting the Dipartimento di Matematica, dell' Università degli Studi di Trento, Povo, Italy, whose hospitality is gratefully acknowledged. They were presented in a series of lectures given by R.T. The authors wish to thank Prof. M. Iannelli and C.N.R.

No boundary conditions need to be imposed at this stage. For the purposes of this paper, the operator $(-\Delta)$ in (1.1) and (1.2) may be replaced by two, possibly different, uniformly elliptic operators of order two, with constant coefficients without effecting the proofs and results. We then introduce

$$E_w(t) \equiv \int_\Omega \left[w_t^2(t) + |\nabla w(t)|^2\right] d\Omega; \; E_z(t) \equiv \int_\Omega \left[z_t^2(t) + |\nabla z(t)|^2\right] d\Omega. \tag{1.5}$$

Our goal in the present paper is twofold:

(i) to establish the energy estimate in Theorem 1.1 below; which reconstructs the energy from the boundary measurements modulo lower order terms;

(ii) to provide a "friendly" and explicit proof of Theorem 1.1.

Theorem 1.1. Let w, z be solutions of Eqns. (1.1) (1.2) in the following class

$$\begin{cases} w, z \in H^1(Q) = L_2(0,T; H^1(\Omega)) \cap H^1(0,T; L_2(\Omega)) & (1.6) \\ w_t, \dfrac{\partial w}{\partial \nu}, z_t, \dfrac{\partial z}{\partial \nu} \in L_2(0,T; L_2(\Gamma)). & (1.7) \end{cases}$$

Let T be the time specified in definition (2.1.4) and Remark 2.1.2 of Section 2. Then

(i) there exists a constant $k_T > 0$ such that the following inequality holds true for all $\epsilon_0 > 0$:

$$k_T[E(T) + E(0)] \le \int_0^T \int_\Gamma \left[w_t^2 + z_t^2 + \left(\frac{\partial w}{\partial \nu}\right)^2 + \left(\frac{\partial z}{\partial \nu}\right)^2\right] d\Gamma\, dt$$
$$+ C_{T,\epsilon_0} \left\{ \|w\|^2_{H^{\frac{1}{2}+\epsilon_0}(Q)} + \|z\|^2_{H^{\frac{1}{2}+\epsilon_0}(Q)} \right\}, \tag{1.8}$$

where

$$E(t) \equiv E_w(t) + E_z(t). \tag{1.9}$$

(ii) Assume further that

$$w|_{\Sigma_0} \equiv 0, \text{ and/or, respectively, } z|_{\Sigma_0} \equiv 0, \quad \Sigma_0 = (0,T] \times \Gamma_0, \tag{1.10}$$

where Γ_0 is the portion of the boundary $\Gamma = \Gamma_0 \cup \Gamma_1$ defined by Eqn. (2.1.13) in Section 2. Then, the right-hand side of (1.8) is refined, in the sense that integration over Γ of the w-terms, and/or, respectively, of the z-terms, is replaced by a corresponding integration over Γ_1 only, so that the right-hand side of (1.8) contains instead

$$\int_0^T \int_{\Gamma_1} \left[w_t^2 + \left(\frac{\partial w}{\partial \nu}\right)^2\right] d\Gamma_1 dt, \text{ and/or, respectively, } \int_0^T \int_{\Gamma_1} \left[z_t^2 + \left(\frac{\partial z}{\partial \nu}\right)^2\right] d\Gamma_1 dt, \tag{1.11}$$

according to whichever of the conditions in (1.10) holds true (possibly both). □

As discussed in detail in the *Comments* below, a main source of difficulty in proving estimate (1.8) is due to the presence of arbitrary *first*-order terms in the original variable as well as in the coupled variable, in each of the two equations (1.1) and (1.2).

Once (1.8) is established, one may further refine it by absorbing the interior lower-order terms $\|w\|_{H^{\frac{1}{2}+\epsilon_0}(Q)}$ and $\|z\|_{H^{\frac{1}{2}+\epsilon_0}(Q)}$ by a compactness/uniqueness argument, and thus obtain the desired final estimate.

Theorem 1.2. Let w, z be solutions of Eqns. (1.1), (1.2) in the class (1.6), (1.7). Let both w and z satisfy the boundary conditions in (1.10) on Σ_0. Assume, further, that the only solution of (1.1), (1.2) subject to the over-determined homogeneous B.C.

$$\begin{cases} w|_\Sigma \equiv 0; \\ \left.\dfrac{\partial w}{\partial \nu}\right|_{\Sigma_1} \equiv 0, \end{cases} \quad \text{and} \quad \begin{cases} z|_\Sigma \equiv 0; \\ \left.\dfrac{\partial z}{\partial \nu}\right|_{\Sigma_1} \equiv 0, \end{cases} \tag{1.12}$$

is the trivial solution $w \equiv z \equiv 0$.

Then, there exists a positive constant $k_T > 0$ such that the following energy estimate holds true:

$$k_T[E(T) + E(0)] \leq \int_0^T \int_{\Gamma_1} \left[w_t^2 + z_t^2 + \left(\frac{\partial w}{\partial \nu}\right)^2 + \left(\frac{\partial z}{\partial \nu}\right)^2 \right] d\Gamma_1 dt. \quad \square \tag{1.13}$$

Remark 1.1. The uniqueness result always holds true with sufficiently smooth coefficients. □

Consequences on Exact Controllability. We now supplement Eqns. (1.1), (1.2) with boundary conditions. We need to consider essentially three cases, where $\Sigma_i = (0, T] \times \Gamma_i$, $i = 0, 1$.

Case 1. (Dirichlet/Dirichlet)

$$\begin{cases} w|_{\Sigma_0} \equiv 0; \\ w|_{\Sigma_1} \equiv u_1, \end{cases} \quad \text{and} \quad \begin{cases} z|_{\Sigma_0} \equiv 0; \\ z|_{\Sigma_1} \equiv u_2. \end{cases} \tag{1.14}$$

Case 2. (Neumann/Neumann)

$$\begin{cases} w|_{\Sigma_0} \equiv 0; \\ \left.\dfrac{\partial w}{\partial \nu}\right|_{\Sigma_1} \equiv u_1, \end{cases} \quad \text{and} \quad \begin{cases} z|_{\Sigma_0} \equiv 0; \\ \left.\dfrac{\partial z}{\partial \nu}\right|_{\Sigma_1} \equiv u_2. \end{cases} \tag{1.15}$$

Case 3. (Dirichlet/Neumann)

$$\begin{cases} w|_{\Sigma_0} \equiv 0; \\ w|_{\Sigma_1} \equiv u_1, \end{cases} \quad \text{and} \quad \begin{cases} z|_{\Sigma_0} \equiv 0; \\ \left.\dfrac{\partial z}{\partial \nu}\right|_{\Sigma_1} \equiv u_2. \end{cases} \tag{1.16}$$

The well-posedness result in each of the three cases is standard. As a consequence of the basic energy estimate in (1.13), we obtain exact controllability at $t = T$ for problem (1.1), (1.2) in each of the three foregoing cases, within the class of $L_2(0,T;L_2(\Gamma_1))$-controls u_1 and u_2 in the space $L_2(\Omega) \times H^{-1}(\Omega)$ in the case of Dirichlet B.C., and in the space $H^1_{\Gamma_0}(\Omega) \times L_2(\Omega)$ in the case of Neumann B.C.

Theorem 1.3. Let the hypotheses of Theorem 1.2 hold true.

(a) (Continuous observability inequalities) Assume first $u_1 \equiv u_2 \equiv 0$ in (1.14)–(1.16). Then, the following inequalities hold true for (1.1), (1.2):

Case 1 (problem (1.1), (1.2), (1.14) with $u_1 \equiv u_2 \equiv 0$):

$$k_T E(0) \leq \int_0^T \int_{\Gamma_1} \left[\left(\frac{\partial w}{\partial \nu} \right)^2 + \left(\frac{\partial z}{\partial \nu} \right)^2 \right] d\Gamma_1 dt. \tag{1.17}$$

Case 2 (problem (1.1), (1.2), (1.15) with $u_1 \equiv u_2 \equiv 0$):

$$k_T E(0) \leq \int_0^T \int_{\Gamma_1} \left[w_t^2 + z_t^2 \right] d\Gamma_1 dt. \tag{1.18}$$

Case 3 (problem (1.1), (1.2), (1.16) with $u_1 \equiv u_2 \equiv 0$):

$$k_T E(0) \leq \int_0^T \int_{\Gamma_1} \left[\left(\frac{\partial w}{\partial \nu} \right)^2 + z_t^2 \right] d\Gamma_1 dt. \tag{1.19}$$

(b) (Exact controllability at $t = T$). By duality, problem (1.1), (1.2) is exactly controllable as specified in each of the following cases:

Case 1: Assume the following initial conditions

$$\{w_0, w_1\} \in L_2(\Omega) \times H^{-1}(\Omega); \quad \{z_0, z_1\} \in L_2(\Omega) \times H^{-1}(\Omega). \tag{1.20}$$

Then, there exist controls

$$\{u_1, u_2\} \in L_2(0,T;L_2(\Gamma_1)) \times L_2(0,T;L_2(\Gamma_1)), \tag{1.21}$$

such that the corresponding solution of (1.1), (1.2), (1.14) satisfies

$$w(T) = w_t(T) = z(T) = z_t(T) = 0. \tag{1.22}$$

Case 2: Assume the following initial conditions

$$\{w_0, w_1\} \in H^1_{\Gamma_0}(\Omega) \times L_2(\Omega); \quad \{z_0, z_1\} \in H^1_{\Gamma_0}(\Omega) \times L_2(\Omega). \tag{1.23}$$

Then, there exist controls $\{u_1, u_2\}$ as in (1.20) such that the corresponding solution of (1.1), (1.2), (1.15) satisfies (1.22).

Case 3: Assume the following initial conditions

$$\{w_0, w_1\} \in L_2(\Omega) \times H^{-1}(\Omega); \quad \{z_0, z_1\} \in H^1_{\Gamma_0}(\Omega) \times L_2(\Omega). \tag{1.24}$$

Then, there exist controls $\{u_1, u_2\}$ as in (1.21) such that the corresponding solution of (1.1), (1.2), (1.16) satisfies (1.22). □

Comments, literature. To put the above estimate (1.8) for the coupled problem (1.1), (1.2) in perspective, let us consider at first only the w-equation (1.1) with no coupling; i.e., the equation

$$w_{tt} = \Delta w + F_1(w) \quad \text{on } Q, \tag{1.25}$$

with F_1 a first-order differential operator satisfying (1.3) [with $P_2 \equiv 0$]. It is well known that the energy (multiplier) method, based on the principal multiplier $h(x) \cdot \nabla w$, $h(x)$ a suitable vector field over $\bar{\Omega}$, permits to establish a number of key inequalities:

(i) the "regularity inequality" in the Dirichlet homogeneous case (the $L_2(\Sigma_T)$-norm of $\frac{\partial w}{\partial \nu}$ is bounded above by $E_w(0)$, for all T), indeed, even in the case of principal part with variable coefficients [L-L-T], [Li.1];

(ii) the reverse "continuous observability inequality," when coupled with the second multiplier w div h, both in the homogeneous Dirichlet case and in the homogeneous Neumann case, however *only when* F_1 is actually a *zero*-order operator. If F_1 is a bonafide first-order operator, the method fails. [Remarkably enough, the method in proving reverse inequalities was first employed successfully in the *more demanding* uniform stabilization problem, with $F_1 \equiv 0$, but, however, with dissipative, non-homogeneous boundary conditions: in [C.1] for the Neumann case, and in [L-T.2] for the technically more difficult Dirichlet case. As a consequence, the first results on exact controllability were obtained, with $L_2(\Sigma_T)$-controls, on the space $H^1_{\Gamma_0}(\Omega) \times L_2(\Omega)$ in the Neumann case [C.1], and on the space $L_2(\Omega) \times H^{-1}(\Omega)$ in the Dirichlet case, [L-T.2].] There are many references within the framework of these multipliers, too numerous to quote. It suffices to refer to the most recent books [Li.2], [K.1], which give an account of this method.

To obtain "continuous observability" reverse inequalities for (1.25), with F_1 of order one, much more sophisticated methods were subsequently introduced:

(a) methods of micro-local analysis (a new micro-local boundary estimate), combined with recent results on geometric optics and propagation of singularities from the boundary, as in the full general work of [B-L-R] for second-order hyperbolic equations, with variable (smooth) coefficients also in the principal part;

(b) pseudo-differential methods to extend Carleman estimates—which were available in the literature for solutions with compact support and, generally, isotropic operators—to the case of domains with boundary and to anisotropic operators, as carried out in the general and unifying work of [Ta.1–3]. Both works have "unfriendly" proofs, not readily accessible outside specialized circles, due to the generality assumed in each case: a general, second-order hyperbolic equation in [B-L-R]; a general, possibly anisotropic operator, with, however, constant coefficient principal part (to assert the existence of a pseudo-convex function) in [Ta].

It would be a problem to just quote or dig out the required estimates from either [B-L-R] or from [Ta.1–3] to prove Theorem 1.1 for the coupled system (1.1), (1.2)—or how to dispense with geometric conditions by appealing to the methods behind these proofs. Moreover, [Ta.1–3], at least in this first effort, takes the control over the entire boundary.

In this paper, we pursue the Carlemann estimate approach proposed by [Ta]. However, by restricting to second-order equations, we proved—first for the single equation (1.25) in Section 2; next for the coupled system (1.1), (1.2) in Section 3—explicit, direct, friendly-to-follow energy computations and estimates, conducted throughout at the differential level with differential multipliers (rather than at the pseudo-differential level with pseudo-differential multipliers as in the general work of [Ta]). In the process, we dispense with geometric conditions (by virtue of our result Theorem 1.1.4) and, moreover, allow the control action to be active only on a portion of the boundary (unlike [Ta.1–3]). Thus, the method given explicitly here reveals itself as a differential multiplier method, with multipliers (which we call Carleman multipliers), which generalize directly but in a non-trivial way the original multipliers $h \cdot \nabla w$ and w div h mentioned above, and used for continuous observability inequalities and in uniform stabilization inequalities, only for canonical models. The method provides a one-parameter family of (Carleman) estimates in terms of a parameter $\tau > 0$, and its virtue in absorbing first-order terms is clearly displayed by using the additional flexibility of the parameter τ, once taken large enough (see Remark 2.2.1, and proof of Theorem 2.2.6). The estimates refer to solutions of the second-order hyperbolic equation (1.25) above (Section 2) and respectively, of the coupled system (1.1) and (1.2) above (Section 3), with no boundary conditions. They are expressed explicitly in terms of boundary traces as well. As a consequence, we derive exact controllability results in the Dirichlet and Neumann cases, i.e., in cases which rely on $H^1(\Omega) \times L_2(\Omega)$-energy level estimates.

2 A-priori P.D.E.'s Estimates for Second-Order Hyperbolic Equations

2.1 Dynamical Model and Statement of Main Results

Dynamical model. Let Ω be an open bounded domain in R^n with sufficiently smooth boundary Γ, say of class C^1. Throughout this section we shall consider the following second-order hyperbolic equation in the unknown $w(t,x)$:

$$w_{tt} = \Delta w + F(w) + f \text{ in } (0,T] \times \Omega \equiv Q, \tag{2.1.1}$$

where $f \in L_2(Q)$ is a forcing term and where $F(w)$ is a linear first-order differential operator in all variables $\{t, x_1, \dots, x_n\}$ on w with $L_\infty(Q)$-coefficients, thus satisfying the following pointwise estimate: there exists a constant $C_T > 0$ such that

$$|F(w)|^2 \leq C_T \left[w_t^2 + |\nabla w|^2 + w^2\right], \qquad \forall\, t, x \in Q; \tag{2.1.2}$$

$$E(t) \equiv \int_\Omega \left[w_t^2 + |\nabla w|^2\right] d\Omega. \tag{2.1.3}$$

Remark 2.1.1. In the analysis below the operator $(-\Delta)$ in (2.1.1) [Laplacian in the space variables] could be replaced by a second-order uniformly elliptic operator with constant coefficients. □

Pseudo-convex function $\phi(x,t)$. Let $\phi : \Omega \times \mathbf{R} \to \mathbf{R}$ be the (pseudo-convex) function defined by

$$\phi(x,t) \equiv |x-x_0|^2 - c\left|t-\frac{T}{2}\right|^2, \tag{2.1.4a}$$

where $x_0 \in \mathbf{R}^n$, $T>0$, $0<c<1$, are selected so that the following two properties are achieved:

(i)

$$cT > 2\max_{y\in\Omega}|y-x_0|; \tag{2.1.4b}$$

(ii) there exists a subinterval $[t_0,t_1] \subset (0,T)$ such that

$$\phi(x,t) > 1 \text{ for } t \in [t_0,t_1];\ x\in\Omega; \tag{2.1.4c}$$

$$\phi(x,0) < -\delta < 0;\ \phi(x,T) < -\delta < 0, \text{ uniformly in } x\in\Omega, \tag{2.1.4d}$$

for a suitable constant $\delta>0$. We observe the following consequences for future use:

$$\nabla\phi(x,t) = 2(x-x_0) \equiv h(x);\quad \text{div } h = 2\dim\Omega; \tag{2.1.5a}$$

$$\phi_t(x,t) = -2c\left(t-\frac{T}{2}\right);\ \phi_{tt} = -2c;\ \phi_t(x,0) = cT;\ \phi_t(x,T) = -cT; \tag{2.1.5b}$$

$$\text{div}(e^{\tau\phi}h) = e^{\tau\phi}[\tau|h|^2 + \text{div } h];\ \nabla(e^{\tau\phi}) = \tau e^{\tau\phi}\nabla\phi. \tag{2.1.5c}$$

Any radial field centered at x_0 is the gradient of a class of pseudo-convex functions as in (2.1.4).

Remark 2.1.2. (Optimal choice of T) First, by appropriate choice of x_0, on may achieve that

$$2\max_{y\in\Omega}|y-x_0| = \text{ diameter of } \Omega = d_\Omega,$$

i.e., diameter of the smallest sphere containing Ω. Next, since the proof below requires of c only the condition $0<c<1$, one may take T to be *any* number strictly greater than the diameter of $\Omega :\ T = d_\Omega + \epsilon$, and still satisfy requirement (2.1.4a). Such choice of T is optimal, via finite speed of propagation argument. Henceforth, this is the value of T taken in definition (2.1.4). □

Main results. The main results of the present section are the following ones. They are listed in the order in which they are proved.

Theorem 2.1.1. (Carleman estimates) Let w be a solution of Eqn. (2.1.1) in the following class:

$$\begin{cases} w \in H^{1,1}(Q) \equiv L_2(0,T;H^1(\Omega)) \cap H^1(0,T;L_2(\Omega)); & (2.1.6) \\ w_t,\ \dfrac{\partial w}{\partial\nu} \in L_2(0,T;L_2(\Gamma)). & (2.1.7) \end{cases}$$

Let $\phi(x,t)$ be the pseudo-convex function defined by (2.1.4). Then, for $\tau>0$ sufficiently large, the following one-parameter family of estimates holds true:

(i)

$$\begin{aligned} &(BT)|_\Sigma + \tfrac{C_T}{\tau}\int_Q e^{\tau\phi}f^2dQ + TC_T \text{ const}_\tau\|w\|^2_{C([0,T];L_2(\Omega))} \\ &\geq \left(1-c-\tfrac{C_T}{\tau}\right)\int_Q e^{\tau\phi}[|\nabla w|^2 + w_t^2]dQ - C_1Te^{-\delta\tau}[E(T)+E(0)], \end{aligned} \tag{2.1.8a}$$

(ii)

$$(BT)|_\Sigma + \tfrac{C_T}{\tau}\int_Q e^{\tau\phi} f^2 dQ + TC_T \text{ const}_\tau \|w\|^2_{C([0,T];L_2(\Omega))}$$
$$\geq \left(1 - c - \tfrac{C_T}{\tau}\right) e^{\tau} \int_{t_0}^{t_1} E(t)dt - C_1 T e^{-\delta\tau}[E(T) + E(0)], \tag{2.1.8b}$$

where the boundary terms $(BT)|_\Sigma$ over $\Sigma = [0,T] \times \Gamma$ are given by (see (1.2.27) below),

$$(BT)|_\Sigma = \int_\Sigma e^{\tau\phi} \frac{\partial w}{\partial \nu}[h \cdot \nabla w - \phi_t w_t] d\Sigma \; + \; \int_\Sigma \frac{\partial w}{\partial \nu} w \left[\frac{1}{2}\mu - (1+c)e^{\tau\phi}\right] d\Sigma$$
$$+ \; \frac{1}{2}\int_\Sigma e^{\tau\phi}[w_t^2 - |\nabla w|^2] h \cdot \nu \, d\Sigma, \tag{2.1.9}$$

with $\mu(x,t)$ the function defined in (1.2.20). The constants c and δ are defined in (2.1.4), while C_T and C_1 are defined in (2.1.2) and (1.2.29). □

The proof of Theorem 2.1.1 will be given in Section 1.2, where additional interesting results are contained.

Theorem 2.1.2. (i) Let w be a solution of Eqn. (2.1.1) in the class (2.1.6), (2.1.7). Then, for $\tau > 0$ sufficiently large, there exists a constant $k_{\phi,\tau} > 0$ such that the following estimates hold true:

(i_1)

$$(\overline{BT})|_\Sigma \; + \; \frac{C_T}{\tau}\int_Q e^{\tau\phi} f^2 dQ + \text{ const}_{\phi,\tau} \int_Q f^2 dQ$$
$$+ \; TC_T \text{ const}_\tau \|w\|^2_{C([0,T];L_2(\Omega))} \geq k_{\phi,\tau} E(0), \tag{2.1.10a}$$

(i_2)

$$(\overline{BT})|_\Sigma \; + \; \frac{C_T}{\tau}\int_Q e^{\tau\phi} f^2 dQ + \text{ const}_{\phi,\tau} \int_Q f^2 dQ$$
$$+ \; TC_T \text{ const}_\tau \|w\|^2_{C([0,T];L_2(\Omega))} \geq \frac{e^{-C_T T}}{2} k_{\phi,\tau}[E(0) + E(T)], \tag{2.1.10b}$$

where the boundary terms $(\overline{BT})|_\Sigma$ over Σ are given by

$$(\overline{BT})|_\Sigma = (BT)|_\Sigma + \text{ const}_{\phi,\tau} \int_\Sigma \left|\frac{\partial w}{\partial \nu} w_t\right| d\Sigma, \tag{2.1.11}$$

with $(BT)|_\Sigma$ defined by (2.1.9).

(ii) Assume, further, that w satisfies the boundary condition

$$w|_{\Sigma_0} \equiv 0; \quad \Sigma_0 = (0,T] \times \Gamma_0, \tag{2.1.12}$$

where we divide the boundary Γ as $\Gamma = \Gamma_0 \cup \Gamma_1$, where

$$\Gamma_0 = \{x \in \Gamma : \; \nabla\phi \cdot \nu(x) \leq 0\}; \tag{2.1.13}$$

$$\Gamma_1 = \{x \in \Gamma : \; \nabla\phi \cdot \nu(x) > 0\} \tag{2.1.14}$$

with $\nu(x)$ = unit outward normal vector at $x \in \Gamma$, and $x_0 \in R^n$ the point entering the definition of ϕ in (2.1.4) so that $2(x - x_0) = h = \nabla\phi$.

Then, estimate (2.1.10) holds true for $\tau > 0$ sufficiently large, with the boundary terms $(\overline{BT})|_\Sigma$ replaced by $(\overline{BT})|_{\Sigma_1}$ i.e., evaluated only on $\Sigma_1 = [0, T] \times \Gamma_1$. □

The proof of Theorem 2.1.2 will be given in Section 2.3. As a consequence of Theorem 2.1.2 and of a uniqueness theorem, we then obtain the desired "continuous observability inequality" for the homogeneous problem

$$\begin{cases} \psi_{tt} = \Delta\psi + F(\psi) & \text{in } (0, T] \times \Omega \equiv Q; & (2.1.15a) \\ \psi(0, \cdot\,) = \psi_0;\ \psi_t(0, \cdot\,) = \psi_1 & \text{in } \Omega; & (2.1.15b) \\ \psi|_\Sigma \equiv 0 & \text{in } (0, T] \times \Gamma \equiv \Sigma, & (2.1.15c) \end{cases}$$

where F is the first-order linear operator in (2.1.2). The reverse "trace regularity inequality" was proved in [L-L-T.1, [Li.1], [L-T,1] for any $T > 0$.

Theorem 2.1.3. Let T be the time constant which enters the definition of ϕ in (2.1.4a–d). Let the homogeneous, over-determined problem defined by (2.1.15a–c), as well as: $\frac{\partial\psi}{\partial\nu}|_{\Sigma_1} \equiv 0$ on $(0, T] \times \Gamma_1 \equiv \Sigma_1$ admit only the unique solution $\psi \equiv 0$. Then, with reference to problem (2.1.15), there exists a positive constant $\text{const}_T > 0$ such that

$$\int_0^T \int_{\Gamma_1} \left(\frac{\partial\psi}{\partial\nu}\right)^2 d\Gamma_1 dt \geq \text{const}_T \|\{\psi_0, \psi_1\}\|^2_{H_0^1(\Omega)\times L_2(\Omega)}. \quad \square \tag{2.1.16}$$

The proof of Theorem 2.1.3 will be given in Section 2.4. The next result 'absorbs' the tangential derivatives $\frac{\partial w}{\partial s} = \nabla_s w$ (tangential gradient) by the normal derivative $\frac{\partial w}{\partial \nu}$ and w_t. It is taken from [L-T.3, Section 7.2].

Theorem 2.1.4. Let w be a solution of Eqn. (2.1.1) in the class (2.1.6), (2.1.7).

(i) Given $\epsilon > 0$ and $\epsilon_0 > 0$ arbitrarily small and given $T > 0$, there exists a constant $C_{\epsilon,\epsilon_0,T} > 0$ such that

$$\int_\epsilon^{T-\epsilon} \int_\Gamma \left(\frac{\partial w}{\partial s}\right)^2 d\Gamma\, dt \leq C_{\epsilon,\epsilon_0,T}\Big\{\int_0^T \int_\Gamma \Big[\left(\frac{\partial w}{\partial \nu}\right)^2 + w_t^2\Big] d\Sigma$$
$$+ \|w\|^2_{L_2(0,T;H^{\frac{1}{2}+\epsilon_0}(\Omega))} + \|f\|^2_{H^{-\frac{1}{2}+\epsilon_0}(Q_T)}\Big\}. \tag{2.1.17}$$

(ii) If, moreover, w satisfies the boundary condition (2.1.12), then (2.1.17) holds true with Γ replaced by Γ_1. □

The final estimate—the main result of the present section—is given next.

Theorem 2.1.5. Let w be a solution of Eqn. (2.1.1) in the class (2.1.6), (2.1.7). Assume, moreover, that w satisfies the boundary condition (2.1.12). Then the following estimate holds true: There exists a constant $k_{\phi,\tau} > 0$, ϕ the pseudo-convex function in (2.1.4) and τ a sufficiently large parameter, such that

(i)

$$\int_0^T \int_{\Gamma_1} \left[\left(\frac{\partial w}{\partial \nu} \right)^2 + w_t^2 \right] d\Gamma_1 dt \; + \; \text{const}_{\phi,T} \int_Q f^2 dQ + C_{T,\epsilon_0} \|w\|^2_{C([0,T];H^{\frac{1}{2}+\epsilon_0}(\Omega))} \geq k_{\phi,T} E(0); \tag{2.1.18a}$$

(ii) or equivalently,

$$\int_0^T \int_{\Gamma_1} \left[\left(\frac{\partial w}{\partial \nu} \right)^2 + w_t^2 \right] d\Gamma_1 dt \; + \; \text{const}_{\phi,T} \int_Q f^2 dQ + C_{T,\epsilon_0} \|w\|^2_{C([0,T];H^{\frac{1}{2}+\epsilon_0}(\Omega))} \geq k_{\phi,T} \{E(T) + E(0)\}. \tag{2.1.18b}$$

The proof of Theorem 2.1.5 combines Theorems 2.1.2 and 2.1.4, and will be given in Section 2.6.

Consequences on exact controllability. By the standard duality between continuous observability and exact controllability [the input-solution operator is surjective $L_2(0,T;U)$ onto Y at time T if and only if its adjoint is bounded below, we obtain exact controllability of problem (2.1.20) below on the space

$$Y \equiv L_2(\Omega) \times H^{-1}(\Omega); \quad U = L_2(\Gamma). \tag{2.1.19}$$

Theorem 2.1.6. (Exact controllability, Dirichlet case) Assume the hypotheses of Theorem 2.1.3. The mixed problem

$$\begin{cases} w_{tt} = \Delta w + F(w) & \text{in } Q; \quad (2.1.20a) \\ w(0,\cdot) = w_0; \; w_t(0,\cdot) = w_1 & \text{in } \Omega; \quad (2.1.20b) \\ w|_{\Sigma_0} \equiv 0 & \text{in } \Sigma_0; \quad (2.1.20c) \\ w|_{\Sigma_1} = u & \text{in } \Sigma_1, \quad (2.1.20d) \end{cases}$$

with F as in (2.1.2), and where Γ_1 is defined by (2.1.14) for a fixed $x_0 \in \mathbf{R}^n$, is exactly controllable (to, or from, the origin) over the space Y in (2.1.19), within the class of $L_2(0,T;L_2(\Gamma_1))$-controls, where T is the time constant entering in the definition of ϕ in (2.1.4) along with the chosen point x_0. Specifically, given $\{w_0, w_1\} \in Y$, and $\{v_0, v_1\} \in Y$, there exists $u \in L_2(0,T;L_2(\Gamma_1))$ such that the corresponding solution of (2.1.20) satisfies $w(T,\cdot) = v_0$, $w_t(T,\cdot) = v_1$. □

Theorem 2.1.5 proves the major estimate responsible for the continuous observability inequality, in the optimal time T, for the following homogeneous problem

$$\begin{cases} \psi_{tt} = \Delta \psi + F(\psi) & \text{in } (0,T] \times \Omega \equiv Q; \quad (2.1.21a) \\ \psi(0,\cdot) = \psi_0; \; \psi_t(0,\cdot) = \psi_1 & \text{in } \Omega; \quad (2.1.21b) \\ \psi|_{\Sigma_0} \equiv 0 & \text{in } (0,T] \times \Gamma_0 \equiv \Sigma_0; \quad (2.1.21c) \\ \left.\frac{\partial \psi}{\partial \nu}\right|_{\Sigma_1} \equiv 0 & \text{in } (0,T] \times \Gamma_1 \equiv \Sigma_1, \quad (2.1.21d) \end{cases}$$

which is equivalent to exact controllability of the non-homogeneous problem

$$\begin{cases} w_{tt} = \Delta w + F(w) & \text{in } (0,T] \times \Omega \equiv Q; & (2.1.22a) \\ w(0,\,\cdot\,) = w_0;\ w_t(0,\,\cdot\,) = w_1 & \text{in } \Omega; & (2.1.22b) \\ w|_{\Sigma_0} \equiv 0 & \text{in } (0,T] \times \Gamma_0 \equiv \Sigma_0; & (2.1.22c) \\ \left.\dfrac{\partial w}{\partial \nu}\right|_{\Sigma_1} \equiv u & \text{in } (0,T] \times \Gamma_1 \equiv \Sigma_1. & (2.1.22d) \end{cases}$$

Theorem 2.1.7. (Continuous observability, exact controllability, Neumann case) Let T be the constant which enters the definition of ϕ in (2.1.4a–d). Let the homogeneous, over-determined problem (2.1.21a,b,d), along with $\psi|_\Sigma \equiv 0$ admit only the unique solution $\psi \equiv 0$. Then, with reference to problem (2.1.21), there exists a positive constant $C_T > 0$ such that the following continuous observability inequality holds true:

$$\int_0^T \int_{\Gamma_1} \psi_t^2 d\Gamma_1 dt \geq C_T E(0). \tag{2.1.23}$$

Equivalently, problem (2.1.22) is exactly controllable (to, from) the origin on the space $H^1_{\Gamma_0}(\Omega) \times L_2(\Omega)$, within the class of $L_2(0,T;L_2(\Gamma_1))$-controls. □

2.2 Proof of Theorem 2.1.1: Carleman Estimates

In order to handle a general *first*-order operator F in Eqn. (2.1.15a) [i.e., at the "energy level"], a major conceptual and technical jump over the energy method used when $F \equiv 0$ described above is called for. Such definitely non-trivial extension is based on definitely more sophisticated (Carleman) multipliers; first among them the multiplier

$$e^{\tau\phi(x,t)}[\nabla\phi(x,t) \cdot \nabla w(t,x) - \phi_t(x,t)w_t(t,x)], \tag{2.2.1a}$$

where ϕ is the pseudo-convex function introduced in (2.1.4a–d), and τ is a positive free parameter of adjustment, to be eventually chosen sufficiently large as to absorb an energy level term with a large negative constant $\left(-\frac{1}{\epsilon}\right)$ in front, which arises due to the fact that $F(w)$ is first order: see the key step described in Remark 2.2.1 below. Additional multipliers are:

$$w\left[\operatorname{div}(e^{\tau\phi}h) - \frac{d}{dt}\left(e^{\tau\phi}\phi_t\right)\right]; \quad we^{\tau\phi}. \tag{2.2.1b}$$

Henceforth, we shall write freely and interchangeably $h(x) = \nabla\phi$ as in (2.1.5a); moreover, c and T will be throughout the constants appearing in the definition of ϕ in (2.1.4).

Step 1. Proposition 2.2.1. Let w be a solution of Eqn. (2.1.1) in the class (2.1.6), (2.1.7). Then, the following identity holds true, where $\Sigma = [0,T] \times \Gamma$; $Q = [0,T] \times \Omega$:

$$\int_\Sigma e^{\tau\phi}\frac{\partial w}{\partial \nu}[h \cdot \nabla w - \phi_t w_t]d\Sigma + \frac{1}{2}\int_\Sigma e^{\tau\phi}[w_t^2 - |\nabla w|^2]h \cdot \nu\, d\Sigma$$

$$= 2\int_Q e^{\tau\phi}|\nabla w|^2 dQ + \frac{1}{2}\int_Q [w_t^2 - |\nabla w|^2] \operatorname{div}\left(e^{\tau\phi}h\right) dQ$$

$$+\frac{1}{2}\int_Q [w_t^2 + |\nabla w|^2]\frac{d}{dt}\left(e^{\tau\phi}\phi_t\right) dQ + \tau\int_Q e^{\tau\phi}(h\cdot\nabla w)^2 dQ$$

$$-2\tau\int_Q e^{\tau\phi} w_t\phi_t h\cdot\nabla w\, dQ + \ell_{0,T} - r_{1;0,T} - \text{R.H.S.}_2, \tag{2.2.2}$$

where we have set for convenience

$$\ell_{0,T} = \left[\int_\Omega e^{\tau\phi} w_t h\cdot\nabla w\, d\Omega\right]_0^T - \frac{1}{2}\left[\int_\Omega e^{\tau\phi}\phi_t w_t^2 d\Omega\right]_0^T$$

$$= \int_\Omega e^{\tau\phi(x,T)} w_t(T) h\cdot\nabla w(T) d\Omega - \int_\Omega e^{\tau\phi(x,0)} w_t(0) h\cdot\nabla w(0) d\Omega$$

$$+\frac{cT}{2}\int_\Omega e^{\tau\phi(x,T)} w_t^2(T) d\Omega + \frac{cT}{2}\int_\Omega e^{\tau\phi(x,0)} w_t^2(0) d\Omega; \tag{2.2.3}$$

$$|\ell_{0,T}| \leq \text{ const } Te^{-\tau\delta}[E(T) + E(0)], \tag{2.2.4}$$

with const. depending only on h, but not on T or τ;

$$r_{1;0,T} = \frac{1}{2}\left[\int_\Omega e^{\tau\phi}\phi_t|\nabla w|^2 d\Omega\right]_0^T$$

$$= -\frac{cT}{2}\int_\Omega e^{\tau\phi(x,T)}|\nabla w(T)|^2 d\Omega - \frac{cT}{2}\int_\Omega e^{\tau\phi(x,0)}|\nabla w(0)|^2 d\Omega; \tag{2.2.5}$$

$$|r_{1;0,T}| \leq \frac{cT}{2} e^{-\tau\delta}[E(T) + E(0)]; \tag{2.2.6}$$

$$\text{R.H.S.}_2 \equiv \int_Q [F(w) + f] e^{\tau\phi}[\nabla\phi\cdot\nabla w - w_t\phi_t] dQ. \tag{2.2.7}$$

Proof. We multiply both sides of Eqn. (2.1.1) by the multiplier in (2.2.1a), i.e., $e^{\tau\phi}[\nabla\phi\cdot\nabla w - \phi_t w_t]$, and integrate over $Q = [0,T]\times\Omega$.

Left-hand side. On the left-hand side (L.H.S.) we obtain recalling (2.2.3),

$$\text{L.H.S.} \equiv \int_0^T\int_\Omega w_{tt} e^{\tau\phi}[\nabla\phi\cdot\nabla w - \phi_t w_t] d\Omega\, dt$$

$$= \frac{1}{2}\int_Q w_t^2\left[\operatorname{div}\left(e^{\tau\phi}h\right) + \frac{d}{dt}\left(e^{\tau\phi}\phi_t\right)\right] dQ$$

$$-\tau\int_Q e^{\tau\phi} w_t\phi_t h\cdot\nabla w\, dQ - \frac{1}{2}\int_\Sigma e^{\tau\phi} w_t^2 h\cdot\nu\, d\Sigma + \ell_{0,T}. \tag{2.2.8}$$

Indeed, integrating by parts in t, and recalling that $\nabla\phi(x,t) \equiv h(x)$ is time-independent from (2.1.5a), we compute

$$\int_\Omega \int_0^T w_{tt} e^{\tau\phi} \nabla\phi \cdot \nabla w \, dt \, d\Omega = \left[\int_\Omega e^{\tau\phi} w_t h \cdot \nabla w \, d\Omega \right]_0^T$$

$$- \; \tau \int_Q e^{\tau\phi} w_t \phi_t h \cdot \nabla w \, d\Omega - \int_Q e^{\tau\phi} h \cdot \nabla w_t w_t \, dQ, \tag{2.2.9}$$

where the last term in (1.2.9) is rewritten, by means of the usual identity [L-T.4, (A.1)], as

$$\begin{aligned} \int_Q e^{\tau\phi} h \cdot \nabla w_t w_t \, dQ &= \frac{1}{2} \int_Q e^{\tau\phi} h \cdot \nabla(w_t^2) dQ \\ &= \frac{1}{2} \int_\Sigma e^{\tau\phi} w_t^2 h \cdot \nu \, d\Sigma - \frac{1}{2} \int_Q w_t^2 \operatorname{div}\left(e^{\tau\phi} h\right) dQ. \end{aligned} \tag{2.2.10}$$

Similarly, integrating by parts in t, we compute

$$\begin{aligned} \int_\Omega \int_0^T w_{tt} w_t e^{\tau\phi} \phi_t dt \, d\Omega &= \frac{1}{2} \int_\Omega \int_0^T e^{\tau\phi} \phi_t \frac{\partial(w_t^2)}{\partial t} dt \, d\Omega \\ &= \frac{1}{2} \left[\int_\Omega e^{\tau\phi} \phi_t w_t^2 d\Omega \right]_0^T - \frac{1}{2} \int_\Omega \int_0^T w_t^2 \frac{\partial\left(e^{\tau\phi}\phi_t\right)}{\partial t} dt \, d\Omega. \end{aligned} \tag{2.2.11}$$

Using (2.2.10) in (2.2.9) and subtracting off (2.2.11) yields (2.2.8), as desired, recalling $\ell_{0,T}$ in (2.2.3).

Right-hand side. Multiplying the first term Δw on the right-hand side (R.H.S.) of Eqn. (2.1.1) by the multiplier in (2.2.1a), $e^{\tau\phi}[\nabla\phi \cdot \nabla w - \phi_t w_t]$, and integrating over $Q = [0,T] \times \Omega$ yields recalling (2.2.5),

$$\begin{aligned} \text{R.H.S.}_{\cdot 1} &\equiv \int_0^T \int_\Omega \Delta w e^{\tau\phi} [\nabla\phi \cdot \nabla w - \phi_t w_t] d\Omega \, dt \\ &= \int_\Sigma e^{\tau\phi} \frac{\partial w}{\partial \nu} [\nabla w \cdot h - \phi_t w_t] \, d\Sigma - \frac{1}{2} \int_\Sigma e^{\tau\phi} |\nabla w|^2 h \cdot \nu \, d\Sigma \\ &\quad + \frac{1}{2} \int_Q |\nabla w|^2 \left\{ \operatorname{div}\left(e^{\tau\phi} h\right) - \frac{d}{dt}\left(e^{\tau\phi}\phi_t\right) \right\} dQ \\ &\quad - 2 \int_Q e^{\tau\phi} |\nabla w|^2 dQ - \tau \int_Q e^{\tau\phi} (h \cdot \nabla w)^2 dQ \\ &\quad + \tau \int_Q e^{\tau\phi} h \cdot \nabla w \phi_t w_t dQ + r_{1;0,T}. \end{aligned} \tag{2.2.12}$$

Indeed, using Green's first identity, we compute since ϕ_t does not depend on x:

$$\begin{aligned}\int_\Omega \Delta w e^{\tau\phi}[\nabla\phi\cdot\nabla w-\phi_t w_t]d\Omega &= \int_\Gamma \frac{\partial w}{\partial\nu}e^{\tau\phi}[\nabla\phi\cdot\nabla w-\phi_t w_t]d\Gamma \\ &- \int_\Omega e^{\tau\phi}\nabla w\cdot\nabla(\nabla\phi\cdot\nabla w)d\Omega+\int_\Omega e^{\tau\phi}\phi_t\nabla w\cdot\nabla w_t\,d\Omega \\ &- \int_\Omega \nabla w\cdot\nabla\left(e^{\tau\phi}\right)\{\nabla\phi\cdot\nabla w-\phi_t w_t\}d\Omega. \end{aligned}\tag{2.2.13}$$

As to the first integral over Ω on the right-hand side of (2.2.13), with $\nabla\phi=h$, we recall the identity

$$\nabla w\cdot\nabla(h\cdot\nabla w)=H\nabla w\cdot\nabla w+\frac{1}{2}h\cdot\nabla(|\nabla w|^2)\tag{2.2.14}$$

from [L-T.4, Appendix A], where in our present case $H=2$ (identity) by (2.1.5a). Thus, by (2.2.14),

$$-\int_\Omega e^{\tau\phi}\nabla w\cdot\nabla(\nabla\phi\cdot\nabla w)d\Omega=-2\int_\Omega e^{\tau\phi}|\nabla w|^2d\Omega-\frac{1}{2}\int_\Omega e^{\tau\phi}h\cdot\nabla(|\nabla w|^2)d\Omega.\tag{2.2.15}$$

Invoking on the second integral of (1.2.15) the standard identity in [L-T.4, (A.1)], we obtain

$$\begin{aligned}-\int_0^T\int_\Omega e^{\tau\phi}\nabla w\cdot\nabla(\nabla\phi\cdot\nabla w)d\Omega\,dt &= -2\int_Q e^{\tau\phi}|\nabla w|^2dQ-\frac{1}{2}\int_\Sigma e^{\tau\phi}|\nabla w|^2h\cdot\nu\,d\Sigma \\ &+\frac{1}{2}\int_Q|\nabla w|^2\text{div}\left(e^{\tau\phi}h\right)dQ.\end{aligned}\tag{2.2.16}$$

As to the second integral term over Ω on the right-hand side of (2.2.13), we compute, integrating by parts and recalling $r_{1;0,T}$ from (2.2.5),

$$\begin{aligned}\int_0^T\int_\Omega e^{\tau\phi}\phi_t\nabla w\cdot\nabla w_t dt\,d\Omega &= \frac{1}{2}\int_\Omega\int_0^T e^{\tau\phi}\phi_t\frac{\partial(\nabla w\cdot\nabla w)}{\partial t}dt\,d\Omega \\ &= r_{1;0,T}-\frac{1}{2}\int_Q|\nabla w|^2\frac{d\left(e^{\tau\phi}\phi_t\right)}{dt}dQ.\end{aligned}\tag{2.2.17}$$

Next, we insert (2.2.16) and (2.2.17) into (2.2.13) after the latter has been integrated over $[0,T]$ and obtain (2.2.12), as desired, using $\nabla\left(e^{\tau\phi}\right)=e^{\tau\phi}\tau\nabla\phi$.

Finally, we combine the L.H.S. = (2.2.8) with the R.H.S.$_1$ = (2.2.12) and take into account R.H.S.$_2$ = (2.2.7), and we thus obtain (2.2.2). The estimates (2.2.4) and (2.2.6) on $\ell_{0,T}$ and $r_{1;0,T}$ use the properties of the function in (2.1.4d), (2.1.5c). □

Step 2. The following lemma will be invoked repeatedly for various suitable choices of the function $m(x,t)$.

Lemma 2.2.2. Let w be a solution of Eqn. (2.1.1) in the class (2.1.6), (2.1.7). Let $m(x,t)$ be a C^1-function defined over Q. Then, the following identity ("kinetic-potential") holds true:

$$\begin{aligned}\int_Q[w_t^2-|\nabla w|^2]m\,dQ &= \int_Q w\nabla w\cdot\nabla m\,dQ-\int_Q ww_tm_tdQ-\int_Q[F(w)+f]w\,m\,dQ \\ &-\int_\Sigma\frac{\partial w}{\partial\nu}w\,m\,d\Sigma+\left[\int_\Omega w_tw\,m\,d\Omega\right]_0^T\end{aligned}\tag{2.2.18}$$

Proof. We multiply both sides of Eqn. (2.1.1) by wm, integrate by parts in t on the left-hand side, apply Green's first identity on the term Δw on the right-hand side, and obtain (1.2.18). $\square$

Proposition 2.2.3. Let w be a solution of Eqn. (2.1.1) in the class (2.1.6), (2.1.7). Then, the following identity holds true:

$$\int_\Sigma e^{\tau\phi}\frac{\partial w}{\partial\nu}[h\cdot\nabla w-\phi_t w_t]d\Sigma+\frac{1}{2}\int_\Sigma\frac{\partial w}{\partial\nu}w\,\mu\,d\Sigma+\frac{1}{2}\int_\Sigma e^{\tau\phi}[w_t^2-|\nabla w|^2]h\cdot\nu\,d\Sigma$$

$$=2\int_Q e^{\tau\phi}|\nabla w|^2dQ-2c\int_Q e^{\tau\phi}w_t^2dQ+\tau\int_Q e^{\tau\phi}[h\cdot\nabla w-w_t\phi_t]^2dQ$$

$$+\frac{1}{2}\int_Q w\nabla w\cdot\nabla\mu\,dQ-\frac{1}{2}\int_Q ww_t\mu_t dQ-\frac{1}{2}\int_Q[F(w)+f]w\mu\,dQ+\alpha_{0,T}$$

$$-\int_Q[F(w)+f]e^{\tau\phi}[\nabla\phi\cdot\nabla w-\phi_t w_t]dQ, \tag{2.2.19}$$

where we have set via (2.1.5b–c)

$$\mu(x,t)\equiv\operatorname{div}\left(e^{\tau\phi}h\right)-\frac{d}{dt}\left(e^{\tau\phi}\phi_t\right)$$

$$=e^{\tau\phi}\left[\tau|h|^2-\tau\phi_t^2+\operatorname{div}h+2c\right]; \tag{2.2.20}$$

$$\alpha_{0,T}\equiv\ell_{0,T}-r_{1;0,T}+\frac{1}{2}\left[\int_\Omega w_t w\,\mu\,d\Omega\right]_0^T; \tag{2.2.21}$$

$$|\alpha_{0,T}|\le C_1Te^{-\delta\tau}[E(T)+E(0)]$$

$$+C_2\tau\,e^{-\delta\tau}\left\{\|w(T)\|^2_{L_2(\Omega)}+\|w(0)\|^2_{L_2(\Omega)}\right\}, \tag{2.2.22}$$

with constants C_1 and C_2 independent of T or τ.

Proof. First, we apply Lemma 2.2.2 with the choice $m=\mu$ as in (2.2.20), and obtain

$$\frac{1}{2}\int_Q[w_t^2-|\nabla w|^2]\operatorname{div}\left(e^{\tau\phi}h\right)dQ=\frac{1}{2}\int_Q[w_t^2-|\nabla w|^2]\frac{d\left(e^{\tau\phi}\phi_t\right)}{dt}dQ$$

$$+\frac{1}{2}\int_Q w\nabla w\cdot\nabla\mu\,dQ-\frac{1}{2}\int_Q ww_t\mu_t dQ$$

$$-\frac{1}{2}\int_\Sigma\frac{\partial w}{\partial\nu}w\,\mu\,d\Sigma+\frac{1}{2}\left[\int_\Omega w_t w\,\mu\,d\Omega\right]_0^T$$

$$-\frac{1}{2}\int_Q[F(w)+f]w\,\mu\,dQ. \tag{2.2.23}$$

Inserting (2.2.23) into the right-hand side of (2.2.2), to replace the second integral term over Q yields after a cancellation

$$\int_\Sigma e^{\tau\phi}\frac{\partial w}{\partial\nu}[h\cdot\nabla w-\phi_t w_t]d\Sigma+\frac{1}{2}\int_\Sigma\frac{\partial w}{\partial\nu}w\,\mu\,d\Sigma+\frac{1}{2}\int_\Sigma e^{\tau\phi}[w_t^2-|\nabla w|^2]h\cdot\nu\,d\Sigma$$

$$= 2\int_Q e^{\tau\phi}|\nabla w|^2dQ+\int_Q w_t^2\frac{d}{dt}\left(e^{\tau\phi}\phi_t\right)dQ$$

$$+\frac{1}{2}\int_Q w\nabla w\cdot\nabla\mu\,dQ-\frac{1}{2}\int_Q ww_t\mu_t dQ$$

$$-2\tau\int_Q e^{\tau\phi}w_t\phi_t h\cdot\nabla w\,dQ+\tau\int_Q e^{\tau\phi}(h\cdot\nabla w)^2dQ$$

$$+\alpha_{0,T}-\frac{1}{2}\int_Q[F(w)+f]w\mu\,dQ-\text{R.H.S.}_{\cdot 2}, \tag{2.2.24}$$

with $\alpha_{0,T}$ and R.H.S.$_{\cdot 2}$ defined by (2.2.21) and (2.2.7), respectively. We next combine the second, the fifth, and the sixth term on the right-hand side of (2.2.24) in a perfect square, as follows:

$$\int_Q w_t^2\frac{d}{dt}\left(e^{\tau\phi}\phi_t\right)dQ-2\tau\int_Q e^{\tau\phi}w_t\phi_t h\cdot\nabla w\,dQ+\tau\int_Q e^{\tau\phi}(h\cdot\nabla w)^2dQ$$

$$= \tau\int_Q e^{\tau\phi}[h\cdot\nabla w-\phi_t w_t]^2dQ-2c\int_Q e^{\tau\phi}w_t^2dQ, \tag{2.2.25}$$

expanding $\frac{d}{dt}\left(e^{\tau\phi}\phi_t\right)=e^{\tau\phi}[\tau\phi_t^2+\phi_{tt}]$, $\phi_{tt}=-2c$ by (2.1.5b). Inserting (2.2.25) into the right-hand side of (2.2.24) yields (2.2.19), as desired, after recalling also R.H.S.$_{\cdot 2}$ from (2.2.7). The estimate (2.2.22) on $\alpha_{0,T}$ follows from (2.2.21), via (2.2.4), (2.2.6), and (2.2.20). □

Step 3. Theorem 2.2.4. [Final identity] Let w be a solution of Eqn. (2.1.1) in the class (2.1.6), (2.1.7). Then the following identity holds true:

$$(BT)|_\Sigma = (1-c)\int_Q e^{\tau\phi}[|\nabla w|^2+w_t^2]dQ+\tau\int_Q e^{\tau\phi}[h\cdot\nabla w-\phi_t w_t]^2dQ$$

$$+ \int_Q ww_t\left[(1+c)\frac{d\left(e^{\tau\phi}\right)}{dt}-\frac{1}{2}\mu_t\right]dQ+\int_Q w\nabla w\cdot\left[\frac{1}{2}\nabla\mu-(1+c)\nabla\left(e^{\tau\phi}\right)\right]dQ$$

$$+ \int_Q[F(w)+f]w\left[(1+c)e^{\tau\phi}-\frac{1}{2}\mu\right]dQ-\int_Q[F(w)+f]e^{\tau\phi}[\nabla\phi\cdot\nabla w-\phi_t w_t]dQ$$

$$+ \beta_{0,T}, \tag{2.2.26}$$

where the function $\mu(x,t)$ is defined by (2.2.20).

Moreover, the boundary terms $(BT)|_\Sigma$ are given as in (2.1.9) by

$$(BT)|_\Sigma = \int_\Sigma e^{\tau\phi}\frac{\partial w}{\partial\nu}[h\cdot\nabla w - \phi_t w_t]d\Sigma + \int_\Sigma \frac{\partial w}{\partial\nu}w\left[\frac{1}{2}\mu - (1+c)e^{\tau\phi}\right]d\Sigma$$
$$+\frac{1}{2}\int_\Sigma e^{\tau\phi}\left[w_t^2 - |\nabla w|^2\right]h\cdot\nu\, d\Sigma, \tag{2.2.27}$$

with μ defined by (2.2.20), and the time end-point terms at $t=0$ and $t=T$ are

$$\beta_{0,T} = \alpha_{0,T} - (1+c)\left[\int_\Omega e^{\tau\phi}w_t w\, d\Omega\right]_0^T \tag{2.2.28}$$

$$|\beta_{0,T}| \le C_1 T e^{-\delta\tau}[E(T)+E(0)] + C_{2T}e^{-\delta\tau}\left\{\|w(T)\|^2_{L_2(\Omega)} + \|w(0)\|^2_{L_2(\Omega)}\right\}. \tag{2.2.29}$$

Proof. We return to the first two integral terms in Q on the right-hand side of identity (2.2.19) and rewrite them as [recall that $0<c<1$ from (2.1.4)]:

$$2\int_Q e^{\tau\phi}|\nabla w|^2 dQ - 2c\int_Q e^{\tau\phi}w_t^2 dQ$$
$$= (1-c)\int_Q e^{\tau\phi}|\nabla w|^2 dQ + (1-c)\int_Q e^{\tau\phi}|\nabla w|^2 dQ$$
$$+2c\int_Q e^{\tau\phi}[|\nabla w|^2 - w_t^2]dQ. \tag{2.2.30}$$

Next, we apply Lemma 2.2.2, Eqn. (2.2.18), with the choice $m = e^{\tau\phi}$, and obtain

$$\int_Q [|\nabla w|^2 - w_t^2]e^{\tau\phi}dQ = \int_Q ww_t\frac{d}{dt}\left(e^{\tau\phi}\right)dQ$$
$$-\int_Q w\nabla w\cdot\nabla\left(e^{\tau\phi}\right)dQ + \int_\Sigma\frac{\partial w}{\partial\nu}we^{\tau\phi}d\Sigma$$
$$-\left[\int_Q w_t w e^{\tau\phi}d\Omega\right]_0^T + \int_Q [F(w)+f]we^{\tau\phi}dQ. \tag{2.2.31}$$

We then use (2.2.31) into (2.2.30) twice and obtain

$$2\int_Q e^{\tau\phi}|\nabla w|^2 dQ - 2c\int_Q e^{\tau\phi}w_t^2 dQ = (1-c)\int_Q e^{\tau\phi}|\nabla w|^2 dQ + (1-c)\int_Q e^{\tau\phi}w_t^2 dQ$$
$$+ (1+c)\,[\text{Right-Hand Side of (2.2.31)}]. \tag{2.2.32}$$

Inserting (2.2.32) into the right-hand side of (2.2.19) produces the desired identity (2.2.26). The estimate (2.2.29) on β_{0T} follows from (2.2.28) via (2.2.22), and again, the properties (2.1.4d) of ϕ at $t=0$ and $t=T$. □

Step 4. Henceforth, we concentrate our analysis on the right-hand side (R.H.S.) of the fundamental identity (2.2.26) of Theorem 2.2.4. So far the parameter $\tau > 0$ has been arbitrary. The next lemma and its proof show the key virtue of the free parameter τ entering the present multipliers (2.2.1a–b) in dealing with a general first-order differential operator $F(w)$ as in (2.1.2): choosing τ sufficiently large permits to absorb a bad energy level term, which arises precisely because $F(w)$ is of order one.

Lemma 2.2.5. Let w be a solution of Eqn. (2.1.1) in the class (2.1.6), (2.1.7). With reference to selected terms on the right-hand side of identity (2.2.26), we have:

(i) For any $\epsilon > 0$, recalling $\nabla\phi = h(x)$ from (2.1.5a):

$$\tau\int_Q e^{\tau\phi}[\nabla\phi\cdot\nabla w - \phi_t w_t]^2 dQ \quad - \quad \int_Q [F(w)+f]e^{\tau\phi}[\nabla\phi\cdot\nabla w - \phi_t w_t]dQ$$

$$\geq \; -\frac{\epsilon}{2}\int_Q [|\nabla w|^2 + w^2 + w_t^2 + f^2]e^{\tau\phi}dQ$$

$$+ \; \left(\tau - \frac{C_T}{2\epsilon}\right)\int_Q e^{\tau\phi}[\nabla\phi\cdot\nabla w - \phi_t w_t]^2 dQ, \tag{2.2.33}$$

where C_T is the constant in (2.1.2);

(ii) for any $\epsilon > 0$ and const_τ = a constant depending on τ:

$$\int_Q ww_t\left[(1+c)\frac{d\left(e^{\tau\phi}\right)}{dt} - \frac{1}{2}\mu_t\right]dQ \; + \; \int_Q w\nabla w\cdot\left[\frac{1}{2}\nabla\mu - (1+c)\nabla\left(e^{\tau\phi}\right)\right]dQ$$

$$+ \; \int_Q [F(w)+f]w\left[(1+c)e^{\tau\phi} - \frac{1}{2}\mu\right]dQ$$

$$\geq \; -\frac{\epsilon}{2}\int_Q e^{\tau\phi}[|\nabla w|^2 + w_t^2 + f^2]dQ$$

$$- \; \frac{T\,\text{const}_\tau}{2\epsilon}\|w\|^2_{C([0,T];L_2(\Omega))}. \tag{2.2.34}$$

Proof. For both (i) and (ii) we use the inequality $2ab \leq \epsilon a^2 + \frac{1}{\epsilon}b^2$, where a denotes "energy level" terms: w_t, $|\nabla w|$, $F(w)$; while b denotes "lower-order terms" (i.e., w). Here we recall (2.1.2) for $F(w)$ and (2.2.20) for μ, as well as (2.1.5b–c). □

Remark 2.2.1. In the second integral over Q on the left-hand side of (2.2.33), both factors $F(w)$ and $[\nabla\phi\cdot\nabla w - \phi_t w_t]$ are energy level, with F a general first-order operator. The virtue of the free parameter τ is seen in the second term on the right-hand side of (2.2.33), in making the coefficient $\tau - \frac{C_T}{2\epsilon} > 0$ after ϵ has been fixed, and dropping that term, see the next result. In part (ii), Eqn. (2.2.36) and (2.2.37), of the next Theorem 2.2.6, we obtain the desired estimates (2.1.8a–b) of Theorem 2.1.1.

Theorem 2.2.6. Let w be a solution of Eqn. (2.1.1) in the class (2.1.6), (2.1.7). Then:

(i) The following inequality holds true, with $\tau = \frac{1}{\epsilon}$, $\epsilon > 0$ as in Lemma 2.2.5:

$$(BT)|_\Sigma + \frac{C_T}{\tau}\int_Q e^{\tau\phi} f^2 dQ + TC_T \text{ const}_\tau \|w\|^2_{C([0,T];L_2(\Omega))}$$

$$\geq \left(1 - c - \frac{C_T}{\tau}\right)\int_Q e^{\tau\phi}[|\nabla w|^2 + w_t^2]dQ$$

$$+\frac{\tau}{2}\int_Q e^{\tau\phi}[\nabla\phi\cdot\nabla w - \phi_t w_t]^2 dQ + \beta_{0,T}, \tag{2.2.35}$$

where C_T is the constant in (2.1.2) and const_τ = constant depending on τ.

(ii) Inequality (2.2.35) may be made more explicit in the following form: for τ sufficiently large,

$$(BT)|_\Sigma + \frac{C_T}{\tau}\int_Q e^{\tau\phi} f^2 dQ + TC_T \text{ const}_\tau \|w\|^2_{C([0,T];L_2(\Omega))}$$

$$\geq \left(1 - c - \frac{C_T}{\tau}\right)\int_Q e^{\tau\phi}[|\nabla w|^2 + w_t^2]dQ$$

$$\beta_{0,T}, \qquad -C_1 T e^{-\delta\tau}[E(T) + E(0)]. \tag{2.2.36}$$

(iii) Recalling (2.1.4a) for ϕ, estimate (2.2.36) implies: for τ sufficiently large,

$$(BT)|_\Sigma + \frac{C_T}{\tau}\int_Q e^{\tau\phi} f^2 dQ + TC_T \text{ const}_\tau \|w\|^2_{C([0,T];L_2(\Omega))}$$

$$\geq \left(1 - c - \frac{C_T}{\tau}\right) e^\tau \int_{t_0}^{t_1} E(t)dt - C_1 T e^{-\delta\tau}[E(T) + E(0)]. \tag{2.2.37}$$

Proof. (i) We use (2.2.33) and (2.2.34) on the right-hand side of the fundamental identity (2.2.26) and obtain inequality (2.2.35) at once, with τ chosen as $\tau = \frac{C_T}{\epsilon}$, $\epsilon > 0$ as in Lemma 2.2.5.

(ii), (iii). On the right-hand side of (2.2.35), we drop the last (positive) integral term [the benefit of having chosen τ large enough]; we invoke estimate (2.2.29) on $\beta_{0,T}$, and thus we get (2.2.36). Finally, we estimate the first integral term as

$$\int_Q e^{\tau\phi}[|\nabla w|^2 + w_t^2]dQ \geq \int_{t_0}^{t_1}\int_\Omega e^{\tau\phi}[|\nabla w|^2 + w_t^2]dQ$$

$$\geq e^\tau \int_{t_0}^{t_1}\int_\Omega [|\nabla w|^2 + w_t^2]d\Omega\, dt = e^\tau \int_{t_0}^{t_1} E(t)dt, \tag{2.2.38}$$

recalling property (2.1.4c) for ϕ. We thus obtain estimate (2.2.37) from (2.2.36). □

2.3 Proof of Theorem 2.1.2

We shall now refine the basic estimate (2.1.8) = (2.2.36) of Theorem 2.1.1 in terms of only $E(T)$ and $E(0)$, by examining the behavior of $E(t)$. This way we shall prove Theorem 2.1.2.

Step 1. Lemma 2.3.1. Let w be a solution of Eqn. (2.1.1) in the class (2.1.6), (2.1.7). Then, the following inequalities hold true for $T \geq t > 0$:

$$e^{-C_T t}E(0) - \Lambda(T) \leq E(t) \leq [E(0) + \Lambda(T)]e^{C_T t}, \tag{2.3.1}$$

where C_T is (essentially) the constant in (2.1.2), and we have set

$$\Lambda(T) = \int_0^T \int_\Omega f^2 d\Omega\, dt + 2\int_0^T \int_\Gamma \left|\frac{\partial w}{\partial \nu} w_t\right| d\Gamma\, dt + C_T \int_0^T \|w\|^2_{L_2(\Omega)} dt. \tag{2.3.2}$$

Proof. We multiply Eqn. (2.1.1) by w_t and integrate over $(s,t) \times \Omega$, by parts in t on the left-hand side, and by Green's first identity on the right-hand side, thus obtaining for all s, t:

$$E(t) = E(s) + 2\int_s^t \int_\Omega [F(w) + f] w_t d\Omega\, d\sigma + 2\int_s^t \int_\Gamma \frac{\partial w}{\partial \nu} w_t d\Gamma\, d\sigma. \tag{2.3.3}$$

By Schwarz inequality, recalling estimate (2.1.2) and (2.3.2), we obtain for $t \geq s \geq 0$:

$$E(t) \leq [E(s) + \Lambda(T)] + C_T \int_s^t E(\sigma) d\sigma; \tag{2.3.4}$$

$$E(s) \leq [E(t) + \Lambda(T)] + C_T \int_s^t E(\sigma) d\sigma. \tag{2.3.5}$$

We apply the classical argument of the Gronwall's inequality to (2.3.4) and (2.3.5), where we note that the terms into the square brackets are independent of t in (2.3.4) and independent of s in (2.3.5). We thus obtain for $t \geq s \geq 0$:

$$E(t) \leq [E(s) + \Lambda(T)]e^{C_T(t-s)}; \quad E(s) \leq [E(t) + \Lambda(T)]e^{C_T(t-s)}. \tag{2.3.6}$$

Setting $s = 0$ and thus $t > 0$ in (2.3.6) yields (2.3.1). □

Step 2. Theorem 2.3.2. Let w be a solution of Eqn. (2.1.1) in the class (2.1.6), (2.1.7). Then, the following inequality holds true, with τ sufficiently large:

$$(\overline{BT})|_\Sigma + \frac{C_T}{\tau}\int_Q e^{\tau\phi} f^2 dQ + \text{ const}_{\phi,\tau} \int_Q f^2 dQ + TC_T \text{ const}_\tau \|w\|^2_{C([0,T];L_2(\Omega))}$$

$$\geq \left\{\left(1 - c - \frac{C_T}{\tau}\right) k_{t_0,t_1} e^\tau - C_1 T e^{-\delta\tau}\left(1 + e^{C_T T}\right)\right\} E(0), \tag{2.3.7}$$

where the boundary terms $(\overline{BT})|_\Sigma$ are given via (2.1.9) by

$$(\overline{BT})|_\Sigma = (BT)|_\Sigma + k_{T,\tau,c,\delta}\int_\Sigma\left|\frac{\partial w}{\partial\nu}w_t\right|d\Sigma \tag{2.3.8a}$$

$$= \int_\Sigma e^{\tau\phi}\frac{\partial w}{\partial\nu}[h\cdot\nabla w-\phi_t w_t]d\Sigma+\int_\Sigma\frac{\partial w}{\partial\nu}w\left[\frac{1}{2}\mu-(1+c)e^{\tau\phi}\right]d\Sigma$$
$$+\frac{1}{2}\int_\Sigma e^{\tau\phi}[w_t^2-|\nabla w|^2]h\cdot\nu\,d\Sigma+k_{T,\tau,c,\delta}\int_\Sigma\left|\frac{\partial w}{\partial\nu}w_t\right|d\Sigma. \tag{2.3.8b}$$

Thus, with $T>0$ and $0<c<1$ fixed by the definition of ϕ in (2.1.4), taking τ sufficiently large makes the coefficient $\{\quad\}$ in front of $E(0)$ positive in (2.3.7).

Proof. Using the left-hand side inequality in (2.3.1), we compute by (2.3.2),

$$\int_{t_0}^{t_1}E(t)dt\geq\int_{t_0}^{t_1}\left[e^{-C_Tt}E(0)-\Lambda(T)\right]dt=k_{t_0,t_1}E(0)-(t_1-t_0)\left\{\int_0^T\int_\Omega f^2dQ\right.$$
$$\left.+2\int_0^T\int_\Gamma\left|\frac{\partial w}{\partial\nu}w_t\right|d\Sigma+C_T\int_0^T\|w\|^2_{L_2(\Omega)}dt\right\}, \tag{2.3.9}$$

with $k_{t_0,t_1}=\int_{t_0}^{t_1}e^{-C_Tt}dt$. Moreover, using the right-hand side inequality in (2.3.1),

$$E(T)+E(0)\leq\left[1+e^{C_TT}\right]E(0)+e^{C_TT}\left\{\int_0^T\int_\Omega f^2dQ+2\int_0^T\int_\Gamma\left|\frac{\partial w}{\partial\nu}w_t\right|d\Sigma\right.$$
$$\left.+C_T\int_0^T\|w\|^2_{L_2(\Omega)}dt\right\}, \tag{2.3.10}$$

by (2.3.2). Inserting (2.3.9) and (2.3.10) on the right-hand side of inequality (2.2.37) = (2.1.8) yields (2.3.7), (2.3.8), as $\left(1-c-\frac{C_T}{\tau}\right)>0$. □

Estimate (2.3.7) of Theorem 2.3.2 proves Theorem 2.1.2, Eqn. (2.1.10a). Using $E(0)\geq e^{-C_TT}E(T)-\Lambda(T)$ from (2.3.1) once more, yields

$$E(0)=\frac{E(0)}{2}+\frac{E(0)}{2}\geq\frac{e^{-C_TT}}{2}[E(0)+E(T)]-\frac{\Lambda(T)}{2}, \tag{2.3.11}$$

which inserted on the right-hand side of (2.3.7) produces estimate (2.1.10b) in Theorem 2.1.2. Thus, Theorem 2.1.2, part (i), is fully proved. □

Corollary 2.3.3. Let w be a solution of Eqn. (2.1.1) in the class (2.1.6), (2.1.7) which, moreover, satisfies the boundary condition:

$$w|_{\Sigma_0}\equiv 0,\ \text{where }\Gamma_0=\{x\in\Gamma:\ \nabla\phi\cdot\nu\leq 0\}\text{ as in (2.1.13).} \tag{2.3.12}$$

Then, the following inequality holds true for τ sufficiently large: there is a positive constant $k_{\phi,\tau}>0$, such that, if $(\overline{BT})|_{\Sigma_1}$ are the boundary terms $(\overline{BT})$ in (2.3.8b) evaluated, however, on $\Sigma_1=[0,T]\times$

Γ_1, then

$$(\overline{BT})|_{\Sigma_1} + \frac{C_T}{\tau}\int_Q e^{\tau\phi} f^2 dQ + \text{const}_{T,\tau,c,\delta}\int_Q f^2 dQ$$
$$+ TC_T \text{ const}_\tau \|w\|^2_{C([0,T];L_2(\Omega))} \geq k_{\phi,\tau} E(0). \tag{2.3.13}$$

Proof. We split the boundary terms $(\overline{BT})$ in (2.3.8b) on Σ_0 and on Σ_1. On Σ_0, the boundary condition (2.3.12) makes all terms vanish, except two of them. Moreover, we have on Σ_0 : $h \cdot \nabla w = \frac{\partial w}{\partial \nu} h \cdot \nu$; $|\nabla w| = \left|\frac{\partial w}{\partial \nu}\right|$, where $h = \nabla \phi$ by (2.1.5a), and thus

$$\begin{aligned}(\overline{BT})|_\Sigma &= (\overline{BT})|_{\Sigma_0} + (\overline{BT})|_{\Sigma_1} \\ &= \int_{\Sigma_0} e^{\tau\phi}\frac{\partial w}{\partial \nu} h \cdot \nabla w \, d\Sigma_0 - \frac{1}{2}\int_{\Sigma_0} e^{\tau\phi}|\nabla w|^2 h \cdot \nu \, d\Sigma_0 + (\overline{BT})|_{\Sigma_1} \\ &= \frac{1}{2}\int_{\Sigma_0} e^{\tau\phi}\left(\frac{\partial w}{\partial \nu}\right)^2 h \cdot \nu \, d\Sigma_0 + (\overline{BT})|_{\Sigma_1} \qquad (2.3.14) \\ &\leq (\overline{BT})|_{\Sigma_1}, \qquad (2.3.15)\end{aligned}$$

recalling $h \cdot \nu \leq 0$ on Γ_0. Inserting (2.3.15) into the left-hand side of (2.1.10a) = (2.3.7) yields (2.3.13), as desired. □

Theorem 2.3.2 and Corollary 2.3.3 prove part (i) and part (ii), respectively, of Theorem 2.1.2.

2.4 Proof of Theorem 2.1.3: Continuous Observability Inequality

Step 1. Proposition 2.3.3. For the solution of the problem

$$\begin{cases} \psi_{tt} = \Delta\psi + F(\psi) & \text{in } (0,T] \times \Omega; \qquad (2.4.1a) \\ \psi(0, \cdot) = \psi_0, \ \psi_t(0, \cdot) = \psi_1 & \text{in } \Omega; \qquad (2.4.1b) \\ \psi|_\Sigma \equiv 0 & \text{in } (0,T] \times \Gamma, \qquad (2.4.1c) \end{cases}$$

where F is the first-order differential operator as in (2.1.2), and where $T > 0$ is the constant that enters the definition of ϕ in (2.1.4), the following estimate holds true: for $\tau > 0$ sufficiently large, there exists a positive constant $C_{\phi,\tau} > 0$ such

$$\int_0^T \int_{\Gamma_1} \left(\frac{\partial \psi}{\partial \nu}\right)^2 d\Gamma_1 dt + TC_T \text{ const}_\tau \|\psi\|^2_{C([0,T];L_2(\Omega))}$$
$$\geq C_{\phi,\tau} \|\{\psi_0, \psi_1\}\|^2_{H_0^1(\Omega) \times L_2(\Omega)}. \tag{2.4.2}$$

Proof. As in the proof of Corollary 2.3.3, Eqn. (2.3.14), we obtain with w there replaced by ψ now,

$$\begin{aligned}(\overline{BT})|_{\Sigma} &= \int_{\Sigma} e^{\tau\phi}\frac{\partial\psi}{\partial\nu}h\cdot\nabla\psi\, d\Sigma - \frac{1}{2}\int_{\Sigma} e^{\tau\phi}|\nabla\psi|^2 h\cdot\nu\, d\Sigma \\ &= \frac{1}{2}\int_{\Sigma} e^{\tau\phi}\left(\frac{\partial\psi}{\partial\nu}\right)^2 h\cdot\nu\, d\Sigma \leq C_{\phi,\tau}\int_0^T\int_{\Gamma_1}\left(\frac{\partial\psi}{\partial\nu}\right)^2 d\Gamma_1 dt, \end{aligned} \tag{2.4.3}$$

recalling, in the last step, the definition of the part of the boundary Γ_0 in (2.1.13) = (2.3.12). Using (2.4.3) on the left-hand side of inequality (2.3.7) yields (2.4.2), as desired. □

Step 2. (Absorption of the lower-order term)

Lemma 2.3.4. Let ψ be a solution of problem (2.3.12) with $\{\psi_0,\psi_1\} \in H_0^1(\Omega)\times L_2(\Omega)$, so that inequality (2.4.2) holds true. Let the homogeneous, over-determined problem defined by (2.1.15a–c), as well as $\frac{\partial\psi}{\partial\nu}|_{\Sigma_1} \equiv 0$ on $(0,T]\times\Gamma_1 \equiv \Sigma_1$ admit only the unique solution $\psi \equiv 0$. Then, there exists a constant $k_T > 0$ such that

$$\|\psi\|^2_{C([0,T];L_2(\Omega))} \leq k_T \int_0^T\int_{\Gamma_1}\left(\frac{\partial\psi}{\partial\nu}\right)^2 d\Gamma_1 dt. \tag{2.4.4}$$

Proof. This follows by a compactness/uniqueness argument which is now standard [L.2], [K.1], [L-T.4]. □

3 Proof of Theorem 1.1

We shall heavily rely on the results proved in the preceding Section 2 for the single Eqn. (2.1.1).

Step 1. Proposition 3.1 (Carleman estimates). Let w, z be solutions of Eqns. (1.1), (1.2), in the class (1.6), (1.7). Let $\phi(x,t)$ be the pseudo-convex function defined by (2.1.4). Then, for $\tau > 0$ sufficiently large, the following one-parameter family of estimates holds true:

$$\begin{aligned}&\left(1-c-\frac{2C_T}{\tau}\right)\int_Q e^{\tau\phi}\left[|\nabla w|^2 + w_t^2 + |\nabla z|^2 + z_t^2\right] dQ - C_1 T e^{-\delta\tau}[E(T)+E(0)] \\ &\quad\leq (BT(w))|_{\Sigma} + (BT(z))|_{\Sigma} + TC_T\ \text{const}_\tau\left\{\|w\|^2_{C([0,T];L_2(\Omega))} + \|z\|^2_{C([0,T];L_2(\Omega))}\right\}, \end{aligned} \tag{3.1}$$

where $E(t)$ is defined by (1.9) and $(BT(w))|_{\Sigma}$ and $(BT(z))|_{\Sigma}$ are the boundary terms over Σ defined as in (2.1.9) [or (2.2.27)] for the variable w and z, respectively [Eqn. (3.1) is the counterpart for system (1.1), (1.2) of Eqn. (2.1.8a) for the single equation (2.1.1).

Proof. We apply the Carleman estimates (2.1.8a) of Theorem 2.1.1 to the w-equation (1.1) with $f = P_1(z)$ and, respectively, to the z-equation (1.2) with $f = P_2(w)$. We thus obtain, respectively,

$$\left(1 - c - \frac{C_T}{\tau}\right) \int_Q e^{\tau\phi} \left[|\nabla w|^2 + w_t^2\right] dQ - C_1 T e^{-\delta\tau}[E_w(T) + E_w(0)]$$

$$\leq \frac{C_T}{\tau} \int_Q e^{\tau\phi} |P_1(z)|^2 dQ + (BT(w))|_\Sigma + TC_T \text{ const}_\tau \|w\|^2_{C([0,T];L_2(\Omega))}; \tag{3.2}$$

$$\left(1 - c - \frac{C_T}{\tau}\right) \int_Q e^{\tau\phi} \left[|\nabla z|^2 + z_t^2\right] dQ - C_1 T e^{-\delta\tau}[E_z(T) + E_z(0)]$$

$$\leq \frac{C_T}{\tau} \int_Q e^{\tau\phi} |P_2(w)|^2 dQ + (BT(z))|_\Sigma + TC_T \text{ const}_\tau \|z\|^2_{C([0,T];L_2(\Omega))}, \tag{3.3}$$

for τ sufficiently large. We next recall the pointwise bounds (1.4) and (1.3) for $P_1(z)$ and $P_2(w)$ in, respectively, (3.2) and (3.3), to obtain

$$\left(1 - c - \frac{C_T}{\tau}\right) \int_Q e^{\tau\phi} \left[|\nabla w|^2 + w_t^2\right] dQ - C_1 T e^{-\delta\tau}[E_w(T) + E_w(0)]$$

$$\leq \frac{C_T}{\tau} \int_Q e^{\tau\phi} [|\nabla z|^2 + z_t^2] dQ + (BT(w))|_\Sigma$$

$$+ \; TC_T \text{ const}_\tau \left\{ \|w\|^2_{C([0,T];L_2(\Omega))} + \|z\|^2_{C([0,T];L_2(\Omega))} \right\}; \tag{3.4}$$

$$\left(1 - c - \frac{C_T}{\tau}\right) \int_Q e^{\tau\phi} \left[|\nabla z|^2 + z_t^2\right] dQ - C_1 T e^{-\delta\tau}[E_z(T) + E_z(0)]$$

$$\leq \frac{C_T}{\tau} \int_Q e^{\tau\phi} [|\nabla w|^2 + w_t^2] dQ + (BT(z))|_\Sigma$$

$$+ \; TC_T \text{ const}_\tau \left\{ \|z\|^2_{C([0,T];L_2(\Omega))} + \|w\|^2_{C([0,T];L_2(\Omega))} \right\}. \tag{3.5}$$

Summing up (3.4) and (3.5) and recalling (1.9) yields (3.1), as desired. □

Step 2. Proposition 3.2. Let w, z be solutions of Eqns. (1.1), (1.2), in the class (1.6), (1.7). Then, the following inequalities hold true for $T \geq t > 0$:

$$e^{-C_T t} E(0) - \bar{\Lambda}(T) \leq E(t) \leq [E(0) + \bar{\Lambda}(T)] e^{C_T t}, \tag{3.6}$$

where $E(t)$ is defined by (1.9) and now

$$\bar{\Lambda}(T) = C_T \int_0^T \int_\Omega [w^2 + z^2] d\Omega \, dt + 2 \int_0^T \int_\Gamma \left| \frac{\partial w}{\partial \nu} w_t + \frac{\partial z}{\partial \nu} z_t \right| d\Gamma \, dt. \tag{3.7}$$

[This result is the counterpart for system (1.1), (1.2) of Eqn. (2.3.1) of Lemma 2.3.1 for the single equation (2.1.1).]

Proof. We return to identity (2.3.3), which we write for the w-equation (1.1) with $f = P_1(z)$ and, respectively, for the z-equation (1.2) with $f = P_2(w)$. We obtain, respectively, for all s, t:

$$E_w(t) = E_w(s) + 2\int_s^t \int_\Omega [F_1(w) + P_1(z)]w_t d\Omega\, d\sigma + 2\int_s^t \int_\Gamma \frac{\partial w}{\partial \nu} w_t d\Gamma\, d\sigma; \tag{3.8}$$

$$E_z(t) = E_z(s) + 2\int_s^t \int_\Omega [F_2(z) + P_2(w)]z_t d\Omega\, d\sigma + 2\int_s^t \int_\Gamma \frac{\partial z}{\partial \nu} z_t d\Gamma\, d\sigma. \tag{3.9}$$

Summing up (3.8) and (3.9) yields by (1.9),

$$E(t) = E(s) + 2\int_s^t \int_\Omega \{[F_1(w) + P_1(z)]w_t + [F_2(z) + P_2(w)]z_t\}\, d\Omega\, d\sigma$$
$$+2\int_s^t \int_\Gamma \left[\frac{\partial w}{\partial \nu} w_t + \frac{\partial z}{\partial \nu} z_t\right] d\Gamma\, d\sigma. \tag{3.10}$$

Using Schwarz inequality in the integral over Ω and recalling the bounds (1.3), (1.4) yields $t \geq s \geq 0$,

$$E(t) \leq [E(s) + \bar{\Lambda}(T)] + C_T \int_s^t E(\sigma)d\sigma \tag{3.11}$$

$$E(s) \leq [E(t) + \bar{\Lambda}(T)] + C_T \int_s^t E(\sigma)d\sigma, \tag{3.12}$$

which are the counterpart of Eqns. (2.3.4), (2.3.5), where now, however, $\bar{\Lambda}(T)$ is defined by (3.7). From here on the proof proceeds as in Lemma 2.3.1 by the classical argument of the Gronwall's inequality and yields (3.6), as desired. □

Step 3. Proposition 3.3. Let w, z be solutions of Eqns. (1.1), (1.2), in the class (1.6), (1.7). Then, the following inequality holds true, with τ sufficiently large:

$$\left\{\left(1 - c - \frac{2C_T}{\tau}\right) k_{t_0,t_1} e^{\tau} - C_1 T e^{-\delta\tau}\left(1 + e^{C_T T}\right)\right\} E(0)$$
$$\leq (\overline{BT}(w,z))|_\Sigma + \text{const}_{T,\tau}\left\{\|w\|^2_{C([0,T];L_2(\Omega))} + \|z\|^2_{C([0,T];L_2(\Omega))}\right\}, \tag{3.13}$$

where taking τ sufficiently large makes the coefficient { } in front of $E(0)$ positive in (3.13), and where we have set

$$(\overline{BT}(w,z))|_\Sigma = (BT(w))|_\Sigma + (BT(z))|_\Sigma + k_{T,\tau,c,\delta}\int_\Sigma \left|\frac{\partial w}{\partial \nu} w_t + \frac{\partial z}{\partial \nu} z_t\right| d\Sigma. \tag{3.14}$$

[Estimate (3.13) is the counterpart for system (1.1), (1.2) of Eqn. (2.3.7) of Theorem 2.3.2 for the single equation (2.1.1).]

Proof. We proceed as in the proof of Theorem 2.3.2. Using the left-hand side inequality in (3.6), we compute by (3.7),

$$\int_{t_0}^{t_1} E(t)dt \geq \int_{t_0}^{t_1} \left[e^{-C_T t}E(0) - \bar{\Lambda}(T)\right] dt$$

$$= k_{t_0,t_1} E(0) - (t_1 - t_0)\left[C_T \int_0^T \int_\Omega [w^2 + z^2]\right] d\Omega\, dt$$

$$+2\int_0^T \int_\Gamma \left|\frac{\partial w}{\partial \nu} w_t + \frac{\partial z}{\partial \nu} z_t\right| d\Gamma\, d\sigma. \tag{3.15}$$

Moreover, recalling property (2.1.4c) for ϕ and (1.9), we estimate with reference to (3.1)

$$\int_Q e^{\tau\phi}\left[|\nabla w|^2 + w_t^2 + |\nabla z|^2 + z_t^2 dQ > e^\tau \int_{t_0}^{t_1} E(t)dt. \tag{3.16}$$

Using now the right-hand side inequality in (3.6), we estimate as in (2.3.10):

$$E(T) + E(0) \leq \left[1 + e^{C_T T}\right] E(0) + e^{C_T T}\Big\{C_T \int_0^T \int_\Omega [w^2 + z^2] d\Omega\, d\sigma$$

$$+2\int_0^T \int_\Gamma \left|\frac{\partial w}{\partial \nu} w_t + \frac{\partial z}{\partial \nu} z_t\right| d\Gamma\, dt. \tag{3.17}$$

We now use (3.15) into (3.16) and substitute the result, along with (3.17), into the left-hand side of (3.1). This way we obtain (3.13). □

Step 4. Proposition 3.4. Let w, z be solutions of Eqns. (1.1), (1.2), in the class (1.6), (1.7). Then

(i) Given $\epsilon > 0$ and $\epsilon_0 > 0$ arbitrarily small, and given $T > 0$, there exists a constant $C_{\epsilon,\epsilon_0,T} > 0$ such that

$$\int_\epsilon^{T-\epsilon} \int_\Gamma \left[\left(\frac{\partial w}{\partial s}\right)^2 + \left(\frac{\partial z}{\partial s}\right)^2\right] d\Gamma\, dt \leq C_{\epsilon,\epsilon_0,T}\Big\{\int_0^T \int_\Gamma \left[\left(\frac{\partial w}{\partial \nu}\right)^2 + w_t^2 + \left(\frac{\partial z}{\partial \nu}\right)^2 + z_t^2\right] d\Gamma\, dt$$

$$+\|w\|^2_{H^{\frac{1}{2}+\epsilon_0}(Q)} + \|z\|^2_{H^{\frac{1}{2}+\epsilon_0}(Q)}\Big\}; \tag{3.18}$$

(ii) if, moreover, w and/or z satisfy the boundary condition (1.10) on Σ_0, then the corresponding integral term for w and/or for z replaces Γ with Γ_1.

[This result is the counterpart for system (1.1), (1.2) of Eqn. (2.1.17) of Theorem 1.1.4 for the single equation (2.1.1).]

Proof. (i) We apply (2.1.17) of Theorem 1.1.4 to the w-equation (1.1) with $f = P_1(z)$ and, respectively, to the z-equation (1.2) with $f = P_2(w)$. We obtain:

$$\int_{\epsilon}^{T-\epsilon}\int_{\Gamma}\left(\frac{\partial w}{\partial s}\right)^2 d\Gamma\, dt \;\le\; C_{\epsilon,\epsilon_0,T}\left\{\int_0^T\int_{\Gamma}\left[\left(\frac{\partial w}{\partial \nu}\right)^2 + w_t^2\right] d\Sigma\right.$$

$$\left. + \|w\|^2_{L_2(0,T;H^{\frac{1}{2}+\epsilon_0}(\Omega))} + \|P_1(z)\|^2_{H^{-\frac{1}{2}+\epsilon_0}(Q)}\right\}; \qquad (3.19)$$

$$\int_{\epsilon}^{T-\epsilon}\int_{\Gamma}\left(\frac{\partial z}{\partial s}\right)^2 d\Gamma\, dt \;\le\; C_{\epsilon,\epsilon_0,T}\left\{\int_0^T\int_{\Gamma}\left[\left(\frac{\partial z}{\partial \nu}\right)^2 + z_t^2\right] d\Sigma\right.$$

$$\left. + \|z\|^2_{L_2(0,T;H^{\frac{1}{2}+\epsilon_0}(\Omega))} + \|P_2(w)\|^2_{H^{-\frac{1}{2}+\epsilon_0}(Q)}\right\}. \qquad (3.20)$$

Next we sum up (3.19) and (3.20) and use

$$\|P_1(z)\|_{H^{-\frac{1}{2}+\epsilon_0}(Q)} \le C\|z\|_{H^{\frac{1}{2}+\epsilon_0}(Q)}; \quad \|P_2(w)\|_{H^{-\frac{1}{2}+\epsilon_0}(Q)} \le C\|w\|_{H^{\frac{1}{2}+\epsilon_0}(Q)}, \qquad (3.21)$$

since P_1 and P_2 are first-order differential operators in all variables. This way we obtain (3.18).
The proof of part (ii) is the same, recalling part (ii) of Theorem 2.1.4. □

We can now complete the proof of Theorem 1.1.

Step 5. Proposition 3.5. Let w, z be solutions of (1.1), (1.2) in the class (1.6), (1.7). Then the following inequality holds true for τ sufficiently large:

(i) there exists a positive constant $k_{\phi,\tau} > 0$ such that

$$k_{\phi,\tau}E(0) \;\le\; \int_0^T\int_{\Gamma}\left[\left(\frac{\partial w}{\partial \nu}\right)^2 + w_t^2 + \left(\frac{\partial z}{\partial \nu}\right)^2 + z_t^2\right] d\Gamma\, dt$$

$$+\text{const}_{T,\tau,\epsilon_0}\left\{\|w\|^2_{H^{\frac{1}{2}+\epsilon_0}(Q)} + \|z\|^2_{H^{\frac{1}{2}+\epsilon_0}(Q)}\right\}, \qquad (3.22a)$$

or equivalently,

$$k_{\phi,\tau}[E(0)+E(T)] \;\le\; \int_0^T\int_{\Gamma}\left[\left(\frac{\partial w}{\partial \nu}\right)^2 + w_t^2 + \left(\frac{\partial z}{\partial \nu}\right)^2 + z_t^2\right] d\Gamma\, dt$$

$$+\text{const}_{T,\tau,\epsilon_0}\left\{\|w\|^2_{H^{\frac{1}{2}+\epsilon_0}(Q)} + \|z\|^2_{H^{\frac{1}{2}+\epsilon_0}(Q)}\right\}, \qquad (3.22b)$$

(ii) If, moreover, w and/or z satisfy the boundary condition (1.10) on Σ_0, then the corresponding integral term for w and/or for z replaces Γ with Γ_1.

[This result is the counterpart for system (1.1), (1.2) of Eqn. (2.1.18) of Theorem 2.1.5 [= Eqn. (2.6.2) of Theorem 2.6.1].

Proof. (i) We proceed as in the proof of Theorem 2.6.1. For fixed $\epsilon > 0$ small we apply estimate (3.13) of Proposition 3.3 over the interval $[\epsilon, T-\epsilon]$, rather than over $[0,T]$.

We obtain, with $k_{\phi,T,\epsilon} > 0$,

$$\begin{aligned} k_{\phi,T,\epsilon}E(\epsilon) &\leq \overline{BT}(w,z)|_{[\epsilon,T-\epsilon]\times\Gamma} \\ &\quad +\text{const}_{T,T,\epsilon}\left\{\|w\|^2_{C([0,T];L_2(\Omega))} + \|z\|^2_{C([0,T]L_2(\Omega))}\right\}. \end{aligned} \tag{3.23}$$

Using the left-hand side inequality in (3.6) with $t = \epsilon$, we have by (3.7),

$$E(\epsilon) \geq e^{-C_T\epsilon}E(0) - \left[C_T\int_0^T\int_\Omega [w^2+z^2]d\Omega\, dt + 2\int_0^T\int_\Gamma\left|\frac{\partial w}{\partial\nu}w_t + \frac{\partial z}{\partial\nu}z_t\right| d\Gamma\, dt\right]. \tag{3.24}$$

We next insert (3.24) into the left-hand side of (3.23), thus obtaining

$$\begin{aligned} k_{\phi,T,\epsilon}e^{-C_T\epsilon}E(0) &\leq \overline{BT}(w,z)|_{[\epsilon,T-\epsilon]\times\Gamma} \\ &\quad +\text{const}_{T,T,\epsilon}\left\{\|w\|^2_{C([0,T];L_2(\Omega))} + \|z\|^2_{C([0,T]L_2(\Omega))}\right\} \\ &\quad +2\int_0^T\int_\Gamma\left|\frac{\partial w}{\partial\nu}w_t + \frac{\partial z}{\partial\nu}z_t\right| d\Gamma\, dt. \end{aligned} \tag{3.25}$$

But, by virtue of estimate (3.18) of Proposition 3.4, we readily see via the definition of $\overline{BT}(w,z)$ over the domain $[\epsilon, T-\epsilon]\times\Gamma$ in (3.14), and the definition of $BT(w)$ or $BT(z)$ in (2.1.9) except on $[\epsilon, T-\epsilon]\times\Gamma$, that

$$\begin{aligned} &\overline{BT}(w,z)|_{[\epsilon,T-\epsilon]\times\Gamma} + 2\int_0^T\int_\Gamma\left|\frac{\partial w}{\partial\nu}w_t + \frac{\partial z}{\partial\nu}z_t\right| d\Gamma\, dt \\ &\leq C_{\epsilon,\epsilon_0,T}\left\{\int_0^T\int_\Gamma\left[\left(\frac{\partial w}{\partial\nu}\right)^2 + w_t^2 + \left(\frac{\partial z}{\partial\nu}\right)^2 + z_t^2\right] d\Gamma\, dt + \|w\|^2_{H^{\frac{1}{2}+\epsilon_0}(Q)} + \|z\|^2_{H^{\frac{1}{2}+\epsilon_0}(Q)}\right\}. \end{aligned} \tag{3.26}$$

Inserting (3.26) into the right-hand side of (3.25) yields (3.22a), as desired.

To prove (3.22b), we write as in (2.3.11),

$$E(0) = \frac{E(0)}{2} + \frac{E(0)}{2} \geq \frac{e^{-C_TT}}{2}[E(0)+E(T)] - \frac{\bar{\Lambda}(T)}{2}, \tag{3.27}$$

recalling $E(0) \geq e^{-C_TT}E(T) - \bar{\Lambda}(T)$ from the right-hand side of inequality (3.6). We then insert (3.27) into (3.22a) already proved, and we then establish (3.22b) via (3.7) for $\bar{\Lambda}(T)$. Part (i) is proved.

For part (ii), we use part (ii) of Theorem 2.1.4 instead. □

References

[B-L-R.] C. Bardos, J. Lebeau, and J. Rauch, Sharp sufficient conditions for the observation, control, stabilization of the wave equation from the boundary, *SIAM J. Control & Optim.* **30** (1992), 1024–1065.

[C.1] G. Chen, Energy decay estimates and exact controllability for the wave equation in a bounded domain, *J. Math. Pures et Appliques* (9) **58** (1979), 249–274.

[L-L-T.1] I. Lasiecka, J. L. Lions, and R. Triggiani, Non homogeneous boundary value problems for second order hyperbolic operators, *J. de Math. Pures et Appl.* **65** (1986), 149–192.

[L-T.1] I. Lasiecka and R. Triggiani, Regularity of hyperbolic equations under $L_2(0,T;L_2(\Gamma))$-Dirichlet boundary terms, *App. Math. & Optim.* **10** (1983), 275–286.

[L-T.2] I. Lasiecka and R. Triggiani, Exponential uniform energy decay rates of the wave equation in a bounded region with $L_2(0,\infty;L_2(\Gamma))$-boundary feedback in the Dirichlet B.C., *J. Diff. Eqns.* **66** (1987), 340–390.

[L-T.3] I. Lasiecka and R. Triggiani, Uniform stabilization of the wave equation with Dirichlet or Neumann feedback control without geometrical conditions, *Appl. Math. & Optim.* **25** (1992), 189–224.

[L-T.4] I. Lasiecka and R. Triggiani, Exact boundary controllability for the wave equation with Neumann boundary control, *Appl. Math. & Optim.* **19** (1989), 243–290. (Also, preliminary version in Springer-Verlag Lecture Notes 100 (1987), 316–371.)

[Li.1] J. L. Lions, *Control of Singular Distributed Systems,* Gauthier-Villar, Paris (1983).

[Li.2] J. L. Lions, *Exact Controllability, Stabilization and Perturbations*, Masson, Paris (1988).

[K.1] V. Komornik, *Exact Controllability and Stabilization. The Multiplier Method,* Masson (1995).

[Ru.1] D. L. Russell, Exact boundary controllability theorems for wave and heat processes in star complemented regions, in *Differential Games and Control Theory,* Roxin-Lin-Sternberg, eds., Marcel Dekker, New York (1974), 291–320.

[Ta.1] D. Tataru, A-priori pseudo-convexity energy estimates in domains with boundary and exact boundary controllability for conservative P.D.E.'s, Ph.D. Thesis, University of Virginia, May 1992.

[Ta.2] D. Tataru, Boundary controllability for conservative P.D.E.'s, *Appl. Math. & Optim.*, **31** (1995), 257–297.

[Ta.3] D. Tataru, A-priori estimates of Carleman type in domains with boundary, *J. Math. Pure et Appl.* **73** (1994), 355–387.

Static and Dynamic Behavior of a Fluid–Shell System

Jean-Paul Marmorat CMA-EMP, Sophia Antipolis, France

Jean-Paul Zolésio Institut Non Linéare de Nice, CNRS, Sophia Antipolis, France, and Centre de Mathematiques Appliquées, INRIA-ENSMP, Sophia Antipolis, France

JEAN-PAUL MARMORAT AND JEAN-PAUL ZOLÉSIO

ABSTRACT. We model the physical system formed by an elastic shell and an incompressible fluid. We first study the quasi-static evolution of the shape of the shell when the volume of the fluid varies slowly and then its small vibrations for a given fixed volume.

1. INTRODUCTION

The presence of liquids in containers on board of spacecraft and their motion give rise to a number of dynamical problems which must be taken into account by the designer to minimize or remove undesirable perturbations which could affect attitude pointing. The problem of sloshing liquids is becoming more important in spacecrafts as larger amounts of fluids (e.g. add period propellant) are being used.

This work is devoted to the modeling of the quasi-steady evolution of a tank with membrane used in space systems. In the absence of a gravity field and in order to keep the liquid located in a given part of the tank, using an elastic membrane make it possible to preserve a constant pressure in the fluid. That pressure is not provided by the tension of the membrane but by a gas which is located on the other side of the membrane. At rest, the surface of the membrane is large enough to contain the fluid when the tank is full.

We make some assumptions. The fluid is taken out of the tank at a very low rate, so that at any time t we consider its volume $v(t)$ to be a quasi-constant. With this assumption we may try to solve separately two problems: a static one and a dynamical one. In the static problem we neglect the dynamics of the membrane and of the fluid and we determine the shape of the membrane at a given constant volume v by minimizing a non-linear expression of the total energy of the system. In the dynamical problem we describe the small oscillations of the coupled membrane-fluid system around its equilibrium shape. The displacements are supposed small and the dynamical equations are linear. The last assumption concerns the geometry: for the sake of simplicity we assume the whole system to be axisymmetric and we only model the axisymmetric displacements and oscillations.

We denote by D the tank that is supposed to be rigid. D_0 is the "horizontal equatorial" disk. Its boundary ∂D_0 is a circle of radius R. S is the membrane attached on ∂D_0 and contained in D. It splits the open set D into two open subsets Ω and Ω^C, $D = \Omega \cup \Omega^c \cup S$. Ω is the domain occupied by the fluid with boundary $S \cup \Sigma$ and Ω^C is the domain occupied by the gas which is maintained at constant pressure p_0.

The equations will be written in an orthonormal axis system attached to the center of D_0. The z axis is a symmetry axis of the tank. The plane $x0y$ contains D_0.

We describe the Quasi-Steady Problem: The volume v of fluid, i.e the volume of Ω will be the parameter governing the quasi-static evolution of the tank. We suppose that v varies very slowly, for example v decreases from v_{max} to v_{min}. At each value v, we suppose that the membrane S is at equilibrium. This membrane is subject to large displacements and we know the initial configuration (S_0, Ω_0) and the current configuration (S, Ω).

We shall solve a succession of small displacements for this membrane with a membrane energy that will integrate the history of the strain and the memory of the initial configuration. For simplicity, we consider $v = v_{max} - t$ for time t increasing from 0 to t_{max}. So we shall speak of the time t which governs that quasi-static evolution. The problem consists in minimizing, for each time t, the energy of the steady system.

That energy is the sum of three terms: the potential energy due to the gravity g which is in the opposite direction of Oz; the tangential membrane energy; the stretch energy that we choose, at each time, as a norm of the main curvature of surfaces S_t and S_0, estimated in the initial location S_0. We make the hypothesis that at each time, S_t is the graph of a mapping $g_t(.)$ with g_t null on D_0 and verifying the admissible constraints. In this paper we consider a fuel tank which is axisymmetric so that the initial position of the membrane being also axisymmetric the evolution process will remain axisymmetric and we shall proceed with the specification of the resulting 2-dimensional problem.

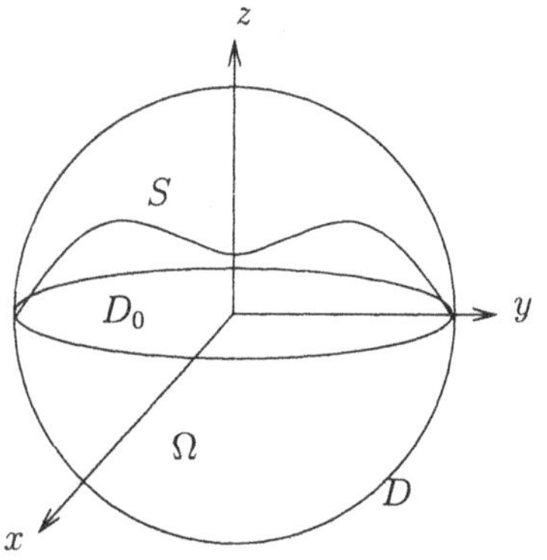

FIGURE 1. Fluid-membrane tank

2. Mathematical Modeling

We consider that during the process the membrane remains an axisymmetric graph . For each volume v occupied by the fluid the surface occupied by the middle membrane is of the following form:

$$S(v) = \{ \, (x, \, z(v)(x)) \mid 0 \leq x \leq 1 \, \} \tag{2.1}$$

We assume the membrane to be fairly inextensible (in practice, the coefficient k_s in the stretch energy E_s defined below will be large enough). Under these conditions the length of any parallel circle remains approximately constant, which means that a material point M_0 with abscissa x_0 on S_0 keeps the same abscissa when the volume v decreases from v_0 (maximal volume corresponding to a full tank and initial position z_0 of the membrane) to v the current volume. In other words the trajectory of any material point of the surface for different values of v is a vertical line

$$M(v) = (\, x, \, z(v)(x) \,) \text{ with } M(v_0) = (\, x, \, z(v_0)(x) \,)$$

We assume that if h is the membrane thickness, the two principal curvatures C_1 and C_2 of the surface at any point remain small compared to h^{-1}. With this assumption the large displacements of the membrane cause small deformations. This approximation is correct at first order (in the small parameter $h||D^2 b_\Omega||_{S(v)}$ where Ω is the set occupied by the fluid and b_Ω is the oriented distance function). The energy to be minimized should take into account the changes in the principal curvatures at each point of the membrane $S(v)$ and also the local variation of the surface element. These two points respectively account for the following energy terms E_c and E_s . In this paper we shall also focus our attention on the existence question for the classically associated minimization problem: the set Z_{ad} being defined as

(2.2) $$Z_{ad} = \{z \in H^1_{z_0}(0,1), \frac{z''}{(1+(z')^2)^{\frac{3}{2}}} \in L^2(0,1), z(1)=0\}$$

we shall consider for each given value v of the volume the subset

$$Z^v_{ad} = \{ \ z \in Z_{ad} \ s.t. \ \int_0^1 z(x)x dx = v \ \}$$

And the variational problem

(2.3) $$Min \ \{ \ E(z) \mid z \in Z^v_{ad} \ \}$$

The admissible family of graphs will take into account the different possible constraints usually encountered: smoothness of the surface considered, given volume, upper bound on some curvature of the boundary. The energy is the sum of three terms : the strain energy related to the curvatures of the surface

(2.4) $$E_c(z) = k_c \int_0^1 \{[\frac{z''}{(1+(z')^2)^{\frac{3}{2}}} - \frac{z_0''}{(1+(z_0')^2)^{\frac{3}{2}}}]^2 + [\frac{z'}{x(1+(z')^2)^{\frac{1}{2}}} - \frac{z_0'}{x(1+(z_0')^2)^{\frac{1}{2}}}]^2\} x(1+(z_0')^2)^{\frac{1}{2}} dx$$

The stretch energy

(2.5) $$E_s(z) = k_s \int_0^1 x[(1+(z')^2)^{\frac{1}{2}} - (1+(z_0')^2)^{\frac{1}{2}}]^2 dx$$

And the potential energy

(2.6) $$E_p(z) = k_p \int_0^1 z^2 x dx$$

$E(z) = E_c(z) + E_s(z) + E_p(z)$

3. Topologies and Different Possibilities for the Set Z_{ad}

The initial shape of the membrane z_0 is assumed to be a smooth function over [0,1] and we consider the usual Sobolev spaces built on the open set]0,1[with measure $\mu(x) = x(1+(z_0')^2)^{\frac{1}{2}}$. We consider the linear set

(3.1) $$L^2_{z_0}(0,1) = \{\varphi \ s.t. \ \varphi x^{\frac{1}{2}}(1+(z_0')^2)^{\frac{1}{4}} \in L^2(0,1)\}$$

And

(3.2) $$H^1_{z_0}(0,1) = \{\varphi \in L^2_{z_0}(0,1) \ s.t. \ (\varphi)' \in L^2_{z_0}(0,1))$$

The energy functional E is defined on the set Z_{ad}. Obviously that set is larger than $\mathcal{H} = \{z \in H^2_{z_0}(0,1), z(1) = 0\}$ (if z belongs to $\mathcal{H}$ then $\frac{|z''|}{(1+(z')^2)^{\frac{3}{2}}} \leq |z''|$ and z belongs to Z_{ad}), that is $\mathcal{H} \subset Z_{ad}$.

The main question is now to minimize the energy functional over Z^v_{ad}. That set is neither linear nor convex . In order to bypass that difficulty we introduce the functions s_n (resp. s) associated with any element z_n (resp. z) in Z_{ad}

$$(3.3) \qquad\qquad s = \frac{u'}{(1+(z')^2)^{\frac{1}{2}}}$$

We have

$$(3.4) \qquad\qquad s' = \frac{z''}{(1+(z')^2)^{\frac{3}{2}}}$$

And the curvature energy can be rewritten as follows:

$$(3.5) \qquad E_c(z) = k_c \int_0^1 \{[(\frac{z'}{(1+(z')^2)^{\frac{1}{2}}})' - (\frac{z_0'}{(1+(z_0')^2)^{\frac{1}{2}}})']^2 + [\frac{z'}{x(1+(z')^2)^{\frac{1}{2}}} - \frac{z_0'}{x(1+(z_0')^2)^{\frac{1}{2}}}]^2\} x(1+(z_0')^2)^{\frac{1}{2}} dx$$

that is also:

$$(3.6) \qquad E_c(z) = k_c \int_0^1 \{[s' - s_0']^2 + [x^{-1}s - x^{-1}s_0]^2\} x(1+(z_0')^2)^{\frac{1}{2}} dx$$

Given an element z_0 in Z_{ad}, the energy functional $E(z)$ is defined for any element z in Z_{ad} . As we have seen , the associated element s is in $H^1_{z_0}(0,1)$ so that the trace $s(0)$ is continuously defined and we set:

$$(3.7) \qquad\qquad Z^v_0 = \{ z \in Z^v_{ad} / s(0) = 0 \}$$

when z is in Z^v_0 we shall write , in view of the definition of s , $z'(0) = 0$ instead of $s(0) = 0$.

Theorem 3.1. *Let there be given z_0 in Z^v_0 and a minimizing sequence z_n of the energy E in the set Z^v_0. Then there exists a subsequence, still denoted z_n converging weakly in $H^1_{z_0}(0,1)$ to an element z of Z^v_0. Moreover the associated sequence s_n weakly converges in $H^1_{z_0}(0,1)$ to the element s associated to the limiting element z. Finally the element z minimizes the energy functional E over Z^v_0.*

Proof.

It is immediate to verify that the sequences z_n and s_n are both bounded in $H^1_{z_0}(0,1)$ (for the boundedness of z_n we refer to the curvature energy which remains bounded while for the boundedness s_n we refer to the stretch energy). We derive the weak-$H^1_{z_0}(0,1)$ convergences of these two sequences (after two successive subsequences extractions) respectively to an element z and to an other element , say r. Then via the compactness of the inclusion mapping of $H^1_{z_0}(0,1)$ into $L^2_{z_0}(0,1)$ we derive the strong-$L^2_{z_0}(0,1)$ convergences of these two sequences. We can also derive that the sequences z_n and s_n converge almost every where in $(0,1)$ respectively to z and r From the strong-$L^2_{z_0}(0,1)$ convergence of the sequence s_n we get that for almost every x in $(0,1)$ we have $s_n(x)$ converge to $r(x)$, but from the very definition of

s_n we have $z_n'(x)^2 = \frac{s_n^2}{1-s_n^2}$ so that from the almost every where convergence of s_n derives the almost everywhere convergence of $z_n'(x)^2$ and then, as $z_n'(x) = s_n(x)(1+(z_n')^2(x))$, the almost everywhere convergence of $z_n'(x)^2$ itself. Now $z_n'(x)^2$ converges both weakly-$H^1_{z_0}(0,1)$ and almost every where. Using Mazur's theorem we know that one can find finite linear-convex combinations , $w_n = \Sigma_{k\in K_n}\lambda_k^n z_k$, each set K_n being finite, with $\Sigma_{k\in K_n}\lambda_k^n = 1$ and the strong convergence in $L^2_{z_0}(0,1)$ of the sequence w_n' to z'. As a first consequence of the previous results of Mazur follows that (after one more subsequence extraction) the sequence w_n converges almost everywhere in $]0,1[$ to the same limit as the sequence z_n itself . Necessarily that limit is $z'(x)$. The two possible limits, weak and almost everywhere for the sequence z_n' are the same. Finally the limiting element r takes the form $r(x) = s(x) = \frac{z'(x)}{1+(z'(x))^2}$.

With the same technique we would prove the following results:

Theorem 3.2. *The set Z_0 is a complete metric space equipped with the following distance function:*

$$d(z_1,z_2) = \int_0^1 (s(z_1) - s(z_2))^2 A_0 dx + \int_0^1 (z_1' - z_2')^2 x dx \tag{3.8}$$

Theorem 3.3. *$L^2(0,1)$ is dense in the set Z_0 equipped with the previous metric*

Theorem 3.4. *let $z \in Z_0$, with $z' \in L^\infty(0,1)$ and $\varphi \in H^2(0,1)$, then for ϵ small enough ($0 \le \epsilon(\varphi)$) we have $z + \epsilon\varphi \in Z_0$*

Proof.

The main point is that Z' and φ' are elements of $C^0([0,1])$ so that we can give a bound to the following term:

$$C = [2z'\varphi' + \epsilon(\varphi')^2]$$

We have to prove that the term

$$D = \frac{(z")^2 + 2\epsilon z"\varphi" + (\epsilon)^2(\varphi")^2}{(1+(z')^2+\epsilon C)^3} \tag{3.9}$$

is an element of $L^1(0,1)$ using the usual expansion

$$(A+\epsilon B)^{-3} = A^{-3} - 3\epsilon\frac{B}{(A+\epsilon\theta B)^4}$$

for some $\theta, 0 \le \theta \le 1$, we get

$$\begin{aligned} D = {} & \frac{z"^2}{(1+Z'^2)^3} + \frac{2\epsilon z"\varphi" + \epsilon^2\varphi"^2}{(1+Z'^2)^3} \\ & -3\epsilon\frac{z"^2(2z'\varphi' + \epsilon\varphi'^2)}{(1+z'^2+\epsilon\theta C)^4} \\ & -2\epsilon\frac{2\epsilon z"\varphi"(2z'\varphi' + \epsilon\varphi'^2)}{(1+z'^2+\epsilon\theta C)^4} \\ & -3\epsilon^3\frac{\varphi"^2(2z'\varphi' + \epsilon'^2)}{(1+z'^2+\epsilon\theta C)^4} \end{aligned}$$

let $E = (2\epsilon z"\varphi" + \epsilon^2\varphi"^2)/(1 + Z'^2)^3$. We get:

$$E = 2\epsilon \frac{z"}{(1+(z')^2)^{\frac{3}{2}}} \frac{\varphi"}{(1+(z')^2)^{\frac{3}{2}}} + F$$

We get easily that $|F| \leq \epsilon^2\varphi"^2 \in L^1(0,1)$ and also each of the two terms in the product are in $L^2(0,1)$.Let us consider now the term:

$$G = -3\epsilon \frac{z"^2(2z'\varphi' + \epsilon\varphi'^2)}{(1 + z'^2 + \epsilon\theta C')^4}$$

We shall make use of the fact that elements of $H^1(0,1)$ are bounded over $[0,1]$.Then there exists m positive such that $C \leq m$. We then get

$$(3.10) \qquad |G| \leq 3\epsilon m \frac{z"^2}{(1 - \epsilon m + z'^2)^4} \leq \frac{3\epsilon m}{1-\epsilon m} \frac{z"^2}{(1-\epsilon m + z'^2)^2}$$

$$\leq \frac{3\epsilon m}{1-\epsilon m} \frac{z"^2}{(\frac{1}{2}(1+z'^2))^2}.$$

In fact, we can improve that result as follows

Theorem 3.5. *let $z \in Z_0$, and $\varphi \in H^2(0,1)$, then for ϵ small enough ($0 \leq \epsilon(\varphi)$) we have $z + \epsilon\varphi \in Z_0$*

Proof.

Consider the set $E = \{x \in [0,1] s.t. |z'(x)| \leq 1\}$ the integrals can be decomposed in two integrals over E and its complementary set E^c. The conclusion derives from the previous result since on E we can bound from below the term

$$(1 + (z' + \epsilon\varphi')^2)^3 \geq (\frac{1}{2}(1 + (z')^2))^3.$$

In order to derive the previous existence result we had to take the initial shape of the membrane to be smooth enough, in particular at point $t = 1$ which does not permit to take $z_0(x) = (1 - x^2)^{\frac{1}{2}}$. We could replace the previous energy minimization by the following constraint problem:

$$(3.11) \qquad Min \{ \ E(z) \mid z \in Z^v_{ad} \ s.t. \int_0^1 z(x)x dx = v,$$

$$|z'(x)| \leq h(|z"(x)|) \ a.e.x \ \}$$

where h is a given lower semi continuous function. For example h could be a constant.

4. Necessary Optimality Conditions

The existence of a (at least one) minimum in the set Z_0^v for the energy functional E leads to the following result :

Theorem 4.1. *Given any element z_0 in the set Z_0 and any v in $]0, v_{max}$ there exists a solution $(z, \alpha) \in (Z)^v \times \mathbb{R}^N$ to the following non linear problem:*

$$z'(0) = 0 \quad , \quad z(1) = 0 \tag{4.1}$$

$$\begin{aligned} \mathbb{A}(z) = k_c\{ \; & 2[\,(\frac{z''}{(1+(z')^2)^{\frac{3}{2}}} - \frac{z_0''}{(1+(z_0')^2)^{\frac{3}{2}}})\frac{1}{(1+(z')^2)^{\frac{3}{2}}}\,]'' + \\ & 6((\frac{z''}{(1+(z')^2)^{\frac{3}{2}}} - \frac{z_0''}{(1+(z_0')^2)^{\frac{3}{2}}})\frac{z'}{(1+(z')^2)^{\frac{5}{2}}})' \\ & -2\,[\,(\frac{z''}{x(1+(z')^2)^{\frac{1}{2}}} - \frac{z_0''}{x(1+(z_0')^2)^{\frac{1}{2}}})\frac{1}{x(1+(z')^2)^{\frac{1}{2}}}\,]' + \\ & +2[\,(\frac{z''}{x(1+(z)^2)^{\frac{1}{2}}} - \frac{z_0''}{x(1+(z_0')^2)^{\frac{1}{2}}})\frac{(z')^2}{x(1+(z')^2)^{\frac{3}{2}}}\,]'\;\} \\ & -2\,k_s\,[\,(x(1+(z')^2)^{\frac{1}{2}} - x(1+(z_0')^2)^{\frac{1}{2}})\frac{z'}{(1+(z')^2)^{\frac{1}{2}}}\,]' \\ & +2\,k_p\,z\,x \;=\; \alpha\,x \end{aligned} \tag{4.2}$$

Theorem 4.2. *The(non linear) operator $\mathbb{A}(.)$ maps Z_0 in $H^{-2}(0,1)$*

Remark. Of course with this non linear problem is not associated a uniqueness result for the solutions.

5. Quasi-Steady Evolution of the Membrane

Using the variational formulation for the position $z(v)$ of the membrane at each volume v we consider now a minimization algorithm for the functional E which is an iterative one .The quasi-steady evolution process consists in assuming that the volume of the domain Ω occupied by the fluid decays so slowly from the initial value v_{max} to the value v that inertial forces can be neglected. For this purpose we introduce a discretization process for the values of the variable v which governs that evolution. Let v_0 be given and also the incremental step δv; we set $v_{n+1} = V_n - \delta v$. Let z_{n+1} be the graph parametrization of the membrane associated with volume v_{n+1}, it will be obtained via minimization of the functional E over the set $Z_0^{v_{n+1}}$. This minimization will be performed through an iterative second order quasi-newton algorithm. This algorithm will generate a sequence denoted by : z_{n+1}^k , $0 \leq k \leq k_{n+1}$ with the initial iteration chosen as

$$z_{n+1}^0 = z_n^{k_n} \tag{5.1}$$

that is to say that the sequence z_{n+1}^k does not start at $k = 0$ with the initial shape of the membrane z_0, but with the last iteration of the previous iterative process associated to the previous volume v_n, this last iteration $z_n^{k_n}$ being considered by the choice of k_n (large enough) as a good approximation for the solution z_n . Finally the initial shape is not only used in the very initialization of the process but mainly during each iterative process since in each energy functional E the element z_0 is preserved. For each value of the parameter v the solution may be non unique. Starting from v_{max} we know that $z(v_{max})$ is uniquely determined as a data.

6. Bifurcation of the Solutions to the Quasi-steady Evolution

At each step in solving the previous evolution problem the little parameter is δv. ΔAt each step the functional to be minimized is not convex. Nevertheless at the first steps we can recover the uniqueness of the solution $z(v)$. For this let us consider the second order derivative of the energy. We put $A = (1 + (z')^2)^{-\frac{1}{2}}$

$$\forall w \in \mathbb{D}(]0,1[), E_c"(z;w,w) = \int_0^1 (w"A^3 - 3z"z'w'A^5)^2\, x(1+(z_0')^2)^{\frac{1}{2}}dx \tag{6.1}$$

$$+\int_0^1 (x^{-1}Aw' - x^{-1}A^3(z')^2w')^2\, x(1+(z_0')^2)^{\frac{1}{2}}dx$$

$$-\int_0^1 (z"A^3 - z_0"A_0^3)(3z'w"w'A^5 + 3w'(z'w" + z"w')A^5\, x(1+(z_0')^2)^{\frac{1}{2}}dx$$

$$-\int_0^1 (x^{-2}A^2(z' - z_0')(w')^2z'(3 + 3(z')^2A^4)\, x(1+(z_0')^2)^{\frac{1}{2}}dx$$

the first two terms are squares hence always positive . In the two last terms the sign cannot be controlled but in the first iterations on the evolution parameter v both terms $(z"A^3 - z_0"A_0^3)$ and $x^{-2}A^2(z' - z_0')$ will be small as $z - z_0$ is small in "Z_0^v norm". The same phenomena occurs in the stretch energy as we have:

$$E_s"(z;w,w) = k_s \int_0^1 (x(Az'w')^2 + x(A - A_0)w'z'(A - A^3w'z'))dx \tag{6.2}$$

Let the gravity g be zero . The classical setting would be to look for solutions (at least one) to the following non linear evolution problem: find the largest T and z such that :

$$t = v_{max} - v\ ,0 \le t \le T,\ T \le v_{max}\quad z \in C^0([0,T[, Z_0)\quad \alpha \in C^0([0,T]) \tag{6.3}$$

$$z(t) \in Z_0^{v_{max}-t}\ i.e.\ \int_0^1 z(t)(x)x dx = v(t) = v_{max} - t \tag{6.4}$$

$$z(0) = z_0 \in Z_0^{v_{max}} \tag{6.5}$$

$$z'(0) = 0\ ,\ z(1) = 0 \tag{6.6}$$

$$k_c\{\ 2[\ (\frac{z"(t)}{(1+(z'(t))^2)^{\frac{3}{2}}} - \frac{z_0"}{(1+(z_0')^2)^{\frac{3}{2}}})\frac{1}{(1+(z'(t))^2)^{\frac{3}{2}}}\ ]" + \tag{6.7}$$

$$6((\frac{z"(t)}{(1+(z'(t))^2)^{\frac{3}{2}}} - \frac{z_0"}{(1+(z_0')^2)^{\frac{3}{2}}})\frac{z'(t)}{(1+(z')^2)^{\frac{5}{2}}})'$$

$$-2\ [\ (\frac{z"(t)}{x(1+(z'(t))^2)^{\frac{1}{2}}} - \frac{z_0"}{x(1+(z_0')^2)^{\frac{1}{2}}})\frac{1}{x(1+(z'(t))^2)^{\frac{1}{2}}}\ ]' +$$

$$+2[\ (\frac{z"(t)}{x(1+(z'(t))^2)^{\frac{1}{2}}} - \frac{z_0"}{x(1+(z_0')^2)^{\frac{1}{2}}})\frac{(z'(t))^2}{x(1+(z'(t))^2)^{\frac{3}{2}}}\ ]'\ \}$$

$$-2\,k_s\,[\,(x(1+(z'(t))^2)^{\frac{1}{2}} - x(1+(z_0')^2)^{\frac{1}{2}}\,)\frac{z'(t)}{(1+(z'(t))^2)^{\frac{1}{2}}}\,]'$$

$$+2\,k_p\,z(t)\,x \;=\; \alpha(t)\,x$$

Remark: We need $g = 0$ in order to be sure that $z_0 = z(0)$. From the previous considerations it follows the existence of a closed interval $[0, T^*]$ on which the previous problem "has unique solution" , more precisely such that all solutions to the previous evolution problem coincide on that interval. We shall denote by $z^* \in C^0([0, T^*], Z_0)$ that unique solution. By definition of T^*, several solutions exist after T^* and we have a bifurcation situation at T^*. It seems numerically that such a global solution (i.e. for which $T = v_{max}$) does not exists. The problem (6.3)-(6.7) characterizes the extremality of the energy term during the quasi-evolution. As we already said at the beginning of that evolution process the solution represents a local minimum of the energy functional but as the volume decreases it may become more and more unstable so that "suddenly" at some step the solution jumps to a stable one for which the energy is lower. It is important to notice that in this modeling such a jump is "energy free" as the quasi steady evolution modeling does not consider the liquid which is in the volume Ω . Of course that jumping of the solution is not a quasi steady one and would increase the energy if we consider the contribution of the kinetic energy of the fluid.

7. Dynamical case

We consider now small displacements of the membrane with little deformation but without neglecting the membrane and fluid inertia as was done in previous sections. Also we consider now a fixed volume v. The unitary normal field out going to the volume Ω containing the liquid is denoted by n while the tangential differential operators on the surface S are denoted by ∇_S for the tangential gradient and D_S for the tangential differential matrix operator If u is a vector field defined on the surface S, $D_S u$ is, at each point $x \in S$, the point matrix whose transposed is derived from the tangential gradients of each component of u as follows: $D_S u^* = (\nabla_S u_1, ..., \nabla_S u_n)$. The tangential deformation tensor is the symmetrical part :

$$\epsilon_S(u) = \frac{1}{2}(D_S u + D_S u^*)$$

At a given volume v of the open domain Ω the boundary S is now the graph $\{(x, z(x)), 0 \leq x \leq 1\}$ of a function z minimizing the static problem. We consider the elastic displacement $u = u(t, x)$. We decompose the field u into its tangential part u_S and normal part w, $u = u_S + wn$ with $u_S.n = 0$. For simplicity we shall neglect the tangential displacement u_S as we are dealing with linear elasticity and consider $w = w(t, x)$ as being the displacement variable in the problem. As we said the volume v is now kept constant, which implies a linear constraint on w;

$$\int_S w\,dS \;=\; 0$$

This dynamical study implies the total energy in the system , in particular the kinematic energy of the membrane as well as the kinematic energy of the fluid lying in Ω. The fluid is assumed to be incompressible and perfect so that its velocity vector field derives from a potential φ, defined up to a constant function, which is harmonic in the domain Ω. Those kinetic energies are respectively given by:

(7.1) $$E_{k,S} = \frac{1}{2}\int_S w_t(t,x)^2 dS, \quad E_{k,\Omega} = \frac{1}{2}\int_\Omega ||\nabla\varphi(t,x)||^2 dx$$

The sum of the potential energy of the fluid and the internal energy stored in the membrane is at a minimum solution of the static problem and the first variation (linear in w) of this sum is null. Thus, the variation of E is quadratic in w and takes the form:

(7.2) $$E_{e,S} = \frac{1}{2}\int_S \epsilon_S w(t,x)..C..\epsilon_S w(t,x)\ dS$$

where $C = C_{i,j,k,l}$ is a four entries surface elasticity tensor.

Fluid pressure on the membrane is given by the linearized Bernoulli equation:

(7.3) $$p = p_0 - \rho\{\frac{\partial\varphi}{\partial t} + gwn.k\ \}$$

So that $p - p_0$ is the force applied on the surface by the fluid and the corresponding dissipated energy W is

(7.4) $$W = \int_S \rho\{\frac{\partial\varphi}{\partial t} + gwn.k\ \}w\ dS$$

The compatibility condition between the membrane S and the potential fluid located in the open domain Ω says that a point x on the surface can be considered either as being in the membrane or in the flow, and can move with the field $w_t n$ as well as with the field $\nabla\varphi$. As a result we get condition

(7.5) $$w_t = \frac{\partial}{\partial n}\varphi \text{ on } S$$

Now we note

$$H^1(\Omega)_* = \{\varphi | \varphi \in H^1(\Omega),\ \int_\Omega \varphi = 0\}$$

$$H_0^1(S)_* = \{w | w \in H_0^1(S),\ \int_S w = 0\}$$

$$L^2(S)_* = \{w | w \in L^2(S),\ \int_S w = 0\}$$

Finally the extremality of the action

$$A = \int_0^T [E_{k,S} + E_{k,\Omega} - E_{e,S} - E_{p,S} - W]dt$$

leads to the following problem for the weak formulation of the fluid structure problem whose unknowns are (φ, w)

(7.6) $$(\varphi, w) \in H^1(\Omega)_* \times H_0^1(S)_*$$

$$\forall(\psi, y)\ \in\ H^1(\Omega)_* \times H_0^1(S)_*$$

$$\int_S \{w_{tt}y + \epsilon_S(wn)..C..\epsilon_S(yn) + (\varphi_t + gwk.n)y\}\ ds = 0$$

and

$$\int_\Omega \nabla\varphi.\nabla\psi\ dx = \int_S \frac{\partial}{\partial t} w\ \psi\ ds \tag{7.7}$$

In [2] was proved the following existence result using a Galerkin approximation

Theorem 7.1. *let S be a C^1 surface, given data*

$$(\varphi_0, w_0, w_1) \in H^1(\Omega)_* \times H_0^1(S)_* \times L^2(S)_*$$

the problem 7.6 has a unique solution (φ, w) verifying:

$$\epsilon_S(wn) \in L^2([0,T] \times S)$$
$$(wn)_t \in L^2([0,T] \times S)$$
$$\nabla\varphi \in L^2([0,T] \times \Omega)$$
$$\varphi \in L^2([0,T] \times \Omega)\ ,\ w \in L^2([0,T] \times S)$$

and the initial conditions:

$$w(0) = w_0\ ,\ w_t(0) = w_1\ ,\ \varphi(0) = \varphi_0$$

We denote by $\mathcal{A}$ and $\mathcal{G}$ the operators from $H_0^1(S)_*$ to $H^{-1}(S)$ defined by

$$(\mathcal{A}w, y) = \int_S \epsilon_S(wn)..C..\epsilon_S(yn)dS$$

and

$$(Gw, y) = \int_S gwyk.ndS$$

and we consider the pseudo differential operator:

$$\mathbb{N}_S \in \mathbb{L}(H_*^{-\frac{1}{2}}(S), H^{\frac{1}{2}}(S)/\mathbb{R}) \text{ defined by } \mathbb{N}_S.Y = \Psi|_S \tag{7.8}$$

where Ψ is one solution of the following boundary value problem

$$\Delta\Psi = 0 \text{ in } \Omega,\ \frac{\partial}{\partial n}\Psi = Y \text{ on } S,\ \frac{\partial}{\partial n}\Psi = 0 \text{ on } \partial\Omega - S, \tag{7.9}$$

This definition of $\mathbb{N}_S$ is independent on the choice of the solution Φ (two solutions differ by a constant function).

In fact, if we note

$$\mathbb{N} \in \mathbb{L}(H_*^{-\frac{1}{2}}(\partial\Omega), H^{\frac{1}{2}}(\partial\Omega)/\mathbb{R})$$

the classical Neumann pseudo differential operator defined by:

$$\Delta\Phi = 0 \text{ in } \Omega,\ \frac{\partial}{\partial n}\Phi = v \text{ on } \partial\Omega \tag{7.10}$$

$$\Delta\Psi = 0 \text{ in } \Omega,\ \frac{\partial}{\partial n}\Psi = v \text{ on } \partial\Omega \tag{7.11}$$

$$(\mathbb{N}u, v) = \int_\Omega \nabla\Phi\nabla\Psi \tag{7.12}$$

by

$$s \in \mathbb{L}(H^{\frac{1}{2}}(\partial\Omega)/\mathbb{R}, H^{\frac{1}{2}}(S)/\mathbb{R})$$

the restriction to S of functions defined on $\partial\Omega$: $\phi \to \phi|_S$ and by

$$\imath \in \mathbb{L}(H_*^{-\frac{1}{2}}(S), H_*^{-\frac{1}{2}}(\partial\Omega))$$

its dual injection: $(\imath u, \phi) = (u, s\phi)$ then we have

$$\mathbb{N}_S = s \circ \mathbb{N} \circ \imath$$

Moreover the operator $\mathbb{N}_S$ is positive:

$$\int_\Omega ||\nabla\Psi||^2 = (\mathbb{N}_S Y, Y)_{L^2(S)}. \geq 0$$

Then we have $\varphi|_S = \mathbb{N}_S w_t$ and w is the weak solution of the following differential equation:

$$(I + N)w_{tt} + (\mathcal{A} + \mathcal{G})w = 0 \tag{7.13}$$

with initial conditions

$$w(0) = w_0 \quad w_t(0) = w_1 \tag{7.14}$$

The stability of the membrane S is related to the real parts of the eigenvalues λ of the following problem: we search solutions of the previous dynamical problem in the form

$$(\varphi(t,.)\,,\, w(t,.)) \;=\; (\Phi(.)\,,\, W(.))\; e^{\lambda\, t} \tag{7.15}$$

the two functions Φ and W being respectively defined on Ω and S. the dynamical problem becomes

$$\forall Y \in H^1(S),\ \forall\Psi \in H^1(\Omega)_*$$

$$\int_S \epsilon_S(Wn)..C..\epsilon_S(Yn) + \lambda^2 \int_S WYds\,\lambda \int_S \Phi Yds \;+ \int_S gWk.nYdS = 0 \tag{7.16}$$

and

$$\int_\Omega \nabla\Phi.\nabla\Psi\ dx \;=\; \lambda \int_S W\ \Psi\ ds \tag{7.17}$$

with the boundary compatibility condition

$$\frac{\partial}{\partial n}\Phi \;=\; \lambda\, W \text{ on } S, \text{ and } \frac{\partial}{\partial n}\Phi \;=\; 0 \text{ on } \partial\Omega\backslash S \tag{7.18}$$

Then we get $\Phi \;=\; \lambda\, \mathbb{N}_S.W$ and the following eigenvalue problem:

$$(\lambda^2\,(I + N) + (\mathcal{A} + \mathcal{G}))W = 0 \quad W \in H_0^1(S)_* \quad W \neq 0 \quad \lambda \in \mathbb{C} \tag{7.19}$$

With the assumptions on operator $\mathcal{G}$, $(I + N)^{-1}$ is positive and bounded, $(\mathcal{G} + \mathcal{A})$ is positive and has a compact resolvant, then we have a spectrum of positive eigenvalues $-\lambda^2$ so the real parts of the λ's are zero as we introduced no dissipation in the modeling of the problem. Such a structural damping would be introduced in considering a viscous fluid of some friction term at the interface S. Using integration by part on the surface (see [3]) S

it can easily be verified that, H being the mean curvature of S, the operator $\mathcal{A}$ takes the following form:

(7.20) $$\mathcal{A}w = -div_S(C..\epsilon_S(w.n)\ ).n\ +\ H < (\ C..\epsilon_S(w.n)\ ).n\ ,\ n >$$

In that study the modeling of the membrane could be replaced by the intrinsic shell modeling developed in [4] or by its asymptotic version as presented in [5].

8. References

[1] M.Delfour, J.P.Zolésio, oriented distance function, J.Func.Anal.1994

[2] N.Clariond, J.P.Zolésio, Hydroelastic Behavior of an Inextensible Membrane. in "Boundary Control and Variation", Lecture Notes in Pure and Applied Mathematics, vol.263, pp.124-135, Marcel Dekker Inc. New-York,1994.

[3] J.Sokolowski, J.P.Zolésio, Introduction to Shape Optimization, computational mathematics, vol. 16, Springer Verlag, 1991.

[4] M.Delfour, J.P. Zolésio. A Boundary Differential Equation for Thin Shells, Journal of Differential Equations, vol. 119, n¡ 2, p. 426-449, July 1, 1995

[5] M.C. Delfour, J.P. Zolésio. Comparison of Shell Modeling in Intrinsic Geometry. To appear in the volume of CRM anniversary (Univ. de Montreal)..

CMA-EMP, BP 207, 06904 Sophia-Antipolis, France
E-mail address: jpmcma.cma.fr

CNRS-INLN, 1361 Route des Lucioles, 06560 Valbonne, France
E-mail address: zolesiocma.cma.fr

Inf–Sup Conditions for an Elliptic Operator in the Spaces $W_0^{1,p}W_0^{1,q}$ Approximated with Lagrange Finite Element

J. Pousin

SWISS FEDERAL INSTITUTE OF TECHNOLOGY
Department of Mathematics
CH 1015 Lausanne, Switzerland
and
INSA LYON
Lab. Modélisation et Calcul Scientifique
URA 740 CNRS, Bât. 403, 20, av. Einstein, F 69621 Villeurbanne Cedex, France

Abstract

Let Ω be a bounded convex plane polygon domain and let $\mathcal{T}_h$ be an affine equivalent quasi-uniform triangulation of $\overline{\Omega}$. We define V_h as the space of continuous functions on $\overline{\Omega}$ which are affine on each triangle κ of $\mathcal{T}_h$, and we introduce the following Banach spaces : $X_h = V_h$ endowed with the $W_0^{1,p}(\Omega)$ semi-norm; $Y_h = V_h$ endowed with the $W_0^{1,q}(\Omega)$ semi-norm with $\frac{1}{p}+\frac{1}{q}=1$. If A denotes the following elliptic operator : $A = \sum_{i,j=1}^{2} \partial_j a_{ij} \partial_i$, where the functions $a_{ij} \in C^1(\overline{\Omega}; IR)$; $a_{ij} = a_{ji}$, then we define the bilinear form $a(\cdot,\cdot)$ associated to A by : $a(\varphi,\psi) = \int_\Omega - \sum_{i,j=1}^{2} a_{ij}\, \partial_i\varphi\, \partial_j\psi\, dx$. We prove the Inf-Sup conditions in $X_h \times Y_h$ for the bilinear form $a(\cdot,\cdot)$ for all $p \in (\frac{p^*}{p^*-1}, p^*)$ where p^* depends on a_{ij}, the size of the inner angles of Ω (see (2.3)) and satisfies $2 < p^* < +\infty$.

1 MOTIVATIONS

Let X and Y be two Banach spaces, Y' denotes the dual of Y, and let $F : X \to Y'$ be a nonlinear C^1-mapping. In a previous work with J. Rappaz (see Pou-Rap [8], Pou-Rap [9]), we proposed a general framework for deriving a priori and a posteriori error estimates for nonlinear problems :

$$F(u) = 0 ; \quad \Leftrightarrow \langle F(u), v\rangle = 0 \quad \forall v \in Y, \tag{1.1}$$

approximated through a Petrov-Galerkin procedure. That is to say : if $X_h \subset X$, $Y_h \subset Y$ are finite dimensional subspaces of X and Y, we define an approximation u_h of $u \in X$ solution to problem (1.1) by looking for $u_h \in X_h$ verifying :

$$\langle F(u_h), v_h \rangle = 0 \quad \forall v_h \in Y_h. \tag{1.2}$$

The error estimates are established if consistency and stability properties for the approximated problem (1.2) hold true.

Consistency is a consequence of approximation of X and Y by X_h and Y_h, whereas stability is a consequence of (discrete) Inf-Sup conditions in $X_h \times Y_h$ for a bilinear form $b(\cdot,\cdot)$ built with $DF(u)$ the derivative of the mapping F at the point u solution to (1.1) :

$$b(\psi, v) = \langle DF(u) \cdot \psi, v \rangle \quad \forall \psi \in X, \quad \forall v \in Y.$$

In order to have the derivability of mapping F, we need to work in general Sobolev's spaces $X = W_0^{1,r}(\Omega)$ where r has to satisfy the condition $r > 2$ (which implies $X \hookrightarrow C^0(\overline{\Omega})$).

Let us mention another example where discrete Inf-Sup conditions are needed for deriving error estimates. When one deals with weakly coupled nonlinear problems such as compressible flows in a $2-D$ domain Ω, a simplified model reads (see for example Pir [7]) :

$$-\Delta u + \nabla p = h(T) \quad \text{in } \Omega; \tag{1.3}$$

$$-\operatorname{div}(k \nabla \theta) = G(\nabla u) \quad \text{in } \Omega, \tag{1.4}$$

where $k \in C^1(\overline{\Omega}; IR)$ is given and where G stands for an homogeneous polynomial function of degree 2 with respect to the derivatives of the components velocity vector u. So, if pressure P is in $L^2(\Omega)$, u should be sought in ($W^{1,2}(\Omega)$) and $G(\nabla u)$ belongs to $L^1(\Omega)$ which is not included in $W^{-1,2}(\Omega)$. The temperature problem (1.4) cannot be solved in $W_0^{1,2}(\Omega)$ but has to be solved in $W_0^{1,r}(\Omega)$ with $1 < r < 2$. This leads to consider the approximation of problem (1.4) with a Petrov-Galerkin method in $W_0^{1,r}(\Omega)$. It is known (see Azi-Bab [1]) that Inf-Sup conditions in approximated spaces and interpolation results for approximated spaces (see Ber [2] for $1 < r < 2$ and Cia [3] for $2 < r < +\infty$) yield error estimates.

The result concerning the Inf-Sup conditions which will be stated may be known but the proof does not seem to appear anywhere.

Now, let us specify the notations which will be used throughout this paper.

Let Ω be a convex bounded plane polygon domain, for $1<r<+\infty$, we denote by $W_0^{1,r}(\Omega)$ the usual Sobolev's spaces of real functions defined on Ω, equipped with the semi-norm $|\cdot|_{1,r}$ which is an equivalent norm (see for example Gri [6]), and the Sobolev's space $W^{2,r}(\Omega)$ will be equipped with the norm $\|\cdot\|_{2,r}$.

If $(Z,\|\cdot\|_Z)$ is a Banach space, Z' will stand for its dual, the norm of which is :

$$\|\ell\|_{Z'} = \sup_{\substack{\varphi\in Z\\ \|\varphi\|_Z=1}} \ell(\varphi). \tag{1.5}$$

If p is given such as $1<p<+\infty$, let us recall that $W^{-1,p}(\Omega)$ stands for the dual space of $W_0^{1,q}(\Omega)$ with $\frac{1}{p}+\frac{1}{q}=1$.

For $1\le i,j\le 2$ let $a_{ij}\in C^1(\overline{\Omega};IR)$; $a_{ij}=a_{ji}$ be given, we define the operator A by :

$$A=\sum_{i,j=1}^{2}\partial_j a_{ij}\,\partial_i, \tag{1.6}$$

and we denote by $a(\cdot,\cdot)$ the bilinear form associated to A defined by :

$$a(\varphi,\psi)=\int_\Omega -\sum_{i,j=1}^{2} a_{ij}\,\partial_i\varphi\,\partial_j\psi\,dx. \tag{1.7}$$

Let $\mathcal{T}_h$ be a triangulation of Ω, and let V_h be the space of continuous functions on $\overline{\Omega}$ which are affine on each triangle κ of $\mathcal{T}_h$:\

$$V_h=\{\varphi_h\in C^0(\overline{\Omega};IR);\ \varphi_h|_\kappa \text{ is affine } \forall\kappa\in\mathcal{T}_h \text{ and } \varphi_h=0 \text{ on } \partial\Omega\}. \tag{1.8}$$

An outline of this paper is as follows : In section 2, we recall some results concerning the operator A, and we give an existence result for an auxiliary problem which will be used in the next section. In section 3, thanks to a result of Ran-Sco [10] we prove that the bilinear form $a(\cdot,\cdot)$ verifies the Inf-Sup conditions on $V_h\times V_h$ equipped respectively with the semi-norms $|\cdot|_{1,p}$ and $|\cdot|_{1,q}$ with $\frac{p^*}{p^*-1}<p<p^*$, $\frac{1}{p}+\frac{1}{q}=1$, where p^* depends on Ω, a_{ij} and satisfies : $2<p^*<+\infty$.

In section 4, some possible extensions to convection diffusion operators are presented.

2 PRELIMINARIES CONCERNING OPERATOR A

We recall that the operator A is an isomorphism from $W_0^{1,p}(\Omega)$ onto $W^{-1,p}(\Omega)$ and we prove a result of existence for an auxiliary problem which will be used in the next section.

We assume that the operator A defined by (1.6) verifies the following strong ellipticity hypothesis :

H1) $$\exists \alpha > 0, \quad \forall x \in \overline{\Omega}, \quad \forall \xi \in IR^2 \quad \sum_{i,j=1}^{2} a_{ij}(x)\, \xi_i\, \xi_j \leq -\alpha |\xi|^2. \tag{2.1}$$

Let p be fixed, such as $1 < p < +\infty$, we define q by $q = \frac{p}{p-1}$ and we set $X = W_0^{1,p}(\Omega)$, $Y = W_0^{1,q}(\Omega)$.

Now, we need to introduce some more notations. We consider the operators A_j deduced from A by freezing the coefficients a_{ij} at the corners x_j of Ω.

$$A_j = \sum_{k,\ell=1}^{2} a_{k\ell}(x_j)\, \partial_k\, \partial_\ell .$$

If we denote by $\mathcal{A}_j$ the symmetric matrix whose entries are $a_{k\ell}(x_j)$, we denote by $\mathcal{T}_j$ the matrix such as :

$$-\mathcal{T}_j^{\mathrm{T}}\, \mathcal{A}_j\, \mathcal{T}_j = I_{IR^2} \text{ (here, } ^{\mathrm{T}} \text{ stands for the transposition),} \tag{2.2}$$

and we denote by $T_j \in \mathcal{L}(IR^2)$ the mapping, the matrix of which is $\mathcal{T}_j$.

For $j, 1 \leq j \leq J$, we define $\omega_j(A)$ as the measure of the angle at point $\mathcal{T}_j(x_j)$ of the image of $\overline{\Omega}$ by T_j. We denote by ω the maximum size of the inner angles $\omega_j(A)$ at the points $\mathcal{T}_j(x_j)$.

Finally we define p^* by :

$$p^* = \begin{cases} \frac{2}{2-\pi/\omega} & \frac{\pi}{2} < \omega < \pi; \\ +\infty & 0 < \omega \leq \frac{\pi}{2}. \end{cases} \tag{2.3}$$

We will assume the following hypothesis holds :

H2) $$0 < \omega < \pi.$$

We have :

Theorem 2.1. Assume hypothesis H1) holds, then A is an isomorphism from X onto Y'. Moreover, if hypothesis H2) holds then A is an isomorphism from $W_0^{1,\overline{p}}(\Omega) \cap W^{2,\overline{p}}(\Omega)$ onto $L^{\overline{p}}(\Omega)$ for all $\overline{p}$ such as $\frac{p^*}{p^*-1} < \overline{p} < p^*$.

Remark 2.1. If for $1 \leq j \leq J$ the mappings T_j are angles preserving mappings, then ω is the maximum size of the inner angles of polygon Ω. This situation appears frequently in applications when the matrices $\mathcal{A}_j$ are a multiple of identity i.e. A has the following structure : $Au = \operatorname{div}(k\nabla u)$ with $k \in C^1(\overline{\Omega};IR)$.

Before proving theorem 2.1, let us recall a result we need in the following.

Lemma 2.1. Let X and Y' be two reflexive Banach spaces and let L belong to $\mathcal{L}(X,Y')$. If the adjoint of L, $L^* \in \operatorname{Isom}(Y,X')$ then $L \in \operatorname{Isom}(X,Y')$.

Proof. Since X and Y' are reflexive Banach spaces, for all $L \in \mathcal{L}(X,Y')$ corresponds a unique adjoint $L^* \in \mathcal{L}(Y,X')$ such as $\|L^*\|_{\mathcal{L}(Y,X')} = \|L\|_{\mathcal{L}(X,Y')}$ (see theorem 4.10 of Rud [11] p. 99). If $L^* \in \operatorname{Isom}(Y,X')$ then we have

$$R(L^*) = X',$$

so we deduce that the kernel of L, $N(L)$ satisfies (see theorem 4.12 of Rud [11] p. 94) :

$$N(L) = {}^{\perp}R(L^*) = \{\varphi \in X \text{ s.a. } \ell(\varphi) = 0 \quad \forall \ell \in R(L^*) = X'\} = \{0\}. \tag{2.4}$$

We know that $L^* \in \operatorname{Isom}(Y,X')$ (theorem 4.15 of Rud [11] p. 97) applies and we get

$$R(L) = Y'. \tag{2.5}$$

Thus, $L \in \mathcal{L}(X,Y')$ is one to one and thus $L^{-1} \in \mathcal{L}(Y',X)$ (see theorem 4.1 of Sch [12] p. 63) that is to say $L \in \operatorname{Isom}(X,Y')$.

#

Proof of Theorem 2.1

<u>First step</u> : Let us prove that : A belongs to $\operatorname{Isom}(X,Y')$, the subspace of $\mathcal{L}(X,Y')$ the elements of which are isomorphisms.

For $p = 2$, this result has been known for a long time as Lax-Milgram's theorem. So let us consider the case $2 < p < +\infty$, the case $1 < p < 2$ will be treated in a second time by means of a transposition argument.

For all $\ell \in Y'$, since Ω is bounded, we remark that $\ell \in W^{-1,2}(\Omega)$, so there exists a unique $u \in W_0^{1,2}(\Omega)$ such as :

$$a(u,v) = \langle \ell, v \rangle \quad \forall v \in W_0^{1,2}(\Omega). \tag{2.6}$$

We have $u \in W_0^{1,2}(\Omega)$, $Au \in Y'$ then since Ω is a plane convex polygon, item (5.5) of Dau [4] p. 241 asserts $u \in X$.

The operator $A \in \mathcal{L}(X,Y')$ and is one to one thus $A^{-1} \in \mathcal{L}(Y',X)$ (see theorem 4.1 of Sch [12] p. 63). That proves $A \in \mathrm{Isom}(X,Y')$.

The case $1 < p < 2$: If we define the operator L as A from X onto Y', due to the property $a_{ij} = a_{ji}$, L^* represents the operator A from Y onto X' ($Y = W_0^{1,q}(\Omega)$ with $2 < q < +\infty$).

We have previously proved that $L^* \in \mathrm{Isom}(Y,X')$; lemma 2.1 applies and we get $L \in \mathrm{Isom}(X,Y')$.

Second step : Let us prove the regularity result : $A \in \mathrm{Isom}\,(W_0^{1,\bar{p}}(\Omega) \cap W^{2,\bar{p}}(\Omega)\,;\, L^{\bar{p}}(\Omega))$, for all $\bar{p}$ such as $\frac{p^*}{p^*-1} < \bar{p} < p^*$, where p^* is given by (2.3).

The domain Ω is a polygon, the coefficients of operator A are C^1, so only in a neighbourhood of the angles of the boundary $\partial\Omega$ of Ω the regularity may fail.

Due to hypothesis H2) we know that the maximum size of the inner angles of the image of $\overline{\Omega}$ by T_j for $1 \le j \le J$ is less than π.

If we set $q^* = \frac{p^*}{p^*-1}$, it is easy to check that :

$$\mathrm{card}\,\left\{m \in \mathbb{Z} \text{ such as } -\frac{2}{q^*} < \frac{m\pi}{\omega_j(A)} < 0\right\} = 0 \quad \text{for } 1 \le j \le J. \tag{2.7}$$

By means of the perturbation method for the Laplacian, we deduce the desired result (see theorem 5.2.2 of Gri [6] p. 266) :

$$A \in \mathrm{Isom}\,(W_0^{1,\bar{p}}(\Omega) \cap W^{2,\bar{p}}(\Omega)\,;\, L^{\bar{p}}(\Omega)) \text{ for all } \bar{p} \text{ satisfying } 2 \le \bar{p} < p^*.$$

The case $\frac{p^*}{p^*-1} < \bar{p} < 2$ is treated by duality argument since $a_{ij} = a_{ji}$, $1 \le i,j \le 2$.

The theorem 2.1 is proved.

#

As consequence of theorem 2.1 we have the following.

For all p such as $1 < p < +\infty$, and for all $\varphi \in Y$ be given, for $i = 1,2$, define ψ_i^φ by :

$$\psi_i^{\varphi} = \begin{cases} |\partial_i\varphi|^{q-2}\,\partial_i\varphi & \text{if } \partial_i\varphi \neq 0; \\ 0 & \text{if } \partial_i\varphi = 0. \end{cases} \tag{2.8}$$

Corollary 2.1. Let $\varphi \in Y$ be given, and for $i = 1,2$ let ψ_i^{φ} be defined by (2.8). Then there exists a unique $u \in X$ and a positive constant c_1 (depending on q and on Ω) such as :

$$a(u,w) = \int_{\Omega} \sum_{i=1}^{2} \psi_i^{\varphi}\,\partial_i w\,dx \quad \forall w \in Y, \tag{2.9}$$

and such as :

$$\|u\|_X \leq c_1 \|\varphi\|_Y. \tag{2.10}$$

Proof. Set $L = A$ considered as an operator from X onto Y' theorem 2.1 asserts $L \in \mathrm{Isom}(X,Y')$. So we just have to check that :

$$\partial_i\,\psi_i^{\varphi} \in Y' \quad \text{for } 1 \leq i \leq 2. \tag{2.11}$$

If we set $p = \frac{q}{q-1}$ then since $\varphi \in Y$ we have $\psi_i^{\varphi} \in L^p(\Omega)$ for $1 \leq i \leq 2$ and we conclude that (2.11) is true.

We set

$$u = L^{-1}\Big(\sum_{i=1}^{2} \partial_i\,\psi_i^{\varphi}\Big); \quad c_1 = \left\|L^{-1}\right\|_{\mathcal{L}(Y',X)},$$

and corollary 2.1 is proved.

#

Remark 2.2. If we assume $a_{ij} \in C^0(\overline{\Omega}; IR)$ then the result $A \in \mathrm{Isom}(X,Y')$ remains valid. This result is not trivial since if $a_{ij} \in L^{\infty}(\Omega)$, ellipticity of operator A is not a sufficient condition and the isomorphism result fails (see example 5.1 of Dau [4] p. 241).

3 DISCRETE INF-SUP CONDITIONS

Before stating the main result of this section let us specify the hypothesis we assume to hold concerning the triangulation $\mathcal{T}_h$.

H3) $\mathcal{T}_h$ is an affine equivalent quasi uniform triangulation of $\overline{\Omega}$ (see Cia [3] p. 132 for a precise definition).

For p given such as $\frac{p^*}{p^*-1} < p < p^*$ where p^* is defined by (2.3) we define respectively X_h and Y_h as the space V_h equipped with the semi-norm $|\cdot|_{1,p}$ and the space V_h equipped with the semi-norm $|\cdot|_{1,q}$ where $\frac{1}{p}+\frac{1}{q}=1$. We have :

Theorem 3.1. Assume hypotheses H1), H2) and H3) are satisfied, then there exists $\beta > 0$ such as :

$$\mathop{Inf}_{\substack{u_h \in X_h \\ |u_h|_{1,p}=1}} \quad \mathop{Sup}_{\substack{v_h \in Y_h \\ |v_h|_{1,q}=1}} \quad a(u_h, v_h) \geq \beta \quad \forall h > 0, \tag{3.1}$$

$$\mathop{Inf}_{\substack{v_h \in Y_h \\ |v_h|_{1,q}=1}} \quad \mathop{Sup}_{\substack{u_h \in X_h \\ |u_h|_{1,p}=1}} \quad a(u_h, v_h) \geq \beta \quad \forall h > 0. \tag{3.2}$$

Proof. For $p = 2$, (3.1) is a trivial consequence of H1) (take $v_h = u_h$). First let us state (3.2) for $2 < p < p^*$, and then we will prove (3.1) by means of a transposition argument. Then the case $\frac{p}{p^*-1} < p < 2$ will be treated. Let $\varphi_h \in Y_h$ be given such as $|\varphi_h|_{1,q} = 1$, and for $i = 1,2$ define $\psi_i^{\varphi_h}$ by :

$$\psi_i^{\varphi_h} = \begin{cases} |\partial_i \varphi_h|^{q-2} \, \partial_i \varphi_h & \text{if } \partial_i \varphi_h \neq 0; \\ 0 & \text{if } \partial_i \varphi_h = 0. \end{cases}$$

Corollary 2.1 ensures the existence of $u \in X$ such as :

$$a(u,v) = \int_\Omega \sum_{i=1}^{2} |\partial_i \varphi_h|^{q-2} \partial_i \varphi_h \, \partial_i v \, dx \quad \forall v \in Y, \tag{3.3}$$

and such as :

$$|u|_{1,p} \leq c_1 |\varphi_h|_{1,q}. \tag{3.4}$$

We set

$$\beta_1 = \frac{1}{c_1} \tag{3.5}$$

and we choose $v = \varphi_h$, then (3.3) and (3.4) yield :

$$a(u, \varphi_h) = |\varphi_h|_{1,q}^q = |\varphi_h|_{1,q} \geq \beta_1 |u|_{1,p}. \tag{3.6}$$

Now we introduce R_h the $W_0^{1,2}(\Omega)$-projector from X into V_h defined by :

$$a(\theta - R_h\theta, w_h) = 0 \quad \forall w_h \in V_h \quad \forall \theta \in X. \tag{3.7}$$

Theorem 2.1 allows us to use the following stability result concerning R_h due to Ran-Sco [10] (see theorem 2 p. 438 and extension remarks p. 439) :

$$\| R_h\theta \|_{1,p} \leq c_2 \| \theta \|_{1,p} \quad \forall \theta \in X, \tag{3.8}$$

where c_2 is independent of h.

From (3.8) and equivalence of semi-norm and norm in $W_0^{1,p}(\Omega)$, it follows :

$$| R_h\theta |_{1,p} \leq c_3 | \theta |_{1,p} \quad \forall \theta \in X, \tag{3.9}$$

with c_3 independent of h.

Combining (3.6) and (3.9) we get :

$$a(R_h u, \varphi_h) \geq \tfrac{\beta_1}{c_3} | R_h u |_{1,p}. \tag{3.10}$$

So we set

$$u_h = \frac{R_h u}{| R_h u |_{1,p}}; \quad \beta = \tfrac{\beta_1}{c_3}, \tag{3.11}$$

which proves item (3.2) for $2 < p < p^*$.

Now, let us establish (3.1) for $2 < p < p^*$. The bilinear form $a(\cdot,\cdot)$ is continuous on $X_h \times Y_h$, then for all $v_h \in Y_h$ we have $a(\cdot, v_h) \in X_h'$ and thus we have defined a linear operator L_h :

$$\begin{aligned} L_h : Y_h &\to X_h' \\ v_h &\mapsto a(\cdot, v_h) \text{ such as } a(\varphi_h, v_h) = \langle \varphi_h, L_h v_h \rangle \quad \forall \varphi_h \in X_h \end{aligned} \tag{3.12}$$

$L_h \in \mathcal{L}(Y_h, X_h')$.

Evaluating $\| L_h v_h \|_{X'}$, the condition (3.2) implies that L_h is injective. Since the spaces X_h and Y_h are finite dimensional Banach spaces their duals X_h' and Y_h' are finite dimensional Banach spaces of same dimension. We have : dim X_h' = dim Y_h and we deduce that since L_h is injective, L_h is also surjective and thus $L_h \in \text{Isom}(Y_h, X_h')$. If we denote by $(L_h^{-1})^*$ the adjoint of L_h^{-1} which belongs to $\text{Isom}(Y_h', X_h)$ we have $(L_h^*)^{-1} = (L_h^{-1})^* \in \text{Isom}(Y_h', X_h)$ and from (3.2) we deduce :

$$\| (L_h^*)^{-1} \|_{\mathcal{L}(Y_h', X_h)} = \| L_h^{-1} \|_{\mathcal{L}(X_h', Y_h)} \leq \tfrac{1}{\beta}. \tag{3.13}$$

We can write :

$$|u_h|_{1,p} = \left|(L_h^*)^{-1} L_h^* u_h\right|_{1,p} \le \frac{1}{\beta} \left\| L_h^* u_h \right\|_{Y'} = \frac{1}{\beta} \sup_{\substack{v_h \in Y_h \\ |v_h|_{1,q}=1}} a(u_h, v_h) \tag{3.14}$$

which proves (3.1) for $2 < p < p^*$.

Since $a(\cdot,\cdot)$ is symmetric we deduce from what we have done before that items (3.1) and (3.2) are valid for $\frac{p^*}{p^*-1} < p < 2$ (for such a p (3.1) is nothing else than (3.2) where the role of p and q are exchanged). Thus the theorem 3.1 is proved.

#

4 EXTENSION TO CONVECTION DIFFUSION OPERATORS

In this section we consider a convection diffusion operator, and we specify how the results of the previous section can be extend to this operator by means of a perturbation technique.

Let $b_i \in C^0(\overline{\Omega}; IR)$ $1 \le i \le 2$ be given, we define the operator B by :

$$B = \sum_{i,j=1}^{2} \partial_j a_{ij} \partial_i + \sum_{i=1}^{2} b_i \partial_i = A + \sum_{i=1}^{2} b_i \partial_i, \tag{4.1}$$

Let us start with a result concerning operator B.

For p such as :

$$\frac{p^*}{p^*-1} < p < p^*, \tag{4.2}$$

where p^* is defined by (2.3), we denote by X the Banach space $W_0^{1,p}(\Omega)$ and by Y the Banach space $W_0^{1,q}(\Omega)$ with $\frac{1}{p} + \frac{1}{q} = 1$.

Lemma 4.1. Assume hypotheses H1), H2) hold. Then the operator B defined by (4.1) satisfies $B \in \text{Isom}(X, Y')$.

Proof. We use the Fredholm's alternative. First we prove that B is injective. Assume there exists $\varphi \in X$ such as :

$$B\varphi = 0. \tag{4.3}$$

Then (4.3) is equivalent to :

$$A\varphi = -\sum_{i=1}^{2} b_i \partial_i \varphi, \tag{4.4}$$

and theorem 2.1 yields :

$$\varphi \in W^{2,p}(\Omega) \cap X. \tag{4.5}$$

Due to Sobolev's injections (see Gri [6]) since Ω is $2-D$, for all $r>1$ we have :

$$W^{2,r}(\Omega) \cap X \hookrightarrow W_0^{1,2}(\Omega),$$

so $\varphi \in W_0^{1,2}(\Omega)$ and satisfies (4.3). The weak maximum principle (see Gil-Tru [5] theorem 8.1 p. 168) implies $\varphi = 0$. This proves the injectivity of B.

We set $T = A^{-1}$, we know that : $T \in \text{Isom}(Y', X)$ and that $T \in \text{Isom}(L^p(\Omega), W^{2,p}(\Omega) \cap X)$ (see theorem 2.1). Moreover, due to the compact injection of $W^{2,p}(\Omega) \cap X$ into X we have : T is a compact operator from $L^p(\Omega)$ into X. We deduce that the operator :

$$T \sum_{i=1}^{2} b_i \partial_i \tag{4.6}$$

is compact as operator from X into X.

Thus the operator $TB = I + T \sum_{i=1}^{2} b_i \partial_i$ is a Fredholm's operator from X into X of zero index since the kernel of $(I + T \sum_{i=1}^{2} b_i \partial_i)$ is $\{0\}$.

The Fredholm's alternative applies and $TB \in \text{Isom}(X, X)$ since B is injective from X into Y'. We conclude that $B \in \text{Isom}(X, Y')$ since $T \in \text{Isom}(Y', X)$.

Lemma 4.1 is proved.

#

Now, we introduce an approximation T_h of operator T by a Petrov-Galerkin procedure, and we give a result of convergence of T_h toward T.

$$T_h : Y' \to X_h$$
$$f \mapsto T_h f \text{ satisfying :}$$

$$a(T_h f, v_h) = \langle f, v_h \rangle \quad \forall v_h \in Y_h. \tag{4.7}$$

The operator T_h is well defined because of theorem 3.1 (see Azi-Bab [1] for example).

When h goes to zero, we have the following :

Lemma 4.2. Assume hypotheses H1) to H3) hold. Then :

$$\lim_{h\to 0} \|T - T_h\|_{\mathcal{L}(L^p(\Omega),X)} = 0. \tag{4.8}$$

Proof. A standard interpolation result combined with the regularity property of T; $T \in \mathrm{Isom}(L^p(\Omega), W^{2,p}(\Omega) \cap X)$ allow us to prove (4.8).

For all $f \in Y'$, by following the classical procedure to evaluate

$$a(T_h f - \varphi_h, v_h) \quad \forall v_h \in Y_h, \tag{4.9}$$

Inf-Sup condition (3.1) yields the estimate :

$$\|T_h f - \varphi_h\|_X \leq \frac{|||a|||}{\beta} \inf_{\varphi_h \in X_h} \|Tf - \varphi_h\|_X, \tag{4.10}$$

where $|||a|||$ stands for the norm of the bilinear form $a(\cdot,\cdot)$.

We denote by π_h the Lagrange's interpolation operator from $C^0(\overline{\Omega})$ into X_h and we have (see Cia [3]) :

$$\|\varphi - \pi_h \varphi\|_X \leq c\, h \|\varphi\|_{2,p} \quad \forall \varphi \in X \cap W^{2,p}(\Omega). \tag{4.11}$$

The triangular inequality, estimate (4.10) and the regularity of T combined with the interpolation result (4.11) lead to :

$$\begin{aligned} \|T_h f - Tf\|_X &\leq (1 + \frac{|||a|||}{\beta})\, c\, h \|T\|_{\mathcal{L}(L^p(\Omega), X \cap W^{2,p}(\Omega))} \|f\|_{L^p(\Omega)} \\ &\leq \tilde{c}\, h \|T\|_{\mathcal{L}(L^p(\Omega), X \cap W^{2,p}(\Omega))} \|f\|_{L^p(\Omega)}. \end{aligned} \tag{4.12}$$

with $\tilde{c} = (1 + \frac{|||a|||}{\beta})\, c$.

Lemma 4.2 is proved.

#

Now, let us define the bilinear form $b(\cdot,\cdot)$ associated to the operator B by :

$$\begin{aligned} b(\cdot,\cdot) \quad & X \times Y \to IR \\ & (u,v) \mapsto b(u,v) = \langle Bu, v \rangle \end{aligned} \tag{4.13}$$

where the brackets $\langle \cdot,\cdot \rangle$ stand for the duality pairing between Y' and Y.

The main result of this section is :

Theorem 4.1. Assume hypotheses H1) to H3) hold. Then there exist $h_0 > 0$ and constant $\beta_1 > 0$ such as for all $h \leq h_0$ we have :

$$\underset{\substack{u \in X_h \\ |u|_{1,p}=1}}{Inf} \quad \underset{\substack{v \in Y_h \\ |v|_{1,q}=1}}{Sup} \quad b(u,v) \geq \beta_1 \tag{4.14}$$

$$\underset{\substack{v \in Y_h \\ |v|_{1,q}=1}}{Inf} \quad \underset{\substack{u \in X_h \\ |u|_{1,p}=1}}{Sup} \quad b(u,v) \geq \beta_1 \tag{4.15}$$

Proof. Let us prove item (4.14), then (4.15) will be proved by a transposition technique in the same way as in theorem 3.1.

We have :

$$T B \varphi = \varphi + T \sum_{i=1}^{2} b_i \, \partial_i \varphi \quad \forall \varphi \in X, \tag{4.16}$$

and from theorem 2.1 and lemma 4.1, we deduce :

$$T B \in \text{Isom}(X, X). \tag{4.17}$$

It is not difficult to check that :

$$T_h B \varphi = \varphi + T_h \sum_{i=1}^{2} b_i \, \partial_i \varphi \quad \forall \varphi \in X_h. \tag{4.18}$$

We can write :

$$I + T_h \sum_{i=1}^{2} b_i \, \partial_i = I + T \sum_{i=1}^{2} b_i \, \partial_i + (T_h - T) \sum_{i=1}^{2} b_i \, \partial_i \varphi. \tag{4.19}$$

We choose h_0 such as :

$$h_0 < \frac{1}{\tilde{c} \max\limits_{1 \leq i \leq 2} \| b_i \|_{L^\infty(\bar{\Omega})} \| T \|_{\mathcal{L}(L^p(\Omega), X \cap W^{2,p}(\Omega))}}, \tag{4.20}$$

where the constant $\tilde{c}$ is defined in (4.12). It is then straightforward that

$$\sup_{\substack{\varphi \in X \\ |\varphi|_{1,p}}} \| (T - T_h) \sum_{i=1}^{2} b_i \, \partial_i \, \varphi \|_X < \| [I + T \sum_{i=1}^{2} b_i \, \partial_i]^{-1} \|_{\mathcal{L}(X,X)}^{-1}$$

which implies that $I+T_h \sum_{i=1}^{2} b_i \partial_i$ is invertible from X into itself with a uniformly bounded inverse with respect h (see Sch [13] theorem T2 XIV,7,1 p. 178) :

$$\| [I+T_h \sum_{i=1}^{2} b_i \partial_i]^{-1} \|_{\mathcal{L}(X,X)} \leq \frac{1}{\| [I+T \sum_{i=1}^{2} b_i \partial_i]^{-1} \|_{\mathcal{L}(X,X)}^{-1} - \| (T-T_{h_0}) \sum_{i=1}^{2} b_i \partial_i \|_{\mathcal{L}(X,X)}}$$

$$\forall h \leq h_0. \tag{4.21}$$

The definition of T_h (see (4.7)), the definition of B (see (4.1)), the definition of $b(\cdot,\cdot)$ (see (4.13) and theorem 3.1) provide :

$$\beta |T_h B\varphi|_{1,p} |v|_{1,q} \leq a(T_h B\varphi, v) = \langle B\varphi, v\rangle = b(\varphi, v) \quad \forall v \in Y_h. \tag{4.22}$$

Combining the following inequality :

$$\beta |\varphi|_{1,p} \leq \beta \left| [T_h B]^{-1} [T_h B] \varphi \right|_{1,\varphi} \tag{4.23}$$

with (4.18) and (4.21), if we set :

$$\beta_1 = \frac{\beta}{\| [I+T \sum_{i=1}^{2} b_i \partial_i]^{-1} \|_{\mathcal{L}(X,X)}^{-1} - \| (T-T_{h_0}) \sum_{i=1}^{2} b_i \partial_i \|_{\mathcal{L}(X,X)}},$$

we get :

$$\beta_1 \leq \inf_{\substack{\varphi \in X_h \\ |\varphi|_{1,p}=1}} \sup_{\substack{v \in Y_h \\ |v|_{1,q}=1}} b(\varphi, v) \quad \forall h \leq h_0.$$

Theorem 4.1 is proved.

#

Remarks 4.1. Theorem 4.1 allows us to get a priori error estimates in $|\cdot|_{1,p}$ norm for convection diffusion problem defined with operator B (see Azi-Bab [1]). Theorem 4.1 combined with the Aubin-Nitche procedure lead to a priori error estimates in $L^r(\Omega)$ norms.

Theorem 4.1 can be extend to operator $\tilde{B}$ with 0 order terms :

$$\tilde{B} = \sum_{i,j=1}^{2} \partial_j a_{ij} \partial_i + \sum_{i=1}^{2} b_i \partial_i + b_0$$

with $b_0 \in L^\infty(\Omega)$, as long as b_0 is not a generalized eigenvalue of B. We then get a priori error estimates for non coercive problems in various norms.

REFERENCES

[1] Aziz I.K., Babuska I.
Survey lecture on the mathematical foundations of the finite element method with applications to partial differential equations. Edited by Aziz, Academic Press, New York and London 1972.

[2] Bernardi C.
Optimal finite element interpolation on curved domains. SIAM J. Numer. Anal. 26 (1989), p. 1212-1240.

[3] Ciarlet P.G.
The finite element method for elliptic problems. Studies in Mathematics and its Applications, North Holland 1978.

[4] Dauge M.
Neumann and mixed problems on curvilinear polyhedra. Integral equations and operator theory, Vol. 15, N° 2, 1992.

[5] Gilbarg D., Trudinger N.S.
Elliptic partial differential equations of second order. A serie of comprehensive studies in Mathematics, Springer-Verlag 1983.

[6] Grisvard P.
Elliptic problems in non smooth domains. Monograph and studies in Mathematics, Pitman 1985.

[7] Pironneau O.
Méthode des éléments finis pour les fluides. Collection RMA 7, Masson 1988.

[8] Pousin J., Rappaz J.
Consistance, stabilité, erreurs a priori et a posteriori pour des problèmes non linéaires. C.R. Acad. Sci. Paris, t. 312, série I, p. 699-703, 1991.

[9] Pousin J., Rappaz J.
Consistency, stability, a priori and a posteriori errors for Petrov-Galerkin methods applied to nonlinear problems. Preprint Dept of Math., Swiss Federal Institute of Technology (1992).

[10] Rannacher R., Scott R.
Some optimal error estimates for piecewise linear finite element approximations. Mathematics of computations, Vol. 38, N° 158, p. 437-665, 1982.

[11] Rudin W.
Functional Analysis. Mac Graw Hill, Series in Higher Mathematics 1973.

[12] Schechter M.
Principles of functional analysis. Academic Press 1971.

[13] Schwartz L.
Topologie générale et analyse fonctionnelle. Hermann 1970.

Numerical Method for Shape Identification Problems

Jean R. Roche and Jan Sokołowski*
Université de Nancy I
Departement de Mathematiques, B.P. 239
U.R.A-C.N.R.S. 750
Projet NUMATH INRIA Lorraine
54506 Vandoeuvre-Les-Nancy Cedex, France

1. Introduction

In the present paper a class of shape identification problems is considered. Such problems arise, e.g. in applications to non-destructive identification of cracks, inclusions or voids in solids. The available information (data) are given on a part of the geometrical boundary of the solid (geometrical domain) and the inclusion is to be determined in the interior of the geometrical domain.

In the present paper we restrict ourselves to the class of problems associated with the stationary heat conduction in 2-D. We refer the reader to [1] for a description of the problem and interesting results on the identifiability of cracks on the basis of boundary measurements.

An existance result is established for the class of problems under considerations by an application of a recent result by V. Sverak [6].

The material derivative method [2] is used in order to derive the first order optimality conditions for the shape identification problem. Finally, the boundary element method is applied to numerical solution of the resulting shape optimization problem.

* Systems Research Institute of the Polish Academy of Sciences, ul. Newelska 6, 01-744 Warszawa, Poland. This author gratefully ackowledges the support of the State Committee for Scientific Research of the Republic of Poland, grant # 2 1207 9101

2. Shape identification

Given a bounded domain $D \subset \mathbb{R}^2$ with the boundary ∂D. Denote

$$D_\rho = \{x \in D | \text{dist}(x, \partial D) > \rho\}$$

and define family of open sets Ω of the form

$$\mathcal{U}_{ad} = \{\Omega \subset D | \Omega = D \setminus S, \ S \subset D_\rho\}$$

for $\rho > 0$, ρ small enough.

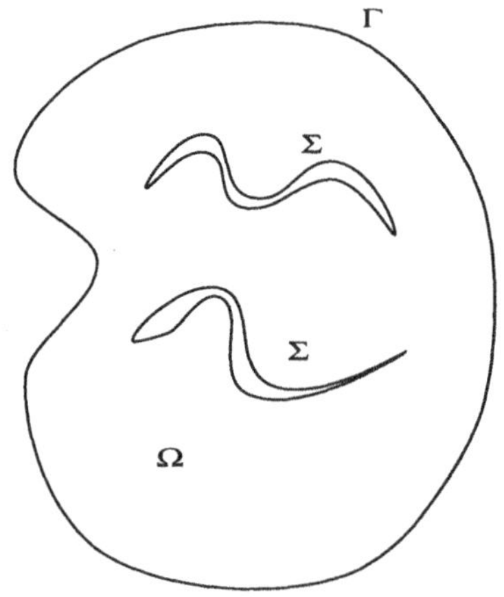

Fig. 1. Domain with inclusions.

Consider the following cost functional

$$J(S) = \int_\Omega |\nabla(w - z)|^2 + \alpha \|w - z\|^2_{H^{\frac{1}{2}}(\Gamma)}$$

where $\alpha \geq 0$ is a constant, $w, u \in H^1(D)$ are given as the unique solutions to the following elliptic equations,

(I) Dirichlet problem

$$\Delta w = 0 \quad \text{in } \Omega$$
$$w = g \quad \text{on } \partial D$$
$$w = 0 \quad \text{on } S$$

(II) Dirichlet–Neumann problem

$$\Delta u = 0 \quad \text{in } \Omega$$
$$\frac{\partial u}{\partial n} = f \quad \text{on } \partial D$$
$$u = 0 \quad \text{on } S$$

Here $g, f \in C(\partial D)$ are given functions, i.e. data of the identification problem under considerations. In particular, such a shape optimization problem can be obtained for identification of cracks, voids or inclusions in solids using as data the temperature and flux distributions on the boundary of solids.

For the shape optimization problem the following results can be established.

(i) The existence of a solution to the problem.

(ii) The first order necessary optimality conditions using the material derivative method.

(iii) Numerical methods of solution by using the integral equations on $\partial\Omega$, i.e. BEM.

3. Existence of an optimal domain.

Consider the following family of admissible domains

$$\mathcal{O}_\ell = \{\Omega \in \mathcal{U}_{ad} | \#S \leq \ell, \quad S = D \setminus \Omega\}$$

where $\#S$ denotes the number of connected components of compact S.

Theorem 1 *For $\alpha > 0$ and any finite ℓ there exists a solution $\Sigma \in \mathcal{O}_\ell$ to the shape identification problem under considerations, i.e.,*

$$J(\Sigma) = \min_{S \in \mathcal{O}_\ell} J(S)$$

Remark. The proof of this theorem is based on the properties of harmonic functions in $\mathbb{R}^2$ and uses the same argument as given in [6] for a slightly different shape optimization problem. In particular the method cannot be directly used in $\mathbb{R}^N$ for $N \geq 3$.

Proof. Let $\{S_i\}$, $i = 1, 2, \ldots$, be a minimizing sequence for the problem under considerations.

Since the cost functional is bounded by a constant, $J(S_i) \leq J(S_1)$, it follows that the corresponding sequences of solutions $\{u_i\}$, $\{w_i\}$, are bounded, i.e.

$$\|u_i\|_{H^1(D)} \leq C$$
$$\|w_i\|_{H^1(D)} \leq C$$

There exists a function $W \in H^1(D)$ such that $w_i - W \in H_0^1(\Omega_i)$. On the other hand, denote by $\tilde{U}_i \in H^1(D \setminus D_{\frac{\ell}{2}})$ a solution to the following elliptic equation

$$\Delta \tilde{U}_i = \Delta \eta u_i + 2\nabla \eta \cdot \nabla u_i \text{ in } D \setminus D_{\frac{\ell}{2}}$$
$$\frac{\partial \tilde{U}_i}{\partial n} = f \text{ on } \partial D$$
$$\tilde{U}_i = 0 \text{ on } \partial D_{\frac{\ell}{2}}$$

and denote

$$U_i = \begin{cases} \tilde{U}_i & \text{in } D \setminus D_{\frac{\varrho}{2}} \\ 0 & \text{in } D_{\frac{\varrho}{2}} \end{cases}$$

where the function $\eta \in C_0^\infty(\mathbb{R}^2)$ satisfies the following conditions

$$\begin{aligned} &0 \le \eta(x) \le 1 \text{ in } \mathbb{R}^2 \\ &\eta(x) = 0 \text{ in } D_{\frac{\varrho}{3}} \\ &\eta(x) = 1 \text{ in } D \setminus D_{\frac{\varrho}{6}} \end{aligned}$$

therefore $\Delta\eta u_i + 2\nabla\eta \cdot \nabla u_i \in L^2(\Omega_i)$ for $i = 1, 2, \ldots$. Then, for $f \in H^{\frac{1}{2}}(\Gamma)$, it follows that

$$\begin{aligned} &\tilde{U}_i \in H^2(D \setminus D_{\frac{\varrho}{2}}) \text{ and} \\ &U_i \in H^1(D) \text{ for } i = 1, 2, \ldots \\ &U_i \rightharpoonup \overline{U} \text{ in } H^1(D) \\ &u_i - U_i \in H_0^1(\Omega_i) \text{ for } i = 1, 2, \ldots \end{aligned}$$

The result of V. Sverak [6] can be formulated in the following way. There exists a subsequence, still denoted by $\{S_i\}$, which converges to $\overline{S}$ in the Hausdorff metric, such that the sequence of metric projections

$$\mathcal{P}_i \; : \; H_0^1(D) \mapsto H_0^1(\Omega_i)$$

converges strongly to the metric projection

$$\overline{\mathcal{P}} \; : \; H_0^1(D) \mapsto H_0^1(D \setminus \overline{S})$$

Therefore, we can pass to the limit on both sides of the following equalities

$$\begin{aligned} w_i - W &= \mathcal{P}_i(w_i - W) \\ u_i - U_i &= \mathcal{P}_i(u_i - U_i) \end{aligned}$$

and we obtain for the weak limits

$$\begin{aligned} &w_i - W \rightharpoonup \overline{w} - W \text{ in } H^1(D) \\ &u_i - U_i \rightharpoonup \overline{u} - \overline{U} \text{ in } H^1(D) \end{aligned}$$

the following equality

$$\begin{aligned} \overline{w} - W &= \overline{\mathcal{P}}(\overline{w} - W) \\ \overline{u} - \overline{U} &= \overline{\mathcal{P}}(\overline{u} - \overline{U}) \end{aligned}$$

which completes the proof of Theorem 1.

4. Optimality conditions

In order to derive the first order necessary optimality conditions using the shape derivatives (instead of material derivatives) we assume that an optimal solution $\Sigma \in \mathcal{O}_\ell$ is a domain with the Lipschitz boundary $\partial\Sigma$. In the case of a crack given by a Lipschitz curve i.e., with the Lebesgue measure $|\Sigma| = 0$, the optimality conditions can be obtained by an application of the material derivative method, taking into account the singularity coefficients at the tips of the crack.

Theorem 2 *Assume that $\alpha = 0$ and $\ell = 1$. If $\Sigma \in \mathcal{O}_1$ is an optimal solution, then*

$$\frac{\partial u}{\partial n} = \frac{\partial w}{\partial n} \quad \text{on } \partial\Sigma$$

The proof of the theorem is based on the fact that for $\Sigma \subset \Omega_\rho$ the shape derivative of the cost functional $J(\cdot)$ is given in the following form

$$dJ(\Sigma; V) = \int_{\partial\Sigma} \left[\left(\frac{\partial u}{\partial n}\right)^2 - \left(\frac{\partial w}{\partial n}\right)^2 \right] V \cdot n d\sigma$$

for any vector field $V(\cdot,\cdot)$ with the compact support in D, n denotes a unit normal vector on $\partial\Omega$ directed outside of Ω.

Indeed, the cost functional takes the form

$$J(S) = \inf_{\varphi \in H_0^1(\Omega)+g} \int_\Omega |\nabla\varphi|^2 dx - \inf_{\varphi \in H_S^1(\Omega)} \left[\int_\Omega |\nabla\varphi|^2 dx - 2\int_\Gamma f\varphi d\Gamma(x) \right] - 2\int_\Gamma g f d\Gamma(x)$$

where

$$\begin{aligned}
&\Omega = D \setminus S \\
&\partial\Omega = \partial D \cap \partial S, \ \partial D = \Gamma \\
&H_S^1(\Omega) = \{\varphi \in H^1(D) | \varphi = 0 \text{ on } S\} \\
&H_0^1(\Omega) + g = \{\varphi \in H_S^1(\Omega) | \varphi = g \text{ on } \Gamma\}
\end{aligned}$$

therefore, the form of the shape derivative $dJ(\Sigma; V)$ is obtained in a standard way for the energy type shape functionals.

5. Integral equations

For the Dirichlet problem (I) we have the following integral equation

$$\int_{\partial\Omega} (g(x) - g(y)) \frac{\partial \ln|x-y|}{\partial n_x} d\sigma(x) = \int_{\partial\Omega} \frac{\partial w_{\text{int}}}{\partial n_x} \ln|x-y| d\sigma(x)$$

with the unknown function $\frac{\partial w_{\rm int}}{\partial n}$.
For the Dirichlet–Neumann problem (II) the integral equations take the following form:

(i) For $y \in \Gamma$,

$$\frac{\partial u_{\rm int}}{\partial n_y} - \frac{1}{\pi}\int_\Gamma \frac{\partial u_{\rm ext}}{\partial n_x}\frac{\partial \ln|x-y|}{\partial n_y}d\sigma(x) + \frac{1}{\pi}\int_\Sigma \frac{\partial u_{\rm int}}{\partial n_x}\frac{\partial \ln|x-y|}{\partial n_y}d\sigma(x) = h(y)$$

with

$$h(y) = -f(y) - \frac{1}{\pi}\int_\Gamma f(x)\frac{\partial \ln|x-y|}{\partial n_y}d\sigma(x)$$

and the unknown functions $\frac{\partial u_{\rm int}}{\partial n}, \frac{\partial u_{\rm ext}}{\partial n}$.

(ii) For $y \in \Sigma$,

$$\frac{1}{\pi}\int_\Gamma \frac{\partial u_{\rm ext}}{\partial n_x}\frac{\partial \ln|x-y|}{\partial n_y}d\sigma(x) - \frac{1}{\pi}\int_\Sigma \frac{\partial u_{\rm int}}{\partial n_x}\frac{\partial \ln|x-y|}{\partial n_y}d\sigma(x) = m(y)$$

where

$$m(y) = \frac{1}{\pi}\int_\Gamma f(x)\frac{\partial \ln|x-y|}{\partial n_y}d\sigma(x)$$

with the unknown functions $\frac{\partial u_{\rm int}}{\partial n}, \frac{\partial u_{\rm ext}}{\partial n}$. Assuming that the function u is continuous across the boundary $\partial\Omega$, the single layer representation can be used to evaluate the unknown solution u on Γ, see the following section for the explicit formulae.

6. Numerical method

The numerical exemple is defined in the following way. Let us denote by

$$\Omega = \{x \in \mathbb{R}^2 | \|x-y\| < 0.9, y \in \mathbb{R}^2\}$$

a ball in $\mathbb{R}^2$ of the radius $r = 0.9$ and the center at $y \in \mathbb{R}^2$. We assume that the inclusion (void) takes the following form,

$$S = \{x \in \mathbb{R}^2 | (\frac{x_1}{a+b})^2 + (\frac{x_2}{a-b})^2 = 1\}\ .$$

That means that we consider a small ellipse as the inclusion.

A harmonic function $u = u(x)$, $x \in \mathbb{R}^2 \setminus S$, can be constructed in such a way that $u = 0$ on $\Sigma = \partial S$. We briefly describe the construction.

If Φ is an conformal mapping that maps the ellipse S onto the unit circle and the exterior of the ellipse S onto the exterior of the circle, then $u(x,y) = \ln|\Phi^{-1}(\xi_1,\xi_2)|$ is a harmonic function such that $u_{|\Sigma} = 0$ and the $\frac{\partial u}{\partial n}$ is given on $\Gamma \cup \Sigma$ in an explicit form.

An approximation of the continuous problem is introduced in order to obtain numerical results, therefore Ω_h is an approximation of Ω with the piecewise linear exterior boundary Γ_h. In the same way S_h with piecewise linear boundary Σ_h is defined for the inclusion S.

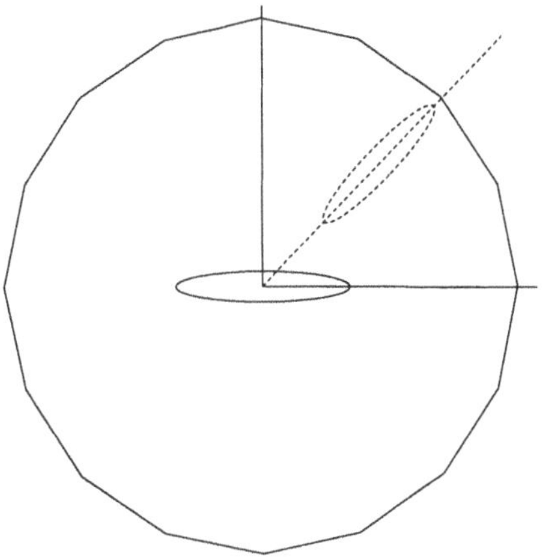

Fig. 2. Approximation of geometrical domain.

Each linear segment of the boundary $\partial\Omega_h \equiv \Gamma_h \cup \Sigma_h$ can be considered as a geometrical finite element and parametrized on the reference segment [0,1].

The numerical method consists in constructing of a minimizing sequence, denoted by $\{S_h^k, u_h^k, w_h^k)\}_{k=1}^{\infty}$, for the discrete shape functional $J_h(S_h)$. The existence of a solution to the shape optimization problem is shown provided that the inclusion is strictly included in Ω, therefore we introduce an additional constraint for the discrete problem to assure that the distance from the inclusion S_h to the boundary Γ_h is less or equal to ρ, ρ is given.

First, an inclusion S_h of the given form is considered, thus only the position of S_h is to be determined. Then, in fact the minimizing sequence of inclusions S_h^k is determined by a sequence in $\mathbb{R}^3$, since S_h^k depends on three parameters, the center $a^k = (a_1^k, a_2^k)$ of the ellipse S_h^k and the angle θ_k of the principal axes of the ellipse.

The minimizing sequence is obtained by locally optimal displacements of the nodes of the boundary Σ_h^k according to a given vector field V^k which defines a rotation combined with a translation. In the following we denote for simplicity e.g., Σ_h for Σ_h^k, i.e. the index k will be omitted.

For the vector field parametrized by θ, a_1, a_2 the continuos gradient takes the following form:

$$
\begin{aligned}
dJ(\Sigma; V) &= \int_{\Sigma} \left[\left(\frac{\partial u}{\partial n}\right)^2 - \left(\frac{\partial w}{\partial n}\right)^2 \right] V \cdot n d\sigma \\
&= C(a_1, a_2, \theta, w, u)
\end{aligned}
$$

To compute a solution of the Euler equation $C(a_1, a_2, \theta, w, u) = 0$ a projected gradient method combined with the Wolfe line search algorithm is used.

At each iteration we evaluate functions $\frac{\partial w}{\partial n}$, $\frac{\partial u}{\partial n}$ on Σ_h and $\frac{\partial w}{\partial n}$, u on Γ_h in order to compute the continuos gradient and the value of the cost funtional,

$$J_h(S_h) = \int_{\partial\Omega_h} (g(x)\frac{\partial w(x)}{\partial n} + f(x)u(x) - 2g(x)f(x)]d\sigma(x)$$

The integral equation associated to the Dirichlet problem takes the following form, for $y \in \Gamma_h$,

$$\int_{\partial\Omega_h} (g(x) - g(y))\frac{\partial \ln|x-y|}{\partial n_x} d\sigma(x) = \int_{\partial\Omega_h} q(x)\ln|x-y| d\sigma(x)$$

and for $y \in \Sigma_h$,

$$\int_{\partial\Omega_h} g(x)\frac{\partial \ln|x-y|}{\partial n_x} d\sigma(x) = \int_{\partial\Omega_h} q(x)\ln|x-y| d\sigma(x)$$

In order to solve the above system of two integral equations, the Galerkin algorithm [2], [3] is used. The following variational fomulation is obtained,

$$b(q, q') = l(q')$$

with

$$b(q, q') = \int_{\partial\Omega_h}\int_{\partial\Omega_h} q(x)q'(x)\ln|x-y| d\sigma(x)d\sigma(x)$$

and

$$l(q') = \int_{\Gamma_h}\int_{\Gamma_h} (g(x) - g(y))\frac{\partial \ln|x-y|}{\partial n_x} d\sigma(x) \ .$$

Now, $q_h(x)$ is a piecewise constant aproximation of $q(x)$. Then, if we denote by L_k each segment of the boundary Γ_h and N_k denotes a finite element on the boundary Σ_h, the following linear system is derived,

$$\begin{pmatrix} A & B \\ & \\ C & D \end{pmatrix} \begin{pmatrix} q_1^1 \\ \cdot \\ q_i^1 \\ \cdot \\ \cdot \\ q_i^2 \\ \cdot \\ q_M^2 \end{pmatrix} = \begin{pmatrix} f_1 \\ \cdot \\ f_i \\ \cdot \\ \cdot \\ f_i \\ \cdot \\ f_{N+M} \end{pmatrix}$$

where

$$a_{ij} = \int_{L_i}\int_{L_j} \ln|x-y| d\sigma(x)d\sigma(y)$$

$$b_{ij} = \int_{L_i}\int_{N_j} \ln|x-y| d\sigma(x)d\sigma(y)$$

$$c_{ij} = \int_{N_i}\int_{L_j} \ln|x-y|d\sigma(x)d\sigma(y)$$

$$d_{ij} = \int_{N_i}\int_{N_j} \ln|x-y|d\sigma(x)d\sigma(y)$$

and

$$f_i = \begin{cases} \int_{L_i}\int_{\Gamma_h}(g(x)-g(y))\frac{n_x\cdot(x-y)}{|x-y|}d\sigma(x)d\sigma(y) & 1 \le i \le N \\ \int_{N_i}\int_{\Gamma_h}(g(x)-g(y))\frac{n_x\cdot(x-y)}{|x-y|}d\sigma(x)d\sigma(y) & N \le i \le N+M \end{cases}$$

The above linear system is symmetric and the associated matrix is positive definite. We use a conjugate gradient method to solve the system.

For the mixed Dirichlet-Neumann problem we solve the following system of integral equations,
for $y \in \Gamma_h$,

$$q(y) - \frac{1}{\pi}\int_{\Gamma_h} q(x)\frac{\partial \ln|x-y|}{\partial n_y}d\sigma(x) + \frac{1}{\pi}\int_{\Sigma_h} q(x)\frac{\partial \ln|x-y|}{\partial n_y}d\sigma(x) = h(y)$$

with

$$h(y) = -f(y) - \frac{1}{\pi}\int_{\Gamma_h} f(x)\frac{\partial \ln|x-y|}{\partial n_y}d\sigma(x)$$

otherwise, for $y \in \Sigma_h$

$$\frac{1}{\pi}\int_{\Gamma_h} q(x)\frac{\partial \ln|x-y|}{\partial n_y}d\sigma(x) - \frac{1}{\pi}\int_{\Sigma_h} l(x)\frac{\partial \ln|x-y|}{\partial n_y}d\sigma(x) = m(y)$$

where

$$m(y) = \frac{1}{\pi}\int_{\Gamma_h} f(x)\frac{\partial \ln|x-y|}{\partial n_y}d\sigma(x) .$$

To solve the latter system of integral equations we introduce again a piecewise constant approximation of unknown functions $q(x)$, $l(x)$ on Γ_h and Σ_h, respectively. Then, with the same notation as for the previous equations we obtain the following linear system,

$$\begin{pmatrix} A & B \\ & \\ C & D \end{pmatrix} \begin{pmatrix} q_1 \\ \cdot \\ q_i \\ \cdot \\ \cdot \\ l_i \\ \cdot \\ l_M \end{pmatrix} = \begin{pmatrix} h_1 \\ \cdot \\ h_i \\ \cdot \\ \cdot \\ m_i \\ \cdot \\ m_M \end{pmatrix}$$

where

$$a_{ij} = \int_{L_i} d\sigma(y) - \frac{1}{\pi}\int_{L_i}\int_{L_j} \frac{n_y\cdot(x-y)}{|x-y|^2}d\sigma(x)d\sigma(y)$$

$$b_{ij} = \frac{1}{\pi}\int_{L_i}\int_{N_j} \frac{n_y \cdot (x-y)}{|x-y|^2} d\sigma(x) d\sigma(y)$$

$$c_{ij} = -\frac{1}{\pi}\int_{N_i}\int_{L_j} \frac{n_y \cdot (x-y)}{|x-y|^2} d\sigma(x) d\sigma(y)$$

$$d_{ij} = \frac{1}{\pi}\int_{N_i}\int_{N_j} \frac{n_y \cdot (x-y)}{|x-y|^2} d\sigma(x) d\sigma(y)$$

and the right-hand side is given by:

$$h_i = -\int_{L_i} f(y) d\sigma(y) - \frac{1}{\pi}\int_{L_i}\int_{\Gamma_h} f(x)\frac{n_y \cdot (x-y)}{|x-y|^2} d\sigma(x) d\sigma(y)$$

and

$$m_i = \frac{1}{\pi}\int_{N_i}\int_{\Gamma_h} f(x)\frac{n_y \cdot (x-y)}{|x-y|^2} d\sigma(x) d\sigma(y)$$

Once, $q_h = (q_1, ..., q_N)$ and $l_h = (l_1, ..., l_M)$ computed we can evaluate an approximation of the solution $u(x)$ using the following integral formulae.

$$u(y) = -\frac{1}{2\pi}\Big(\int_{\Gamma_h} (f(x) - q_h(x)) \ln|x-y| d\sigma(x) + \int_{\Sigma_h} l_h(x) \ln|x-y| d\sigma(x)\Big)$$

$$u(y) = -\frac{1}{2\pi}\Big(\sum_{i=1}^{i=N}\int_{L_i} (f(x) - q_i) \ln|x-y| d\sigma(x) + \sum_{i=1}^{i=M} l_i \int_{N_i} \ln|x-y| d\sigma(x)\Big)$$

This approximation of the solution u is given with precision of order h, the length of the finite element.

The numerical results will be reported in a forthcoming paper.

References

1. A. Ben Abda, Sur quelques problèmes inverses géométriques via des équations de conduction elliptiques : Etude théorique et numérique. THESE, Ecole Nationale d'Ingénieurs de Tunis, 1993.
2. R. Dautray and J.L. Lions, Mathematical Analysis and Numerical Methods for Science and Technology Vol. 4, Springer Verlag, Berlin, 1984.
3. R. Kress, Linear Integral Equations. Springer Series in Applied Mathematical Sciences Vol. 82, Berlin, 1989.
4. F. Santosa and M. Vogelius, A Computational Algorithm to Determine Cracks from Electrostatic Boundary Measurements. Int. J. Engng. Sci. Vol. 29, 1991, 917–937.
5. J. Sokołowski and J.P. Zolesio, Introduction to Shape Optimization. Shape sensitivity analysis. Springer Series in Computational Mathematics Vol. 16, New York, 1992.
6. On optimal shape design. J. Math. Pures Appl. 72, 1993, 537–551.

Partial Regularity of Weak Solutions to Certain Parabolic Equations

Giulia Sargenti
Dipartimento di Matematica, Università di Roma La Sapienza
Piazzale Aldo Moro 2, 00185 Roma, Italy
sargenti@mat.uniroma1.it

Vincenzo Vespri
Dipartimento di Matematica Pura e Applicata, Università dell'Aquila
Via Vetoio, 67010 Coppito (AQ), Italy
vespri@vxscaq.aquila.infn.it

1 Introduction

In this note we study the partial regularity of bounded weak solutions of nonlinear parabolic equations of the type

$$(\beta(u))_t = \text{Div}\, a(x,t,u,\nabla u) \quad \text{in } D'(\Omega_T) \tag{1.1}$$

$$u \in L^2_{\text{loc}}(0,T;L^2_{\text{loc}}(\Omega)). \tag{1.2}$$

Here Ω is a domain in $\mathbb{R}^N$, $N \geq 3$. For $T > 0$ we denote by Ω_T the cylindrical domain $\Omega_T = \Omega \times (0,T)$.

Assume $\beta(\cdot)$ be a maximal monotone graph in $\mathbb{R} \times \mathbb{R}$ satisfying:

$$\beta(s_1) - \beta(s_2) \geq c_0(s_1 - s_2) \quad \forall s_1, s_2 \in \mathbb{R} \tag{1.3}$$

for some given constant $c_0 > 0$,

$$\sup_{|u|\leq M} |\beta(u)| < \infty \tag{1.4}$$

for every $M > 0$. Furthermore we say that a point s is of singularity for β if there β exhibits a jump or a growth more than polynomial. We assume:

β has only a finite numbers of singular points in any finite interval (1.5)

The function $a : \Omega_T \times \mathbb{R}^{N+1} \to \mathbb{R}^N$ is only assumed to be measurable and satisfying the structure conditions

$$a(x,t,u,\nabla u)\cdot\nabla u \geq c_1|\nabla u|^2 - \varphi_0(x,t) \tag{1.6}$$

$$|a(x,t,u,\nabla u)| \leq c_2|\nabla u| + \varphi_1(x,t) \tag{1.7}$$

a.e.$(x,t) \in \Omega_T$. Here c_i, $i = 1,2$ are positive constants and φ_i $i = 0,1$ are nonnegative functions defined in Ω_T, subjected to this condition:

$$\varphi_0, \varphi_1^2 \in L^{\hat r}(0,T;L^{\hat q}(\Omega_T)) \tag{1.8}$$

where

$$\frac{1}{\hat r} + \frac{N}{2\hat q} = 1 - \kappa_1$$

and

$\hat q \in [\frac{N}{2(1-\kappa_1)}, \infty]$, $\hat r \in [\frac{N}{1-\kappa_1}, \infty]$ and $\kappa_1 \in (0,1)$.

We also assume u locally bounded in Ω_T, i.e.

$$\|u\|_{\infty,\Omega_T} \leq M \quad \text{for some constant } M > 0 \tag{1.9}$$

Lastly we assume

u is the limit of a sequence of local smooth solutions of approximating problems (1.10)

By a weak solution of (1.1) we mean a function u which satisfies (1.2) and such that if w is a function defined in Ω_T, $w \subset \beta(u)$, then w and u satisfy

$$\int_\Omega w\varphi(x,\tau)dx\,\Big|_{t_0}^{t} + \int_{t_0}^{t}\int_\Omega \{-w\varphi_t + a(x,\tau,u,\nabla u)\cdot\nabla\varphi\}\,dx\,dt = 0$$

for all $\varphi \in W^{1,2}_{loc}(0,T;L^2_{loc}(\Omega)) \cap L^2_{loc}(0,T;W^{1,2}_0(\Omega))$ and all intervals $[t_0,t] \subset (0,T]$.

In order to state our main theorem we recall the definition of the Hausdorff measure (for more details, see for instance [8] and [10]).

Let G be a subset of $\mathbb{R}^N$ and let F be the set of the balls centered in a point P belonging to G. Define

$$\Phi_{k,\varepsilon}(G) = \inf_{F_h \subset F}\left\{\sum_h 2^{-k}(\text{diam } F_h)^k \mid \bigcup_h F_h \supset G \ \text{diam} F_h = \varepsilon\right\}$$

Lastly define

$$H^k(G) = \liminf_{\varepsilon \to 0} \Phi_{k,\varepsilon}(G).$$

We recall that the Hausdorff dimension of a set G is

$$\mathrm{Dim}_H(G) = \inf\{ k \in \mathbb{R}^+ : H^k(G) = 0 \}$$

Fix $\varepsilon > 0$ and a compact set $K \subset \Omega$. Define

$\Sigma = \{x \in K$: there exists $t \in (\varepsilon, T]$ such that (x,t) is a point of discontinuity for u.

In what follows we say that $c = c(data)$ if c is a constant that can be determinated a priori only in terms of N, M, c_i and the distance between $K \times (\varepsilon, T)$ and the parabolic boundary of Ω_T. Now we can state our main result.

Main Theorem *Let* (1.3)–(1.10) *hold and let* u *be a locally bounded weak solution of (1.1),(1.2). Then* Σ *is a closed set. Moreover there is a* $\gamma > 0$ *such that*

$$\int\int_{\Omega_T} |\nabla u(x,t)|^2 \, dx \, dt \geq c \sum_{n=1}^{\infty} \Phi_{N-2,\gamma^n}(\Sigma) \qquad (1.11)$$

where $c = c(data)$ *is a positive constant.*

We stress that our result implies a quantitative estimate concerning the singular set and we feel that is a necessary step in order to understand better this kind of nonlinear phenomena.

Before concluding this section we recall that these equations arise naturally in several physical phenomena: for instance in the case of transition of phase and in the Buckley–Leverett model of two immiscible fluids in a porous medium (see for instance [1], [3], [5] and [11]).

In the early '80s the continuity of local bounded solutions was proved assuming only a point of singularity for β (see [1], [2], [4], [12] and [13]).

The case of a more general β was faced in [7] where the regularity result was proved in the case $N = 2$ for general operators, while in the case $N \geq 3$ the continuity was proved only for local solutions to equation of the type

$$(\beta(u))_t = \Delta u$$

2 Notation and preliminary results

Let $B(x_o, R)$ the ball of radius R and center x_0 in $\mathbb{R}^N$. For every $P_0 = (x_0.t_0) \in \Omega_T$, we denote with $Q_R(P_0)$ the cylinder of vertex at P_0, height R^2 and cross section $B(x_0, R)$,

$$Q_R(P_0) = B(x_0, R) \times (t_0 - R^2, t_0).$$

Assume that $Q_R(P_0) \subset \Omega_T$ and set

$$\mu^+ = \text{ess sup}_{Q_R(P_0)} u \quad \mu^- = \text{ess inf}_{Q_R(P_0)} u$$

$$\omega \geq \text{ess osc}_{Q_R(P_0)} u = \mu^+ - \mu^-.$$

Now, we state some results useful for the sequel.

Proposition 2.1 *The set Σ is a closed set.*

Proof

The graph β has a finite number of singular points. Hence there is a $\delta_0 > 0$ such that the distance between any two singular points is greater than $2\delta_0$. Let

$$K_0 = \{x_0 \in K_0 : \forall t_0 \in [\varepsilon, T] \text{ there exists } R > 0 : \text{ess osc}_{Q_R(x_0,t_0)} u \leq \delta_0\}$$

Note that K_0 is an open set and $K_0^C \subseteq \Sigma$. In order to prove $K_0^C = \Sigma$ we show that for each $x_0 \in K_0$ and $t_0 \in [\varepsilon, T]$, u is continuous in (x_0, t_0). Actually, by the definition of K_0 there is $Q_R(x_0, t_0)$ where the oscillation of u is less than δ_0. But in such a range β has only a point of singularity. Hence by [1], [2], [4], [12] and [13] u is continuous in $Q_{\frac{R}{2}}(x_0, t_0)$ ■

Moreover we need the following proposition that is a weak version of Besichovitch covering theorem (for the proof see Lemma 2.1, Chapter IV of [9]).

Proposition 2.2 *Let A be a bounded set in $\mathbb{R}^N$ and let $r > 0$. Then there exists a sequence of points $x_i \in A$, $i = 1, \cdots, n_r$ such that*

$$B(x_i, r) \bigcap B(x_j, r) = \emptyset \quad \forall i \neq j$$

$$\bigcup_{i=1}^{n_r} B(x_i, 3r) \supseteq A.$$

The following two lemmas come from [7]:

Proposition 2.3 *For each $x_0 \in K$, for each $0 < r < 1$*

$$\int_0^T \int_{B(x_0,r)} |\nabla u|^2 dx dt \leq k_1 r^{N-2} \tag{2.1}$$

where k_1 is a constant depending only upon the data.

Proof

Let θ be a cutoff function such that $0 \leq \theta \leq 1$, $\theta = 1$ in $B(x_0, r)$, $\theta = 0$ outside $B(x_0, 2r)$ and $|\nabla\theta| \leq \frac{c}{r}$. Let ξ be a cutoff function such that $0 \leq \xi \leq 1$, $\theta = 1$ in (ε, T), and $|\xi_t| \leq \frac{c}{\varepsilon}$. Inequality (2.1) follows multipying (1.1) by $\theta(x)\xi(t)u(x,t)$ and repeating an argument similar to the one of Proposition 2.1 of [7] (for more details see also chapter 2 of [7]). ∎

The following result is proved in [7] Proposition 9.1 (to which we refer the reader for more details).

Proposition 2.4 *There is γ_0 such that for each $x_0 \in \Sigma$, for each $0 < r < 1$*

$$\int\int_\Gamma |\nabla u|^2 dx dt \geq k_2 r^{N-2} \tag{2.2}$$

where Γ is a siutable path piecewise parallel to the coordinate axes strictly contained in the annulus $\{\gamma_0 r \leq \|x - x_0\| \leq r\} \times [0, T]$ and k_2 is a constant depending only upon the data.

3 Proof of the main theorem

By the covering proposition 2.2 for each $r \leq 1$ we can find a finite numbers of balls $B(x_i, r)$ such that

$$B(x_i, \frac{r}{3}) \bigcap B(x_j, \frac{r}{3}) = \emptyset \quad \forall i \neq j$$

$$\bigcup_{i=1}^{n_r} B(x_i, r) \supseteq \Sigma.$$

Obviously

$$\Phi_{N-2,r}(\Sigma) \leq n_r r^{N-2} \tag{3.1}$$

On the other hand by proposition 2.4

$$\int_0^T \int_\Omega |\nabla u|^2 dx dt \geq \sum_{i=1}^{n_r} \int\int_{\Gamma_i^r} |\nabla u|^2 dx dt \geq n_r k_2 r^{N-2} \tag{3.2}$$

where Γ_i^r is the suitable path contained in the annula $\{\gamma_0 r \leq \|x - x_i\| \leq r\} \times [0, T]$ introduced in proposition 2.4.

Note that the previous inequalities imply that

$$\int_0^T \int_\Omega |\nabla u|^2 dx dt \geq k_2 \Phi_{N-2,r}(\Sigma)$$

Let $\gamma = \min(\gamma_0, \frac{k_2}{k_1 3})$ where γ_0 is the constant introduced in proposition 2.4. Let $\Gamma_{\gamma^{n_r}} = \bigcup_{i=1}^{n_{\gamma^{n_r}}} \Gamma_i^{\gamma^{n_r}}$. By the proof of proposition 9.1 of [7] we get

$$\int_0^T \int_{B(x_i,r)} |\nabla u|^2 dx dt \geq k_2 \frac{\text{meas}(\Gamma_{\gamma^{n_r}} \cap \Gamma_i^r)}{\text{meas}\Gamma_i^r} \gamma^{-n} r^{N-2}.$$

On the other hand by proposition 2.3 we get

$$k_2 \frac{\text{meas}(\Gamma_{\gamma^{n_r}} \cap \Gamma_i^r)}{\text{meas}\Gamma_i^r} \gamma^{-n} r^{N-2} \leq k_1 r^{N-2}$$

Therefore the previous inequalities imply that

$$\text{meas}(\Gamma_{\gamma^{n_r}} \cap \Gamma_i^r) \leq 3^{-n} \text{meas}\Gamma_i^r. \tag{3.3}$$

Let

$$D_n = \Gamma_{\gamma^{n_1}} \cap (\cup_{m>n} \Gamma_{\gamma^{m_1}})^C$$

By the proof of proposition 9.1 of [7] we have

$$\int\int_{D_m} |\nabla u|^2 dx dt \geq \frac{1}{3} n_{\gamma^m} k_2 \gamma^{m(N-2)} \geq \frac{1}{3} k_2 \Phi_{N-2,\gamma^m}(\Sigma).$$

On the other hand, as D_m are disjoint sets, we get

$$\int\int_{\Omega_T} |\nabla u|^2 dx dt \geq \sum_{i=1}^{\infty} \int\int_{D_i} |\nabla u|^2 dx dt \geq \frac{1}{3} k_2 \sum_{i=1}^{\infty} \Phi_{N-2,\gamma^i}(\Sigma)$$

i.e. the statement .

References

[1] H.W.ALT AND E.DIBENEDETTO, *Non steady flow of water and oil through inhomogeneous porous media*, Ann. Sc. Norm. Sup. Pisa Ser IV, Vol XII (1985) 335-392

[2] L.CAFFARELLI AND L.C.EVANS, *Continuity of the temperature in the two Stefan problem*, Arch. Rat. Mech. Anal. 81 (1983), 199–220.

[3] G.CHAVENT AND J.JAFFRÈ, *Mathematical models and finite elements methods for reservoir simulation*, North-Holland, 1986.

[4] E.DI BENEDETTO, *Continuity of weak solutions to certain singular parabolic equations*, Ann. Mat. Pura Appl. (4) CXXI (1982), 131–176.

[5] E.DI BENEDETTO, *The flow of two immiscible fluids through a porous medium: regularity of the saturation*, IMA Vol.5. Eds Ericksen and Kinderlherer. Theory and applications of liquid crystals. Springer Verlag. New York (1987) 123-141.

[6] E.DIBENEDETTO, *Degenerate parabolic equations*, Springer Verlag Series Universitext, New York, 1993.

[7] E.DI BENEDETTO AND V.VESPRI, *On the singular equation* $(\beta(u))_t = \Delta u$, Arch. Rat. Mech.to appear.

[8] H.FEDERER, *Geometric measure theory*, Berlin – Heildeberg – New York, Springer Verlag, 1969.

[9] M.GIAQUINTA, *Multiple integrals in the calculus of variations and non linear elliptic systems*, Princeton University Press, 1983.

[10] M.GIAQUINTA AND M.STRUWE, *On the partial regularity of weak solutions of non linear parabolic systems*, Math. Zeitschriff, 178 (1982), 437–451.

[11] S.N. KRUZKOV AND S.M. SUKORJANSKI *Boundary value problems for systems of equations of two phase porous flow type : statement of the problems, question of solvability, justification of approximate methods*, Mat. Sbornik 44 (1977) 62-80.

[12] P.SACHS, *The initial and boundary value problem for a class of degenerate parabolic equations*. Comm. Part. Diff. Equs. 8 (1983) 693-734.

[13] W.ZIEMER, *Interior and boundary continuity of weak solutions of degenerate parabolic equations*, Trans. Amer. Math. Soc., 271 (2) (1982), 733–747.

Local Regularity Properties of the Minimum Time Function

Carlo Sinestrari

Dipartimento di Matematica, Università di Roma "Tor Vergata"
Via della Ricerca Scientifica, 00133 Roma, Italy
e-mail: sinestrari@mat.utovrm.it

Abstract

We study the time optimal control problem with a nonlinear state equation and a general target set. We give conditions for the Lipschitz continuity and the semiconcavity of the minimum time function in a neighbourhood of a given point of the controllable set.

1 Introduction

Consider the system

$$\begin{cases} y'(t) = f(y(t), u(t)), & t \geq 0, \\ y(0) = x, & \end{cases} \tag{1.1}$$

where $x \in \mathbb{R}^n$, $f : \mathbb{R}^n \times U \to \mathbb{R}^n$, and $u : [0, \infty[\to U$ are given, U being a compact subset of $\mathbb{R}^m$.

We assume that a nonempty closed set $\mathcal{K} \subset \mathbb{R}^n$ (called the *target*) is given. We denote by $\mathcal{R}$ the set of all points $x \in \mathbb{R}^n$ such that the solution of (1.1) reaches $\mathcal{K}$ in finite time for some choice of the control function u. Then we define the *minimum time function* $T : \mathcal{R} \to [0, +\infty[$ as the infimum over all $u : [0, \infty[\to U$ of the time taken for the solution to reach $\mathcal{K}$.

We are interested here in the regularity properties of T. It is well known that T is not everywhere differentiable, in general. Differentiability results for T have been proved for linear systems if ∂U is smooth, see [4], while Hölder continuity results have been obtained under weaker controllability assumptions, see [10].

Partially supported by the Italian National Project MURST 40% "Problemi nonlineari...".

There is a satisfactory characterization of the systems whose minimum time function is locally Lipschitz continuous. In fact, it has been proved (see [15], [3], [16]) that $T \in Lip_{loc}(\mathcal{R})$ if and only if there exists $\mu > 0$ such that

$$\min_{u \in U} f(\bar{x}, u) \cdot \nu \leq -\mu |\nu| \tag{1.2}$$

for any $\bar{x} \in \partial \mathcal{K}$ and any vector ν perpendicular to $\mathcal{K}$ at $\bar{x}$. If $\partial \mathcal{K}$ is not a smooth manifold the notion of perpendicular vector is to be understood in the generalized sense of [8], which will be recalled in Section 3. For instance, if $\mathcal{K} = \{0\}$, it can be checked that (1.2) holds if and only if the convex hull of $f(0, U)$ is a neighbourhood of the origin. In this special case this condition was first given by Petrov in [14].

In a recent joint work with Cannarsa ([6]) the author has investigated the semiconcavity and semiconvexity properties of T. Such properties are intermediate between Lipschitz continuity and continuous differentiability. According to the usual definition, a function is called semiconvex (semiconcave) if it is the sum of a convex (concave) function with a smooth function. Semiconcave functions are a very useful class for studying optimal control problems with smooth data, see for instance [5]. They have also been used to obtain uniqueness results for weak solutions of Hamilton–Jacobi equations, see [11], [12]. The differential properties of such functions have been studied in [1], [2] and Hausdorff estimates on the sets of non differentiability of a semiconcave function have been obtained.

It is usually more difficult to obtain semiconcavity results for the value function of an optimal control problem with exit time. For instance, in the minimum time problem, this kind of regularity may fail even for a linear system with target $\mathcal{K} = \{0\}$ (see [6]). Therefore some restrictions on the system and on the target are needed.

In [6] we showed that T is semiconcave in $\mathcal{R} \setminus \mathcal{K}$ provided system (1.1) satisfies, in addition to (1.2), the following conditions:

$$\text{the distance function from } \mathcal{K} \text{ is semiconcave in } \mathbb{R}^n \setminus \mathring{\mathcal{K}}, \tag{1.3}$$

$$f(\cdot, u) \text{ is of class } C^2. \tag{1.4}$$

Assumption (1.3) is satisfied if $\partial \mathcal{K}$ is a C^2 manifold. It does not hold instead if $\mathcal{K}$ is a polyhedron or a point.

The aim of this paper is to prove a local version of these regularity results for systems which do not satisfy the above assumptions. For instance, let us suppose that (1.2) is satisfied only on some part of $\partial \mathcal{K}$, and that (1.3) fails because $\partial \mathcal{K}$ has some corners. One cannot expect Lipschitz continuity or semiconcavity of T in the whole domain of definition in this case; nevertheless, we can show that these properties hold in certain subregions of $\mathcal{R}$. In fact, let $x_0 \in \mathcal{R}$ be given and let A be a neighbourhood of the points of $\mathcal{K}$ reached by optimal trajectories starting from x_0. In Section 3 we show (see Corollary 3.3) that, if (1.2) holds for $\bar{x} \in \partial \mathcal{K} \cap A$ then T is Lipschitz continuous in a neighbourhood of x_0. In Section 4 (see Corollary 4.6) we prove that T is semiconcave near x_0 even if the distance function from $\mathcal{K}$ is semiconcave only in $A \setminus \mathring{\mathcal{K}}$. This assumption is certainly satisfied if the corners of $\partial \mathcal{K}$ lie outside A.

We use in this paper the definition of semiconcavity given for instance in [5] and [1], which is weaker than the one recalled above. This is convenient for various reasons. On one hand, we can relax assumption (1.4) and require $f(\cdot, u)$ to be of class C^1. On the other hand, many interesting properties of semiconcave functions hold with the weaker definition as well (see Theorem 4.3).

2 Preliminaries

We first introduce some notations. For any $r > 0$ and $x \in \mathbb{R}^n$ we set

$$B_r(x) = \{y \in \mathbb{R}^n \,:\, |x - y| < r\}$$

and we abbreviate $B_r = B_r(0)$. Given $Q \subset \mathbb{R}^n$ and $\varepsilon > 0$ we set

$$Q_\varepsilon = \{x \in \mathbb{R}^n \,:\, \mathrm{dist}(x, Q) < \varepsilon\}. \tag{2.1}$$

We also denote by Q^c the complement of Q.

Suppose now that a closed set $\mathcal{K} \subset \mathbb{R}^n$ (called the *target*), a compact set $U \in \mathbb{R}^m$ (called the *control set*), and a measurable locally bounded function $f : \mathbb{R}^n \times U \to \mathbb{R}^n$ are given. We assume that f satisfies

$$|f(x, u) - f(y, u)| \leq L|x - y|, \qquad x, y \in \mathbb{R}^n,\ u \in U \tag{2.2}$$

for some $L > 0$. For fixed $x \in \mathbb{R}^n$ we consider the system

$$\begin{cases} y'(t) = f(y(t), u(t)) \\ y(0) = x. \end{cases} \tag{2.3}$$

A *control* for system (2.3) is a measurable function $u : [0, \infty[\to U$. We denote by $y(\cdot; x, u)$ the corresponding solution of (2.3) and set

$$\theta(x, u) = \min\{t \geq 0 : y(t; x, u) \in \mathcal{K}\}.$$

Then, $\theta(x, u) \in [0, +\infty]$, and we note that $\theta(x, u)$ is the time taken for the trajectory $y(\cdot; x, u)$ to reach $\mathcal{K}$. If $\theta(x, u) < \infty$ we define

$$y_f(x, u) = y(\theta(x, u); x, u).$$

The set

$$\mathcal{R} = \{x \,:\, \theta(x, u) < +\infty \text{ for some control } u\}$$

is called the *controllable set.* The minimum time optimal control problem consists in minimizing $\theta(x, u)$ over all measurable $u : \mathbb{R}_+ \to U$.

Definition 2.1 *The* minimum time function $T : \mathcal{R} \to [0, +\infty[$ *is defined by*

$$T(x) = \inf\{\theta(x, u) \,:\, u : \mathbb{R}_+ \to U\}. \tag{2.4}$$

A control u and the corresponding trajectory $y(\cdot;x,u)$ are called *optimal* for x if $T(x) = \theta(x,u)$. We recall a classical condition which ensures the existence of optimal controls (see e.g. [13]).

Theorem 2.2 *Suppose that $f(x,U)$ is a convex set for any $x \in \mathbb{R}^n$. Then there is an optimal control for any $x \in \mathcal{R}$. In addition, the uniform limit of optimal trajectories is an optimal trajectory.*

It is useful to generalize the above definitions to the case when we prescribe that the trajectories reach the target in a certain part of its boundary.

Definition 2.3 *Let an open set $A \subset \mathbb{R}^n$ be given. We set*

$$\theta_A(x,u) = \begin{cases} \theta(x,u) & \text{if } y_f(x,u) \in \mathcal{K} \cap A \\ +\infty & \text{otherwise.} \end{cases}$$

The minimum time function associated with A is defined by

$$T_A(x) = \inf\{\theta_A(x,u) \ : \ u : \mathbb{R}_+ \to U\}. \tag{2.5}$$

and the controllable set by

$$\mathcal{R}_A = \{x : T_A(x) < +\infty\}.$$

Definition 2.4 *Given $x \in \mathcal{R}_A$ and $\varepsilon \geq 0$ we define*

$$\mathcal{F}_{\varepsilon,A}(x) = \{\bar{x} \in \partial\mathcal{K} \cap A \ : \ \exists u \text{ such that } \theta(x,u) \leq T_A(x) + \varepsilon, \bar{x} = y_f(x,u)\}.$$

We set also

$$\mathcal{F}(x) = \mathcal{F}_{0,\mathbb{R}^n}(x).$$

In this paper we prove some regularity results for the functions of the form T_A introduced in Definition 2.3. Such results apply in particular to the minimum time function, since we have $T = T_{\mathbb{R}^n}$. On the other hand, the study of the local properties of T can be reduced to the study of T_A for a suitable choice of A, as shown by the next result.

Theorem 2.5

(i) *Let $Q \subset \mathcal{K}^c$ be given and let the open set A satisfy*

$$\mathcal{F}_{a,\mathbb{R}^n}(x) \cap A \neq \emptyset, \qquad \forall a > 0, \ \forall x \in Q. \tag{2.6}$$

Then $T = T_A$ in Q.

(ii) *Suppose that $f(x,U)$ is convex for any $x \in \mathbb{R}^n$. Given $x_0 \in \mathcal{K}^c$, let A be an open set containing $\mathcal{F}(x_0)$. Then $T = T_A$ in a neighbourhood of x_0.*

Proof. Part (i) is a direct consequence of the definitions, while part (ii) follows from (i) and Theorem 2.2. ∎

Throughout the paper we shall denote with $d_{\mathcal{K}}(\cdot)$ or simply with $d(\cdot)$ the distance function from the target $\text{dist}(\cdot,\mathcal{K})$.

3 Lipschitz continuity

It has been observed by several authors (see e.g., [15], [3], [16]) that the Lipschitz continuity of the minimum time function is related to a controllability condition which was first formulated by Petrov in the case $\mathcal{K} = \{0\}$ (see [14]). Here we state this condition in the form which is suitable to the study of the localized minimum time function T_A introduced in the previous section. To this purpose we need the following definition from nonsmooth analysis (see [8]).

Definition 3.1 *Let $C \subset \mathbb{R}^n$ be a closed set. Given $\bar{x} \in \partial C$ and $\nu \in (\mathbb{R}^n \setminus \{0\})$, we say that ν is a* perpendicular *(or* proximal normal*) to C at $\bar{x}$ if, for some $\lambda > 0$, we have*

$$dist\,(\bar{x} + \lambda\nu, C) = \lambda|\nu|$$

or, equivalently,

$$B_{\lambda|\nu|}(\bar{x} + \lambda\nu) \cap C = \emptyset.$$

We denote by $\Pi_C(\bar{x})$ the set of perpendiculars to C at $\bar{x}$.

It is easily seen that $\Pi_C(\bar{x})$ reduces to the half line parallel to the usual outer normal to C at $\bar{x}$ if ∂C is a $(n-1)$–manifold of class C^2.

Theorem 3.2 *Let an open set $A \subset \mathbb{R}^n$ be given. Suppose that for any $C \subset\subset A$ there exists $\mu > 0$ such that*

$$\min_{u \in U} f(\bar{x}, u) \cdot \nu \leq -\mu|\nu| \tag{3.1}$$

for any $\bar{x} \in C \cap \partial\mathcal{K}$, $\nu \in \Pi_{\mathcal{K}}(\bar{x})$. Then

(i) *for any $C \subset\subset A$ there exist $c > 0$, $\delta > 0$ such that (see 2.1)*

$$T_A(x) \leq c d_{\mathcal{K}}(x), \qquad \forall\, x \in (\partial\mathcal{K} \cap C)_\delta; \tag{3.2}$$

(ii) *the controllable set $\mathcal{R}_A$ is open and $T_A \in Lip_{loc}(\mathcal{R}_A)$.*

Conversely, given $C \subset A$, if (3.2) holds for some c, δ then

$$\min_{u \in U} f(\bar{x}, u) \cdot \frac{\nu}{|\nu|} \leq -\frac{1}{c} \tag{3.3}$$

for any $\bar{x} \in C \cap \partial\mathcal{K}$, $\nu \in \Pi_{\mathcal{K}}(\bar{x})$.

Proof. Let α be a positive constant such that $C_\alpha \subset\subset A$. Then there exists μ such that (3.1) holds for all $\bar{x} \in C_\alpha \cap \partial\mathcal{K}$. We fix a constant M such that

$$\mu/2 \leq M, \qquad |f(x,u)| \leq M, \quad \forall\,(x,u) \in C_\alpha \times U. \tag{3.4}$$

Let us also set

$$k = \sqrt{1 - \left(\frac{\mu}{4M}\right)^2}, \tag{3.5}$$

$$c = \frac{2}{\mu}(1+k), \qquad \delta = \min\left\{\frac{\mu}{2L}, \frac{\alpha}{2(1+cM)}\right\}, \tag{3.6}$$

where L is the Lipschitz constant of $f(\cdot, u)$. Observe that

$$c(1-k) = \frac{\mu}{8M^2}. \tag{3.7}$$

We claim that (3.2) holds with the constants δ, c defined above. To see this, let $x_0 \in (\partial\mathcal{K} \cap C)_\delta$ be given. Then we construct inductively a sequence $\{x_j\}$ with the following properties:

$$\text{dist}(x_j, \partial\mathcal{K} \cap C) \le \delta + \frac{\mu}{8M}\sum_{h=1}^{j-1} k^h d(x_0), \tag{3.8}$$

$$d(x_j) \le k^j d(x_0). \tag{3.9}$$

For $j = 0$ the above properties are satisfied. To obtain x_{j+1} from x_j we proceed as follows. We first observe that, by (3.8) and (3.7),

$$\begin{aligned} \text{dist}(x_j, \partial\mathcal{K} \cap C) &< \delta + \frac{\mu}{8M}\frac{1}{1-k}d(x_0) = \delta + cMd(x_0) \\ &\le (1+cM)\delta \le \frac{\alpha}{2}. \end{aligned} \tag{3.10}$$

Let us choose $\bar{x}_j \in \mathcal{K}$ such that $d(x_j) = |x_j - \bar{x}_j|$. Inequality (3.10) implies that $\bar{x}_j \in \partial\mathcal{K} \cap C_\alpha$. If we define

$$\nu_j = x_j - \bar{x}_j$$

then ν_j is a perpendicular to $\mathcal{K}$ at $\bar{x}_j$. Let u_j be an element of U for which the minimum in (3.1) is achieved (with $x = \bar{x}_j$ and $\nu = \nu_j$). We define

$$x_{j+1} = y(t_j; x_j, u_j), \qquad j \ge 0,$$

where

$$t_j = \frac{\mu}{8M^2}d(x_j). \tag{3.11}$$

We claim that x_{j+1} satisfies (3.8) and (3.9). Let us set

$$y_j(t) = y(t; x_j, u_j), \qquad 0 \le t \le t_j.$$

By (3.10) we have $y_j(0) = x_j \in C_{\alpha/2}$. Furthermore, by (3.4), $|y_j'| \le M$ as long as $y_j(t) \in C_\alpha$ and, by (3.7) and (3.9),

$$Mt_j \le \frac{\mu d(x_0)}{8M} < cMd(x_0) < \frac{\alpha}{2}.$$

It follows

$$y_j(t) \in C_\alpha, \qquad \forall 0 \le t \le t_j.$$

Thus, by (3.1), (2.2), (3.4), (3.6) and (3.11),

$$\begin{aligned}\frac{1}{2}\frac{d}{dt}|y_j(t)-\bar{x}_j|^2 &= \langle f(y_j(t),u_j), y_j(t)-\bar{x}_j\rangle = \langle f(\bar{x}_j,u_j), x_j-\bar{x}_j\rangle \\ &\quad +\langle f(y_j(t),u_j)-f(\bar{x}_j,u_j), x_j-\bar{x}_j\rangle + \langle f(y_j(t),u_j), y_j(t)-x_j\rangle \\ &\le -\mu|x_j-\bar{x}_j| + L|y_j(t)-\bar{x}_j|\,|x_j-\bar{x}_j| + M|y_j(t)-x_j| \\ &\le -\mu\, d(x_j) + (L\, d(x_j)+M)|y_j(t)-x_j| + L\, d(x_j)^2 \\ &\le (-\mu + L\, d(x_j))\, d(x_j) + (L\, d(x_j)+M)Mt \\ &\le -\frac{\mu}{2}d(x_j) + 2M^2 t \le -\frac{\mu}{4}\, d(x_j),\end{aligned}$$

for all $0 \le t \le t_j$. Therefore

$$d^2(y_j(t_j)) \le |y_j(t_j)-\bar{x}_j|^2 \le d^2(x_j) - \frac{\mu}{2}\, d(x_j)t_j = k^2 d^2(x_j),$$

which shows that x_{j+1} satisfies (3.9). In addition

$$\begin{aligned}\operatorname{dist}(x_{j+1}, \partial\mathcal{K}\cap C) &\le \operatorname{dist}(x_j, \partial\mathcal{K}\cap C) + |x_{j+1}-x_j| \\ &\le \delta + \frac{\mu}{8M}\sum_{h=1}^{j-1} k^h d(x_0) + Mt_j \\ &= \delta + \frac{\mu}{8M}\sum_{h=1}^{j} k^h d(x_0).\end{aligned}$$

Hence the sequence $\{x_j\}$ has the desired properties. Now, from (3.9) it follows

$$d(x_j) \to 0 \text{ as } j \to +\infty. \tag{3.12}$$

Moreover, by (3.11), (3.9) and (3.6),

$$\sum_{j=0}^{\infty} t_j \le \frac{\mu}{8M^2}\frac{1}{1-k}d(x_0) = c\, d(x_0). \tag{3.13}$$

Let us consider a control $\bar{u} : [0, +\infty) \to U$ such that

$$\begin{cases} \bar{u}(t) = u_h, & \text{if } \sum_{j=0}^{h-1} t_j \le t < \sum_{j=0}^{h} t_j \text{ for some } h \ge 0, \\ \bar{u}(t) \text{ arbitrary} & \text{if } t \ge \sum_{j=0}^{\infty} t_j. \end{cases}$$

Then

$$y\left(\sum_{h=0}^{j} t_h; x_0, \bar{u}\right) = x_{j+1}, \qquad j \ge 0.$$

In addition $y(t) \in C_\alpha \subset A$ for $t \le \sum_{j=0}^{\infty} t_j$. Thus, by (3.12),

$$T_A(x_0) \le \theta_A(x_0, \bar{u}) = \sum_{j=0}^{\infty} t_j \le c\, d(x_0),$$

which proves (3.2).

Part (ii) is an easy consequence of (i). In fact, let us fix $x_0 \in \mathcal{R}$. For any $\varepsilon > 0$ there exists a control u_ε such that $\theta_A(x_0, u_\varepsilon) \leq T_A(x_0) + \varepsilon$. Let us set for simplicity $\theta_\varepsilon = \theta_A(x_0, u_\varepsilon)$. Using Gronwall's inequality we obtain

$$|y(\theta_\varepsilon; x, u) - y(\theta_\varepsilon; x_0, u)| = O(|x - x_0|).$$

Therefore, by (3.2) we find, for x near x_0,

$$\begin{aligned} T_A(x) &\leq \theta_\varepsilon + cd(y(\theta_\varepsilon; x, u)) \\ &\leq \theta_\varepsilon + c|y(\theta_\varepsilon; x, u) - y(\theta_\varepsilon; x_0, u)| \\ &\leq T_A(x_0) + \varepsilon + O(|x - x_0|). \end{aligned}$$

From this we deduce the assertions of part (ii).

We turn to the second part of the theorem. We consider C such that (3.2) holds for some c, δ. Let us take $\bar{x} \in \partial\mathcal{K} \cap C$, $\nu \in \Pi_{\mathcal{K}}(\bar{x})$. It is not restrictive to assume $|\nu| = 1$. Then, for some $\lambda > 0$, we have

$$B_\lambda(\bar{x} + \lambda\nu) \cap \mathcal{K} = \emptyset. \tag{3.14}$$

For $\varepsilon \in]0, \delta[$ there exists by (3.2) a control u_ε such that

$$\theta_A(\bar{x} + \varepsilon\nu, u_\varepsilon) < (1 + \varepsilon)c\varepsilon.$$

Let us set for simplicity

$$x_\varepsilon = \bar{x} + \varepsilon\nu, \qquad \theta_\varepsilon = \theta_A(x_\varepsilon, u_\varepsilon), \qquad y_\varepsilon(\cdot) = y(\cdot; x_\varepsilon, u_\varepsilon).$$

Then, by (3.14)

$$|y_\varepsilon(\theta_\varepsilon) - x_\lambda| \geq \lambda,$$

which implies

$$|y_\varepsilon(\theta_\varepsilon) - x_\lambda|^2 - |y_\varepsilon(0) - x_\lambda|^2 \geq \lambda^2 - (\lambda - \varepsilon)^2 = 2\varepsilon\lambda - \varepsilon^2. \tag{3.15}$$

We call M the supremum of f in $(\partial\mathcal{K} \cap C)_\delta \times U$. If ε is small enough we have $y_\varepsilon(t) \in (\partial\mathcal{K} \cap C)_\delta$ for $t \in [0, \theta_\varepsilon]$. Therefore

$$\begin{aligned} \frac{1}{2}\frac{d}{dt}|y_\varepsilon(t) - x_\lambda|^2 &= \langle y_\varepsilon(t) - x_\lambda, f(y_\varepsilon(t), u_\varepsilon(t))\rangle \\ &= \langle \bar{x} - x_\lambda, f(\bar{x}, u_\varepsilon(t))\rangle + \langle y_\varepsilon(t) - \bar{x}, f(\bar{x}, u_\varepsilon(t))\rangle \\ &\quad + \langle y_\varepsilon(t) - x_\lambda, f(y_\varepsilon(t), u_\varepsilon(t)) - f(\bar{x}, u_\varepsilon(t))\rangle \\ &\leq -\lambda \min_{u \in U} \nu \cdot f(\bar{x}, u) + M|y_\varepsilon(t) - \bar{x}| + L|y_\varepsilon(t) - x_\lambda||y_\varepsilon(t) - \bar{x}| \\ &\leq -\lambda \min_{u \in U} \nu \cdot f(\bar{x}, u) + M(\varepsilon + \theta_\varepsilon M) + L(\lambda + M\theta_\varepsilon)(\varepsilon + \theta_\varepsilon M) \\ &\leq -\lambda \min_{u \in U} \nu \cdot f(\bar{x}, u) + O(\varepsilon) \end{aligned}$$

for $t \in [0, \theta_\varepsilon]$ a.e.. Thus

$$|y_\varepsilon(\theta_\varepsilon) - x_\lambda|^2 - |y_\varepsilon(0) - x_\lambda|^2 \leq 2\theta_\varepsilon \left[-\lambda \min_{u \in U} \nu \cdot f(x, u) + O(\varepsilon) \right].$$

Since (3.15) holds and $\theta_\varepsilon \leq (1 + \varepsilon)c\varepsilon$ we obtain

$$2\varepsilon\lambda - \varepsilon^2 \leq 2c(1 + \varepsilon)\varepsilon[-\lambda \min_{u \in U} \nu \cdot f(\bar{x}, u) + O(\varepsilon)],$$

which implies

$$\min_{u \in U} \nu \cdot f(\bar{x}, u) \leq -\frac{1}{c}.$$

Since ν is an arbitrary unit perpendicular vector at $\bar{x}$, we have proved (3.3). ■

Corollary 3.3 *Suppose that $f(x, U)$ is convex for any $x \in \mathbb{R}^n$. Given $x_0 \in \mathcal{R}$, assume that there exists an open set A containing $\mathcal{F}(x_0)$ such that for any $C \subset\subset A$ inequality (3.1) holds for some $\mu > 0$. Then T is Lipschitz continuous in a neighbourhood of x_0.*

Proof. It follows from Theorems 2.5 and 3.2. ■

4 Local semiconcavity results

We now recall the definition and some properties of semiconcave functions.

Definition 4.1 *Let $v : \Omega \to \mathbb{R}$, where $\Omega \subset \mathbb{R}^n$. We say that v is* semiconcave *if, for any $C \subset\subset \Omega$ there exists a nondecreasing function $\omega : \mathbb{R}_+ \to \mathbb{R}_+$ such that $\lim_{h \to 0} \omega(h) = 0$ and*

$$\lambda v(x) + (1 - \lambda)v(y) - v(\lambda x + (1 - \lambda)y) \leq \lambda(1 - \lambda)|x - y|\omega(|x - y|), \qquad (4.1)$$

for any x, $y \in \Omega$, $\lambda \in [0, 1]$ such that x, y, $\lambda x + (1 - \lambda)y \in C$. The function ω is called a modulus of semiconcavity *of v in C.*

Remark 4.2 The usual definition of semiconcavity (see [11], [12], [6]) is more restrictive than the one above, since it requires the modulus to be of the form $\omega(h) = ch$ for some $c > 0$. It is easy to see that this condition holds if and only if $v(x) - c|x|^2/2$ is concave in C.

We recall now some properties of semiconcave functions (see [9] and [8] for the definitions of D^+v and ∂v).

Theorem 4.3 *Let $\Omega \subset \mathbb{R}^n$ be an open set and let $v : \Omega \to \mathbb{R}$ be semiconcave. Then*

(i) *v is locally Lipschitz in Ω;*

(ii) *the superdifferential $D^+v(x)$ coincides with the Clarke generalized gradient $\partial v(x)$ for any $x \in \Omega$;*

(iii) *if we set for* $k \in \{0, \ldots, n\}$

$$S^k(v) = \{x \in \Omega : \dim D^+v(x) = k\}.$$

then $S^k(v)$ *is countably* $\mathcal{H}^{n-k}$*-rectifiable, that is it can be covered, up to a* $\mathcal{H}^{n-k}$*-negligible set, by the union of a countable family of* C^1*-hypersuperfaces of dimension* $n-k$ *(with* $\mathcal{H}^{n-k}$ *we denote the* $(n-k)$*-dimensional Hausdorff measure).*

Proof. See [7], [1]. ∎

We now turn to the main result of this section, concerning the local semiconcavity of the minimum time function.

Theorem 4.4 *Let an open set* A *be given. Suppose that*

(a) $\nabla_x f$ *exists and is continuous;*

(b) *for any* $C \subset\subset A$ *there exists* $\mu > 0$ *such that condition (3.1) holds;*

(c) $d_{\mathcal{K}}$ *is semiconcave in* $A \setminus \overset{\circ}{\mathcal{K}}$.

Consider $Q \subset\subset \mathcal{R}_A$ *with the following property: there exists* $S \subset\subset A$ *such that*

$$\mathcal{F}_{a,A}(x) \cap S \neq \emptyset$$

for any $x \in Q$, $a > 0$. *Then* T_A *is semiconcave in* $Q \setminus \overset{\circ}{\mathcal{K}}$.

We recall a criterion for the semiconcavity of the distance function given in [6].

Proposition 4.5 *If there exists* $\rho > 0$ *such that*

$$\forall x \in \mathcal{K} \cap A \ \exists x_0 \in \mathcal{K} \ : \ x \in \overline{B_\rho(x_0)} \subset \mathcal{K}, \tag{4.2}$$

then $d_{\mathcal{K}}$ *is semiconcave in* $A \setminus \overset{\circ}{\mathcal{K}}$. *More precisely, for any* $C \subset\subset A \setminus \overset{\circ}{\mathcal{K}}$, *inequality (4.1) holds with* $\omega(h) = k_C h/2$, *where*

$$k_C = (\min\{\rho, dist(C, \partial A)\})^{-1}.$$

Corollary 4.6 *Let* f *satisfy condition (a) of Theorem 4.4 and let* $f(x, U)$ *be convex for any* x. *Given* $x_0 \in \mathcal{R}$, *suppose that there exists an open neighbourhood* A *of* $\mathcal{F}(x_0)$ *such that (b) and (c) hold. Then* T *is semiconcave near* x_0.

Proof. From Theorem 2.2 it follows that $\mathcal{F}(x) \subset A$ for any x in a neighbourhood of x_0. Then the assertion follows from Theorems 2.5 and 4.4. ∎

Before proving Theorem 4.4 we need a Lemma.

Lemma 4.7 *Let hypotheses (b) and (c) of Theorem 4.4 hold and let C be a compact subset of A. Then there exists $\rho > 0$, and a nondecreasing function $\omega : \mathbb{R}_+ \to \mathbb{R}_+$ such that $\lim_{h\to 0}\omega(h) = 0$ and*

$$\lambda T_A(x+h) \le T_A(x+\lambda h) + \lambda(1-\lambda)|h|\omega(|h|)$$

for all $x \in \partial\mathcal{K}\cap C, h \in B_\rho, \lambda \in [0,1]$ such that $x + \lambda h \notin \mathcal{K}$.

Proof. Let α be a positive constant such that $C_\alpha \subset\subset A$. By assumption (c) there exists ω_1 such that

$$\lambda d(x_1) + (1-\lambda)d(x_2) - d(\lambda x_1 + (1-\lambda)x_2) \le \lambda(1-\lambda)|x_1 - x_2|\omega_1(|x_1 - x_2|) \qquad (4.3)$$

for any x_1, x_2, λ such that $x_1, x_2, \lambda x_1 + (1-\lambda)x_2 \in C_\alpha \setminus \overset{\circ}{\mathcal{K}}$. Here and in the following we denote with ω_i, $i \ge 1$, a nondecreasing function $\omega_i : \mathbb{R}_+ \to \mathbb{R}_+$ such that $\lim_{h\to 0}\omega_i(h) = 0$. By Theorem 3.2 there exist c, $\delta > 0$ such that

$$T_A(x) \le cd(x), \qquad x \in (\partial\mathcal{K}\cap C)_\delta. \qquad (4.4)$$

We fix $\rho > 0$ such that

$$(1 + 2Mc)\rho < \min\{\delta, \alpha\} \qquad (4.5)$$

where M is the supremum of $|f|$ in $C_\alpha \times U$.
Let $x \in \partial\mathcal{K}\cap C$, $h \in B_\rho$ and $\lambda \in [0,1]$. For fixed $\varepsilon \in (0, \lambda c|h|)$, there exists a control u such that $\theta_A(x + \lambda h, u) < T_A(x + \lambda h) + \varepsilon$. Let us set $\theta^* = \theta_A(x + \lambda h, u)$ and define

$$\bar{u}(t) = u(\lambda t), \quad \tilde{y}(t) = y(t; x + \lambda h, u), \quad \bar{y}(t) = y(t; x + h, \bar{u}).$$

From (3.2) we deduce

$$\theta^* \le 2c\lambda|h|. \qquad (4.6)$$

Then (4.5) implies that $\tilde{y}(t)$ and $\bar{y}(t)$ stay inside C_α for $t \le \lambda^{-1}\theta^*$. We claim that

$$T_A(x+h) \le \lambda^{-1}(T_A(x+\lambda h) + \varepsilon) + (1-\lambda)|h|\omega(|h|) \qquad (4.7)$$

for a suitable $\omega(\cdot)$, which does not depend on x, h, ε. We may assume that $\lambda\theta_A(x+h, \bar{u}) > \theta^*$ (otherwise (4.7) is satisfied with $\omega = 0$). By the dynamic programming principle,

$$T_A(x+h) \le \lambda^{-1}\theta^* + T_A(\bar{y}(\lambda^{-1}\theta^*)). \qquad (4.8)$$

To estimate $T_A(\bar{y}(\lambda^{-1}\theta^*))$, we first observe that, by (2.2) and (4.6),

$$\begin{aligned}
|\tilde{y}(t) - \bar{y}(\lambda^{-1}t)| &= \left|(\lambda - 1)h + \int_0^t [f(\tilde{y}(s), u(s)) - \lambda^{-1}f(\bar{y}(\lambda^{-1}s), u(s))]\, ds\right| \\
&\le (1-\lambda)|h| + (\lambda^{-1} - 1)\int_0^t |f(\bar{y}(\lambda^{-1}s), u(s))|\, ds \\
&\quad + \int_0^t |f(\tilde{y}(s), u(s)) - f(\bar{y}(\lambda^{-1}s), u(s))|\, ds \\
&\le (1-\lambda)\left(|h| + \frac{\theta^* M}{\lambda}\right) + L\int_0^t |\tilde{y}(s) - \bar{y}(\lambda^{-1}s)|\, ds \\
&\le (1-\lambda)|h|(1 + 2cM) + L\int_0^t |\tilde{y}(s) - \bar{y}(\lambda^{-1}s)|\, ds
\end{aligned}$$

for any $t \le \theta^*$. This implies, by Gronwall's inequality and (4.6),

$$|\tilde{y}(t) - \bar{y}(\lambda^{-1}t)| \le k(1-\lambda)|h| \qquad t \le \theta^*, \tag{4.9}$$

where we have set $k = (1 + 2cM)\exp(2Lc\rho)$. We also find

$$|\tilde{y}(\theta^*) - x| \le M\theta^* + |\tilde{y}(0) - x| \le \lambda(1 + 2cM)|h|. \tag{4.10}$$

If we set

$$\tilde{x} = \tilde{y}(\theta^*) \qquad \bar{x} = \bar{y}(\lambda^{-1}\theta^*),$$

then by our definitions $\tilde{x} \in \partial\mathcal{K} \cap C_\alpha$. Furthermore, by (4.9) and (4.6),

$$\begin{aligned} |\bar{x} - x - \lambda^{-1}(\tilde{x} - x)| &= \left| \int_0^{\lambda^{-1}\theta^*} \bar{y}'(t)\,dt - \lambda^{-1}\int_0^{\theta^*} \tilde{y}'(t)\,dt \right| \\ &= \lambda^{-1}\left| \int_0^{\theta^*} [f(\bar{y}(\lambda^{-1}t), u(t)) - f(\tilde{y}(t), u(t))]\,dt \right| \\ &\le \lambda^{-1}\theta^* Lk(1-\lambda)|h| \le (1-\lambda)2cLk|h|^2. \end{aligned}$$

On the other hand, by (4.3), (4.5) and (4.10),

$$\begin{aligned} \lambda d(x + \lambda^{-1}(\tilde{x} - x)) &= \lambda d(x + \lambda^{-1}(\tilde{x} - x)) + (1-\lambda)d(x) - d(\tilde{x}) \\ &\le (1-\lambda)|\tilde{x} - x|\omega_1(\lambda^{-1}|\tilde{x} - x|) \\ &\le (1-\lambda)\lambda(1 + 2cM)|h|\omega_1((1 + 2cM)|h|). \end{aligned}$$

We deduce

$$\begin{aligned} d(\bar{x}) &\le |\bar{x} - x - \lambda^{-1}(\tilde{x} - x)| + d(x + \lambda^{-1}(\tilde{x} - x)) \\ &\le (1-\lambda)2cLk|h|^2 \\ &\quad +(1-\lambda)(1 + 2cM)|h|\omega_1((1 + 2cM)|h|) \\ &= (1-\lambda)|h|\omega_2(|h|). \end{aligned}$$

Taking into account (3.2) and (4.8), we obtain that (4.7) holds, with $\omega(h) = c\omega_2(|h|)$. Since ε is arbitrary, the conclusion follows. ∎

Proof of Theorem 4.4. We will assume for simplicity that for every $x \in \mathcal{R}_A$ there exists an optimal trajectory starting at x, i.e. a solution of (1.1) reaching $\mathcal{K} \cap A$ at time $T_A(x)$. If no such trajectory exists, one can use an approximation procedure as in the previous proof.

We fix $Q \subset\subset \mathcal{R}_A \setminus \mathcal{K}$, and take x_1, x_2, λ such that $x_1, x_2, \lambda x_1 + (1-\lambda)x_2 \in Q$. We set

$$x_\lambda = \lambda x_1 + (1-\lambda)x_2,$$

$$\bar{y}(t) = y(t; x_\lambda, u), \quad y_1(t) = y(t; x_1, u), \quad y_2(t) = y(t; x_2, u),$$

where u is an optimal control for x_λ. By hypothesis we can choose u such that $y(T_A(x_\lambda)) \in \partial\mathcal{K} \cap S$. Using Gronwall's inequality we find

$$|\bar{y}(t) - y_1(t)| < k_1(1-\lambda)|x_2 - x_1|, \quad |\bar{y}(t) - y_2(t)| < k_1\lambda|x_2 - x_1|, \quad t \le T(x). \tag{4.11}$$

Here and in the following k_i, $i \geq 1$, denotes a positive constant, which depends only on the set Q and not on the choice of x_1, x_2 and λ.
We choose T^* such that $T_A(x) < T^*$ for $x \in Q$ and fix $R > 0$ such that $\bar{y}(t), y_1(t), y_2(t)$ are contained in B_R for any choice of initial points in Q and for $t \leq T^*$. By our assumptions on f there exists ω_1 such that

$$|\nabla_x f(x,u) - \nabla_x f(y,u)| \leq \omega_1(|x-y|), \qquad x, y \in B_R,\ u \in U.$$

As before, we denote with ω_i, $i \geq 1$ a non decreasing function from $\mathbb{R}_+$ to $\mathbb{R}_+$ such that $\omega_i(h) \to 0$ as $h \to 0$. Then we obtain

$$\begin{aligned} &|\lambda f(x,u) + (1-\lambda) f(y,u) - f(\lambda x + (1-\lambda)y, u)| \\ &= \left|\lambda(1-\lambda) \int_0^1 \langle \nabla_x f(\lambda x + (1-\lambda)y + \lambda s(y-x), u) \right. \\ &\qquad \left. - \nabla_x f(x + s(1-\lambda)(y-x), u),\, y - x\rangle ds \right| \\ &\leq \lambda(1-\lambda)|y-x|\omega_1(|y-x|). \end{aligned}$$

Therefore, taking into account (4.11), we find (for simplicity we omit the dependence of f on u)

$$\begin{aligned} &|\lambda y_1(t) + (1-\lambda)y_2(t) - \bar{y}(t)| \\ &= \left| \int_0^t [\lambda f(y_1(s)) + (1-\lambda) f(y_2(s)) - f(\bar{y}(s))]\, ds \right| \\ &\leq \int_0^t |\lambda f(y_1(s)) + (1-\lambda) f(y_2(s)) - f(\lambda y_1(s) + (1-\lambda)y_2(s))|\, ds \\ &\quad + \int_0^t |f(\lambda y_1(s) + (1-\lambda)y_2(s)) - f(\bar{y}(s))|\, ds \\ &\leq \lambda(1-\lambda) \int_0^t |y_1(s) - y_2(s)|\omega_1(|y_1(s) - y_2(s)|)\, ds \\ &\quad + L \int_0^t |\lambda y_1(s) + (1-\lambda) y_2(s) - \bar{y}(s)|\, ds \\ &\leq k_2 \lambda(1-\lambda)|x_2 - x_1|\omega_1(k_1|x_2 - x_1|) + L \int_0^t |\lambda y_1(s) + (1-\lambda)y_2(s) - \bar{y}(s)|\, ds, \end{aligned}$$

for any $t \leq T_A(x_\lambda)$. This implies, by Gronwall's inequality,

$$|\lambda y_1(t) + (1-\lambda)y_2(t) - \bar{y}(t)| < k_3 \lambda(1-\lambda)|x_2 - x_1|\omega_1(k_1|x_2 - x_1|) \qquad t \leq T_A(x_\lambda). \quad (4.12)$$

Next we choose a compact set C such that $S \subset C \subset A$ and $\text{dist}(S, \partial C) > 0$. We can associate with C the constants c, δ, ρ and the function ω given by Theorem 3.2 and Lemma 4.7. By (4.11) we have $y_1(t) \in C$, $y_2(t) \in C$ if $|x_2 - x_1|$ and $|t - T_A(x_\lambda)|$ are small enough. We consider three cases.

(i) Let us first suppose that one of the two paths y_1, y_2 (say for instance y_2) reaches the target at a time $t^* < T_A(x_\lambda)$, i.e. $\theta(x_2, u) = t^*$. We set

$$\bar{x}_1 = y_1(t^*),\ \bar{x}_2 = y_2(t^*),\ \bar{x} = \bar{y}(t^*).$$

Then $\bar{x}_2 \in \partial\mathcal{K} \cap C$ if $|x_2 - x_1|$ is small enough (as we will tacitly assume in the rest of the proof). Moreover, by the dynamic programming principle,

$$T_A(x_2) \le t^*, \qquad T_A(x_1) \le t^* + T_A(\bar{x}_1), \qquad T_A(x_\lambda) = t^* + T_A(\bar{x}). \tag{4.13}$$

By (4.11) and Lemma 4.7 we deduce, with $x = \bar{x}_2$ and $h = \lambda^{-1}(\bar{x} - \bar{x}_2)$,

$$\begin{aligned} \lambda T_A(\bar{x}_2 + \lambda^{-1}(\bar{x} - \bar{x}_2)) - T_A(\bar{x}) &\le (1-\lambda)|\bar{x} - \bar{x}_2|\omega(\lambda^{-1}|\bar{x} - \bar{x}_2|) \\ &\le k_1\lambda(1-\lambda)|x_2 - x_1|\omega(k_1|x_2 - x_1|). \end{aligned}$$

On the other hand, by (4.12) and the local Lipschitz continuity of T_A,

$$\begin{aligned} \lambda|T_A(\bar{x}_2 + \lambda^{-1}(\bar{x} - \bar{x}_2)) - T_A(\bar{x}_1)| &\le k_4|\lambda\bar{x}_1 + (1-\lambda)\bar{x}_2 - \bar{x}| \\ &\le k_3k_4\lambda(1-\lambda)|x_2 - x_1|\omega_1(k_1|x_2 - x_1|). \end{aligned}$$

Hence, by (4.13),

$$\begin{aligned} \lambda T_A(x_1) &+ (1-\lambda)T_A(x_2) - T_A(x_\lambda) \le \lambda T_A(\bar{x}_1) - T_A(\bar{x}) \\ &\le \lambda(1-\lambda)[k_1|x_2 - x_1|\omega(k_1|x_2 - x_1|) + k_3k_4|x_2 - x_1|\omega_1(k_1|x_2 - x_1|)] \\ &= \lambda(1-\lambda)|x_2 - x_1|\omega_2(|x_2 - x_1|), \end{aligned}$$

which proves our conclusion.

(ii) Let us now suppose that neither y_1 nor y_2 reach the target before $\bar{y}$, and that

$$\lambda y_1(T_A(x_\lambda)) + (1-\lambda)y_2(T_A(x_\lambda)) \in \mathcal{K}.$$

Then there exists a time $t^* \le T_A(x_\lambda)$ such that

$$\lambda y_1(t^*) + (1-\lambda)y_2(t^*) \in \partial\mathcal{K}.$$

From (4.11) we deduce

$$\begin{aligned} \lambda d(y_1(t^*)) + (1-\lambda)d(y_2(t^*)) &\le \lambda(1-\lambda)|y_1(t^*) - y_2(t^*)|\omega_3(|y_1(t^*) - y_2(t^*)|) \\ &\le \lambda(1-\lambda)k_1|x_2 - x_1|\omega_3(k_1|x_2 - x_1|). \end{aligned}$$

where ω_3 is the modulus of semiconcavity of d in $C\setminus \mathring{\mathcal{K}}$. We conclude, by (3.2) and the dynamic programming principle,

$$\begin{aligned} &\lambda T_A(x_1) + (1-\lambda)T_A(x_2) - T_A(x_\lambda) \\ &\le t^* + \lambda T_A(y_1(t^*)) + (1-\lambda)T_A(y_2(t^*)) - T_A(x_\lambda) \\ &\le \lambda T_A(y_1(t^*)) + (1-\lambda)T_A(y_2(t^*)) \le c[\lambda d(y_1(t^*)) + (1-\lambda)d(y_2(t^*))] \\ &\le \lambda(1-\lambda)ck_1|x_2 - x_1|\omega_3(k_1|x_2 - x_1|) = \lambda(1-\lambda)|x_2 - x_1|\omega_4(|x_2 - x_1|). \end{aligned}$$

(iii) Finally, suppose that neither y_1 nor y_2 reach the target before $\bar{y}$, and that

$$\lambda y_1(T_A(x_\lambda)) + (1-\lambda)y_2(T_A(x_\lambda)) \notin \mathcal{K}.$$

We define

$$\bar{x}_1 = y_1(T_A(x_\lambda)), \quad \bar{x}_2 = y_2(T_A(x_\lambda)), \quad \bar{x} = \bar{y}(T_A(x_\lambda)).$$

By the semiconcavity of d and (4.11) we have

$$\begin{aligned} & \lambda d(\bar{x}_1) + (1-\lambda)d(\bar{x}_2) - d(\lambda\bar{x}_1 + (1-\lambda)\bar{x}_2) \\ \leq\ & \lambda(1-\lambda)|\bar{x}_1 - \bar{x}_2|\omega_3(|\bar{x}_1 - \bar{x}_2|) \\ \leq\ & k_1\lambda(1-\lambda)|x_2 - x_1|\omega_3(k_1|x_2 - x_1|). \end{aligned}$$

Taking into account inequality (4.12) we obtain

$$\begin{aligned} & \lambda d(\bar{x}_1) + (1-\lambda)d(\bar{x}_2) \\ =\ & \lambda d(\bar{x}_1) + (1-\lambda)d(\bar{x}_2) - d(\bar{x}) \\ \leq\ & \lambda d(\bar{x}_1) + (1-\lambda)d(\bar{x}_2) - d(\lambda\bar{x}_1 + (1-\lambda)\bar{x}_2) + |\lambda\bar{x}_1 + (1-\lambda)\bar{x}_2 - \bar{x}| \\ \leq\ & \lambda(1-\lambda)|x_2 - x_1|[k_1\omega_3(k_1|x_2 - x_1|) + k_3\omega_1(k_1|x_2 - x_1|)] \\ =\ & \lambda(1-\lambda)|x_2 - x_1|\omega_5(|x_2 - x_1|). \end{aligned}$$

Thus, by (3.2) and the dynamic programming principle, we conclude

$$\begin{aligned} & \lambda T_A(x_1) + (1-\lambda)T_A(x_2) - T_A(x_\lambda) \\ \leq\ & \lambda T_A(\bar{x}_1) + (1-\lambda)T_A(\bar{x}_2) \\ \leq\ & c(\lambda d(\bar{x}_1) + (1-\lambda)d(\bar{x}_2)) \leq c\lambda(1-\lambda)|x_2 - x_1|\omega_5(|x_2 - x_1|). \end{aligned}$$

Hence T_A is semiconcave in Q. ∎

Remark 4.8 From the above proofs it is easy to see that the modulus of semiconcavity obtained for T_A is of the form $\omega(h) = c_1\tilde{\omega}(c_2 h)$, where c_1 and c_2 are suitable constants and $\tilde{\omega}$ is the supremum of the modulus of semiconcavity of $d_{\mathcal{K}}$ and the modulus of continuity of $\nabla_x f$. Thus, if $\mathcal{K}$ satisfies the interior sphere condition (4.2) and $\nabla_x f$ is locally Lipschitz, then $\omega(h) = c_3 h$ and T_A is semiconcave even with respect to the more restrictive definition recalled in Remark 4.2.

References

[1] Alberti G., Ambrosio L., Cannarsa P., On the singularities of convex functions. Manuscripta Math. **76**, 421-435 (1992).

[2] Ambrosio L., Cannarsa P., Soner H.M., On the propagation of singularities of semi–convex functions. Ann. Scuola Norm. Sup. Pisa Ser. IV **20** 597-616 (1993).

[3] Bardi M., Falcone M., An approximation scheme for the minimum time function. SIAM J. Control Optim. **28**, 950-965, (1990).

[4] Bressan A., On two conjectures by Hájek. Funkcial. Ekvac. **23**, 221-227, (1980).

[5] Cannarsa P., Frankowska H., Some characterizations of optimal trajectories in control theory. SIAM J. Control Optim. **29**, 1322-1347 (1991).

[6] Cannarsa P., Sinestrari C., Convexity properties of the minimum time function. To appear in Calculus of Variations and Partial Differential Equations.

[7] Cannarsa P., Soner H.M., Generalized one-sided estimates for solutions of Hamilton-Jacobi equations and applications. Nonlinear Anal. **13**, 305-323 (1989).

[8] Clarke F., Optimization and nonsmooth analysis. John Wiley, New York, 1983.

[9] Crandall M.G., Evans L.C., Lions P.-L., Some properties of viscosity solutions of Hamilton-Jacobi equations. Trans. Amer. Math. Soc. **282**, 487-502 (1984).

[10] Evans L.C., James M.R., The Hamilton–Jacobi–Bellman equation for time optimal control. SIAM J. Control Optim. **27**, 1477-1489 (1989).

[11] Kruzhkov S.N., Generalized solutions of the Hamilton-Jacobi equations of the eikonal type I. Math. USSR Sb. **27**, 406-445 (1975).

[12] Lions P.L., Generalized solutions of Hamilton-Jacobi equations. Pitman, Boston, 1982.

[13] Lee E.B., Markus L., Foundations of optimal control theory. John Wiley, New York, 1968.

[14] Petrov N.N., Controllability of autonomous systems. Differential Equations **4**, 311-317, (1968).

[15] Petrov N.N., On the Bellman function for the time-optimal process problem. J. Appl. Math. Mech **34**, 785-791, (1970).

[16] Veliov V.M., On the Lipschitz continuity of the value function in optimal control. To appear in J. Optim. Theory Appl.

Some Remarks on the Detectability Condition for Stochastic Systems

Gianmario Tessitore
Dipartimento di Matematica Applicata "G.Sansone"
via di Santa Marta 3, 50139 Firenze.

1 Setting of the problem

The optimal cost of the control problem given by a stochastic linear differential equation with multiplicative noise, as:

$$\begin{cases} d_t y(t) = Ay(t)dt + Bu(t)dt + Cy(t)d\beta_t \\ y(0) = x \end{cases} \tag{1.1}$$

and by a quadratic, infinite horizon, cost functional as:

$$J_\infty(x,u) = \mathbb{E}\int_0^\infty \left(|\sqrt{S}y(s)|^2 + |u(s)|^2\right) ds \tag{1.2}$$

(where β is a standard brownian motion, S is a non-negative self-adjoint $N \times N$-matrix and A, B, C are $N \times N$-matrices) is given by $\mathbb{E}\langle Xx, x\rangle$, X being the minimal non-negative solution of the following Algebraic Riccati Equation:

$$A^*X + XA - XBB^*X + C^*XC + S = 0. \tag{1.3}$$

It is therefore important to find conditions that guarantee the uniqueness of the non-negative solution of the above equation. In that case the control problem is solved whenever one is able to find a non negative solution of (1.3), otherwise one has to show that the solution he has found is minimal. In [5] it is shown that if the "feedback operator" F is defined by: $F = A - BB^*X$ and $\overline{y}_v$, $\forall v \in \mathbb{R}^N$, is the solution of the following "closed loop" equation:

$$\begin{cases} d_t\overline{y}_v(t) = F\overline{y}_v(t)dt + C\overline{y}_v(t)d\beta_t \\ \overline{y}_v(0) = v \end{cases} \tag{1.4}$$

then the following condition:

$$\mathbb{E}\int_0^\infty |\overline{y}_v(s)|^2 ds < \infty \text{ for all } v \in \mathbb{R}^N \tag{1.5}$$

implies that X is the unique solution of (1.3).
Moreover in [2] it has been introduced the following definition, extending the well known "deterministic" detectability condition (see [1]), under which (1.5) holds:

Definition 1.1 (Detectability) *We say that* $(A, \sqrt{S}, C)$ *is detectable if there exists* $K \in \mathcal{L}(\mathbb{R}^N)$ *such that letting* $A_K = A + K\sqrt{S}$ *and* $\xi_K(\cdot, x)$, $x \in \mathbf{L}^2(\Omega, \mathcal{F}_s, \mathbb{P}, \mathbb{R}^N)$, *be the solution of:*

$$\begin{cases} d_t\xi_K(t,x) = A_K\xi_K(t,x)dt + C\xi_K(t,x)d\beta_t \\ \xi_K(0,x) = x \end{cases} \tag{1.6}$$

then:

$$\mathbb{E}\int_0^\infty |\xi_K(t,x)|^2\, dt < +\infty \quad \forall x \in L^2(\Omega, \mathcal{F}_0, \mathbb{P}, \mathbb{R}^N) \tag{1.7}$$

In the deterministic case ($C = 0$ in (1.1)) the above condition reduces to the stabilizability (relatively to the identity) of the following dual system:

$$\begin{cases} \nu'(t) = A^*\nu(t) + \sqrt{S}\eta(t) \\ \nu(0) = v \end{cases} \tag{1.8}$$

(ν being the state and η the control, see (2.2)) and is particularly easy to check in the concrete cases thank to the spectral "Hautus Condition" (see [3] or Remark 2.3 below).
In the general stochastic case definition (1.1) although it has a simple mathematical formulation is not immediately connected to a control problem and seems to be difficult to apply.

In this work we try to give simple ways to decide if $(A, \sqrt{S}, C)$ is detectable. In §2 prove some initial results and then we deduce, by simple considerations on the Lyapunov Equation similar to the ones developed, in relation to stabilizability, in [6], a necessary condition for the detectability of $(A, \sqrt{S}, C)$. In §3 we follow the technique based on duality relations that works in the deterministic case. A basic toll is consequently represented by backward stochastic differential equations, see [8]. First we show that a "Datko theorem" holds for the above backward stochastic differential equations and then we are able to give a sufficient condition for the detectability of $(A,\sqrt{S},C)$ (see 3.3) implying, for instance, that if $\mathrm{Ker}(S) \subset \mathrm{Ker}(C)$ then $(A,\sqrt{S},C)$ is detectable if and only if $(A,\sqrt{S},0)$ is detectable (this last relation refers to a deterministic system and is of immediate test). Moreover we characterize the detectable triplets $(A,\sqrt{S},C)$ when the sate space is $\mathbb{R}^2$.

Finally let us specify some assumptions and notations:

- $(\Omega,\mathcal{E},\mathcal{F},\mathbb{P})$ is a standard stochastic base,
- $\beta : [0,+\infty)\times\Omega \to \mathbb{R}$ is a one-dimensional Brownian motion. Moreover we assume that $\mathcal{F}_t$ is the filtration generated by β; that is:

$$\mathcal{F}_t = \sigma\{\beta_\tau : \tau \le t\}$$

- A, B and C are $N\times N$ matrices and $S\in\Sigma^+(\mathbb{R}^N)$ where by $\Sigma^+(\mathbb{R}^N)$ we denote the set of bounded non negative self-adjoint $N\times N$ matrices.
- by $\mathbf{M}^2_{\mathcal{P}}(0,T,\mathbb{R}^N)$ with $0\le T\le+\infty$ we denote the closed subspace of $\mathbf{L}^2(\Omega\times[0,T],\mathcal{E}\otimes\mathcal{B}([0,T]),\mathbb{P}\otimes\mu,\mathbb{R}^N)$ (where μ is the Lebesgue's measure) given by all equivalence classes that contain a predictable process with respect to the filtration $\mathcal{F}$. Moreover by $\mathbf{C}_{\mathcal{P}}(0,T,\mathbf{L}^2(\Omega,\mathbb{R}^N))$ we denote the subspace of $\mathbf{M}^2_{\mathcal{P}}(0,T,\mathbb{R}^N)$ given by all mean square continuous processes.
- the initial data x of (1.1) belongs to $\mathbf{L}^2(\Omega,\mathcal{F}_0,\mathbb{P},H)$

(the finite dimensionality of the state space will be, for simplicity, assumed throughout the paper but it is is essential only in the proof of Corollary 3.4).

2 General Results

First we notice that if S is invertible the question is trivial

Remark 2.1 If S is invertible then $(A,\sqrt{S},C)$ is detectable. In fact letting $K=-\lambda\sqrt{S^{-1}}$ we get $A_K = A-\lambda I$ and, consequently, $\xi_K(t,x) = e^{-\lambda t}\xi_0(t,x)$. Therefore it is easy to show that for λ large enough $\xi_K(\cdot,x)$ belongs to $\mathbf{M}^2_{\mathcal{P}}(0,\infty,\mathbb{R}^N)$

Then we recall the following stochastic version of Datko Theorem (see [4]):

Remark 2.2 Relation (1.7) is equivalent to the existence of $M \geq 1$ and $\alpha > 0$ such that:

$$\mathbb{E}|\xi_K(t)|^2 \leq M^2 e^{-2\alpha t}\mathbb{E}|x|^2 \quad \forall x \in L^2(\Omega, \mathcal{F}_0, \mathbb{P}, \mathbb{R}^N) \tag{2.1}$$

This section is based on the following results dealing with a deterministic system:

Remark 2.3 In the deterministic case ($C = 0$ in (1.1)) the triplet $(A, \sqrt{S}, 0)$ is detectable (we will say that $(A, \sqrt{S})$ is detectable) if and only if there exists $K \in \mathcal{L}(\mathbb{R}^N)$ such that the exponential of matrix A_K is stable. Clearly this is equivalent to the stability of the exponential of matrix $A_K^* = A^* + \sqrt{S}K^*$ and by a well known result in deterministic linear quadratic control (see [1]) this last statement is equivalent to the following one (normally expressed saying that $(A^*, \sqrt{S})$ is stabilizable relatively to the identity):

$$\forall v \in \mathbb{R}^N \ \exists \eta \in \mathbf{L}^2(0, \infty, \mathbb{R}^N) \text{ such that } \int_0^\infty |\nu(s)|^2 ds < \infty \tag{2.2}$$

where ν is the solution of (1.8).
Finally the couples $(A, \sqrt{S})$ that are stabilizable relatively to the identity can be characterized by the spectral condition due to M. L. J. Hautus (see [1] and [3]). Finally we obtain:

$$(A, \sqrt{S}) \text{ is detectable } \Leftrightarrow \ \mathrm{Ker}(\widetilde{A} - \varsigma I) \cap \mathrm{Ker}(\widetilde{S}) = \{0\} \quad \forall \varsigma \in \mathbb{C}, \quad \mathrm{Re}\,\lambda \geq 0 \tag{2.3}$$

where $\widetilde{A}$ and $\widetilde{S}$ are the complexification of A and S. □

By a standard result in stochastic linear quadratic control (see [5]) $(A, \sqrt{S}, C)$ is detectable if and only if there exists $K \in \mathcal{L}(\mathbb{R}^N)$, $L \in \Sigma^+(\mathbb{R}^N)$ with $L > 0$ and a mild solution $X \in \Sigma^+(\mathbb{R}^N)$ of the following "Algebraic Lyapunov Equation":

$$A_K^* X + X A_K + C^* X C + L = 0 \tag{2.4}$$

This allows us to give the following basic necessary condition.

Proposition 2.4 *If $(A, \sqrt{S}, C)$ is detectable then $(A^*, \sqrt{S})$ detectable too the.*

Proof: The claim follows remarking that if $X \in \Sigma^+(\mathbb{R}^N)$ is a solution of (2.4) then

$$A_K^* X + X A_K + \widehat{L} = 0.$$

where $\widehat{L} = C^*XC + L > 0$. Now it is enough to notice that the above is the Lyapunov equation relative to the deterministic equation obtained letting $C = 0$ in (1.6). □

Corollary 2.5 (Commutative case) *Assume that C and C^* commutate with S and with A then a sufficient condition for $(A, \sqrt{S}, C)$ to be detectable is that $(A^* + \frac{1}{2}C^*C, \sqrt{S})$ is stabilizable relatively to the identity.*

Proof: If $(A^* + \frac{1}{2}C^*C, \sqrt{S})$ is stabilizable then the there exists $K \in \mathcal{L}(\mathbb{R}^N)$ such that the exponential of matrix $A_K + \frac{1}{2}C^*C$ is stable. Moreover K can be chosen so that it commutates with C and C^* (see ([1]). Therefore letting:

$$X = \int_0^{\infty} e^{(A_K + \frac{1}{2}C^*C)^*t} e^{(A_K + \frac{1}{2}C^*C)t} dt$$

then $X \in \Sigma^+(\mathbb{R}^N)$ and solves:

$$(A_K + \frac{1}{2}C^*C)^*X + X(A_K + \frac{1}{2}C^*C)^* + I = 0$$

moreover X commutates with C and C^*. The above relation can therefore be written:

$$A_K^*X + XA_K + C^*XC + I = 0$$

therefore $(A, \sqrt{S}, C)$ is stabilizable. □

As in [6] we can introduce a class of stochastic systems that are equivalent by the detectability point of view:

Proposition 2.6 *Fix any $\lambda \in \mathbb{R}$ then $(A, \sqrt{S}, C)$ is detectable if and only if $(A - \lambda C - \frac{1}{2}\lambda^2 I, \sqrt{S}, C + \lambda I)$ is detectable.*

Proof: Just remark that equation (2.4) can be written:

$$(A_K - \lambda C - \frac{1}{2}\lambda^2 I)^*X + X(A_K - \lambda C - \frac{1}{2}\lambda^2 I) + (C + \lambda I)^*X(C + \lambda I) + I = 0$$

This clearly proofs our claim. □

The following proposition is the main result of this section

Proposition 2.7 (Necessary Condition) *Let, as in Remark 2.3, $\widetilde{A}$, $\widetilde{C}$, $\widetilde{S}$ be the complexification of A, C and S. Then if $(A, \sqrt{S}, C)$ is detectable the following holds:*

$$\operatorname{Ker}(\widetilde{A} - \lambda\widetilde{C} - \frac{1}{2}\lambda^2 I - \zeta I) \cap \operatorname{Ker}(\widetilde{S}) = \{0\} \quad \forall\lambda \in \mathbb{R},\ \forall\zeta \in \mathbb{C} \textit{ with } \operatorname{Re}(\zeta) \geq 0.$$

Proof: Applying Propositions 2.4 and 2.6 we obtain that if $(A, \sqrt{S}, C)$ is detectable, then $(A - \lambda C - \frac{1}{2}\lambda^2 I, \sqrt{S})$ is detectable $\forall \lambda \in \mathbb{R}$. The proof can now be concluded by (2.3). □

Example 2.1 Let $N = 2$ and:

$$A = \begin{pmatrix} a_{1,1} & a_{1,2} \\ a_{2,1} & a_{2,2} \end{pmatrix}, \quad C = \begin{pmatrix} c_{1,1} & c_{1,2} \\ c_{2,1} & c_{2,2} \end{pmatrix}, \quad S = \begin{pmatrix} 0 & 0 \\ 0 & 1 \end{pmatrix}$$

Then if $c_{2,1} \neq 0$ the necessary condition given by Corollary (2.7) is:

$$a_{1,1} + \frac{a_{2,1}c_{1,1}}{c_{2,1}} - \frac{1}{2}\left(\frac{a_{2,1}}{c_{2,1}}\right)^2 < 0$$

Otherwise, if $c_{2,1} = 0$, the necessary condition is:

$$a_{2,1} \neq 0 \quad \text{or} \quad \left(a_{2,1} = 0 \quad \text{and} \quad a_{1,1} + \frac{1}{2}c_{1,1}^2 < 0\right).$$

3 Detectability and Backward Equations

Fixed any $K \in \mathcal{L}(\mathbb{R}^N)$, let $(R_K(\cdot, T)\varrho, \Phi_K(\cdot, T)\varrho)$ belonging to the space $\mathbf{C}_\mathcal{P}(0, T, \mathbf{L}^2(\Omega, \mathbb{R}^N)) \times \mathbf{M}^2_\mathcal{P}(0, T, \mathbb{R}^N)$ be, for all $\varrho \in L^2(\Omega, \mathcal{F}_T, \mathbb{P}, \mathbb{R}^N)$, the unique solution of the following backward stochastic equation (see [8]):

$$\begin{cases} d_t R_K(t,T)\varrho = (-A_K^* R_K(t,T)\varrho - C^*\Phi_K(t,T)\varrho)dt + \Phi_K(T,t)\varrho d\beta_t \\ R_K(T,T)\varrho = \varrho \end{cases} \tag{3.1}$$

Then:

$$\mathbb{E}\langle R_K(0,T)\varrho, x\rangle = \mathbb{E}\langle \varrho, \xi_K(T,x)\rangle \tag{3.2}$$

Therefore (2.1) is equivalent to:

$$\mathbb{E}|R_K(0,T)\varrho|^2 \leq M^2 e^{-2\alpha T}\mathbb{E}|\varrho|^2 \quad \forall \varrho \in L^2(\Omega, \mathcal{F}_T, \mathbb{P}, \mathbb{R}^N) \tag{3.3}$$

Lemma 3.1 (Datko Theorem for Backward Stochastic Equations) *Fix $K \in \mathcal{L}(\mathbb{R}^N)$ then relation (3.3) is equivalent to the existence of a constant c verifying, for all $T>0$, and all $\varrho \in \mathbf{L}^2(\Omega, \mathcal{F}_T, \mathbb{P}, \mathbb{R}^N)$:*

$$\mathbb{E}\int_0^T |R_K(t,T)\varrho|^2 dt \leq c\,\mathbb{E}|\varrho|^2 \tag{3.4}$$

Proof: First let us remark that the semigroup property holds in this case. Namely it is easy to verify that:

$$R(t,T)\varrho = R(t,\tau)R(\tau,T)\varrho \quad \forall \varrho \in \mathbf{L}^2(\Omega, \mathcal{F}_T, \mathbb{P}, \mathbb{R}^N) \ \forall \tau \in [t,T].$$

Moreover by (3.2) there exists $M_0 \geq 1$ and $\alpha_0 \in \mathbb{R}$ such that:

$$\mathbb{E}|R_K(s,T)\varrho|^2 \leq M_0^2 e^{2\alpha_0(T-s)}\mathbb{E}|\varrho|^2 \quad \forall \varrho \in L^2(\Omega, \mathcal{F}_T, \mathbb{P}, \mathbb{R}^N).$$

The proof is now similar to the one working in the standard case, see [4] and [7]. For all $t \leq T-1$, we have:

$$\begin{aligned}
\mathbb{E}|R(t,T)\varrho|^2 &= \int_t^{t+1} \mathbb{E}|R(t,\tau)R(\tau,T)\varrho|^2\,d\tau \\
&\leq M_0^2 e^{2\alpha_0} \int_t^{t+1} \mathbb{E}|R(\tau,T)\varrho|^2\,d\tau \leq M_0^2 e^{2\alpha_0} c\mathbb{E}|\varrho|^2
\end{aligned}$$

Therefore letting $c_1 = M_0^2 e^{2\alpha_0} c$ it holds:

$$\mathbb{E}|R(t,T)\varrho|^2 \leq c_1 \mathbb{E}|\varrho|^2 \tag{3.5}$$

Moreover:

$$(T-t)\mathbb{E}|R(t,T)\varrho|^2 = \int_t^T \mathbb{E}|R(t,\tau)R(\tau,T)\varrho|^2\,d\tau \leq c_1 \int_t^T \mathbb{E}|R(\tau,T)\varrho|^2\,d\tau \leq c_1 c$$

We can now conclude our argument. Let $L = ecc_1$, by the above inequality we get that:

$$\mathbb{E}|R(t,T)\varrho|^2 \leq e^{-1}\mathbb{E}|\varrho|^2 \quad \text{whenever} \quad T-t \geq L.$$

So if $nL \leq (T-t) < (n+1)L$ for some $n \in \mathbb{N}$, applying n times the above relation, we obtain:

$$\mathbb{E}|R(t,T)\varrho|^2 \leq c_1 e^{-n}|\varrho|^2 \leq ec_1 e^{-(T-t)L^{-1}}\mathbb{E}|\varrho|^2 \leq M_1^2 e^{-2\alpha(T-t)}\mathbb{E}|\varrho|^2$$

where $\alpha = (2L)^{-1} > 0$, $M_1 = \sqrt{ec_1}$ (for the last inequality see also [4]). Combining this last inequality (holding whenever $T-t \geq L$) and inequality (3.5) our proof is completed. □

Corollary 3.2 *$(A, \sqrt{S}, C)$ is detectable if and only if there exists $K \in \mathcal{L}(\mathbb{R}^N)$ and a constant $c \in \mathbb{R}$ such that (3.4) hold for all $T > 0$.* □

The next Proposition studies the asymptotic stability of the solutions of equation (3.1) by a kind of Algebraic Riccati Equation with the wrong sign in front of the quadratic term:

Proposition 3.3 *Assume that there exist L and X in $\Sigma^+(\mathbb{R}^N)$ such that $L > 0$, $X \geq I$ and X is a mild solution of:*

$$A_K X + X A_K^* + X C^* C X + L = 0$$

then $(A, \sqrt{S}, C)$ is detectable.

Proof: Let us compute by Itô rule:

$$\begin{aligned}
&d_t\langle X R_K(t,T)\varrho, R_K(t,T)\varrho\rangle = \\
&\quad = 2\langle X R_K(t,T)\varrho, d_t R_K(t,T)\varrho\rangle + \langle X\Phi_K(t,T)\varrho, \Phi_K(t,T)\varrho\rangle dt = \\
&\quad = \left|\sqrt{L} R_K(t,T)\varrho\right|^2 dt + |C X R_K(t,T)\varrho|^2 dt + \left|\sqrt{X}\Phi_K(t,T)\varrho\right|^2 dt + \\
&\quad -2\langle C X R_K(t,T)\varrho, \Phi_K(t,T)\varrho\rangle dt + 2\langle X R_K(t,T)\varrho, \Phi_K(t,T)\varrho\rangle d\beta_t
\end{aligned}$$

Integrating between 0 and T and than taking the mean value we get by easy computations:

$$\mathbb{E}\langle X\varrho, \varrho\rangle = \mathbb{E}\langle X R_K(t,T)\varrho, R_K(t,T)\varrho\rangle + \mathbb{E}\int_0^T \left|\sqrt{L} R_K(t,T)\varrho\right|^2 dt +$$
$$+\mathbb{E}\int_0^T |C X R_K(t,T)\varrho - \Phi_K(t,T)\varrho|^2 dt + \mathbb{E}\int_0^T \left|\sqrt{X-I}\Phi_K(t,T)\varrho\right|^2 dt$$

Therefore

$$\mathbb{E}\int_0^T |R_K(t,T)\varrho|^2 dt \leq |(\sqrt{L})^{-1}|\mathbb{E}\int_0^T \left|\sqrt{L} R_K(t,T)\varrho\right|^2 dt \leq \mathbb{E}\langle X\varrho, \varrho\rangle$$

and this is what we wanted to show. □

Corollary 3.4 *Assume that* $\mathrm{Ker}(S) \subset \mathrm{Ker}(C)$ *then:*

$$(A, \sqrt{S}, C) \text{ is detectable} \iff (A, \sqrt{S}) \text{ is detectable.}$$

Proof: First notice that $\mathrm{Ker}(S) \subset \mathrm{Ker}(C)$ implies that $\exists d > 0$ such that $d^2 S \geq C^* C$. Moreover, $\forall d \neq 0$, $(A, \sqrt{S}, C)$ is detectable if and only if $(A, \sqrt{d^2 S}, C)$ is detectable, therefore we can assume without loss of generality that $S - C^* C \geq 2^{-1} S$.

So when $(A, \sqrt{S})$ is detectable $(A, \sqrt{S - C^*C})$ is detectable too, or, equivalently, $(A^*, \sqrt{S - C^*C})$ is stabilizable relatively to the identity. Therefore there exists $X \in \Sigma^+(\mathbb{R}^N)$ solving:

$$AX + XA^* - X(S - C^*C)X + I = 0.$$

Moreover by the above relation $\mathrm{Ker}(X) = \{0\}$ and consequently $X \geq \varepsilon I$ for some $\varepsilon \in [0, 1]$ (the finite dimensionality of the state space is essential here). Now notice that if $Y \stackrel{\text{def}}{=} \varepsilon^{-1}X$ then Y solves:

$$AY + YA^* - Y(S - C^*C)Y + L = 0$$

where $L = (1 - \varepsilon)Y(S - CC^*)Y + \varepsilon^{-1}I$. Rewriting the above relation as:

$$\left(A - \frac{1}{2}YS\right)Y + Y\left(A - \frac{1}{2}YS\right)^* + YC^*CY + L = 0$$

the claim follows by Proposition 3.3 letting $K = -\frac{1}{2}Y$. □

Example 3.1 Let $\mathbb{R}^N = \mathbb{R}^2$ and A, C, S as in Example (2.1) we want to show that the necessary condition stated in Example (2.1) is sufficient too.

If $c_{2,1} = 0$ we apply Proposition (2.6) with $\lambda = -c_{1,1}$ obtaining that $(A, \sqrt{S}, C)$ is detectable if and only if $(A + c_{1,1}C - \frac{1}{2}c_{1,1}^2 I, \sqrt{S}, C - c_{1,1}I)$ is detectable. Then we notice that $\mathrm{Ker}(S) \subset \mathrm{Ker}(C - c_{1,1}I)$ therefore $(A, \sqrt{S}, C)$ is detectable if and only if $(A^* - c_{1,1}C^* - \frac{1}{2}c_{1,1}^2 I, \sqrt{S})$ is stabilizable. By the deterministic Hautus condition this last condition is equivalent to:

$$a_{2,1} \neq 0 \quad \text{or} \quad \left(a_{2,1} = 0 \quad \text{and} \quad a_{1,1} + \frac{1}{2}c_{1,1}^2 < 0\right).$$

On the contrary if $c_{2,1} \neq 0$ we can, by Proposition (2.6) with $\lambda = a_{2,1}c_{2,1}^{-1}$, reduce ourselves to the case in which $a_{2,1} = 0$. So we have to show that if $c_{2,1} \neq 0$, $a_{2,1} = 0$ and $a_{1,1} < 0$ then $(A, \sqrt{S}, C)$ is detectable.
Now we notice that the matrices A_K with K belonging to $\mathcal{L}(\mathbb{R}^2)$ are all the matrices of the form:

$$A_K = \begin{pmatrix} a_{1,1} & h \\ 0 & k \end{pmatrix} \quad \text{with } h,\ k \text{ in } \mathbb{R}$$

If we choose an $X \in \Sigma^+(\mathbb{R}^N)$ of the following form:

$$X = \begin{pmatrix} 1 & \chi \\ \chi & \varsigma \end{pmatrix} \quad \text{with } \varsigma > \chi^2$$

we easily see (since the vectors $(1,\chi)$ and (χ,ζ) are linearly independent) that for all z and w in $\mathbb{R}$ there exist h and k in $\mathbb{R}$ such that:

$$A_K^* X + X A_K = \begin{pmatrix} 2a_{1,1} & z \\ z & w \end{pmatrix}$$

It is then clear that if we prove that for a suitable choice of χ and ζ with $\zeta > \chi^2$

$$\langle \begin{pmatrix} 1 \\ 0 \end{pmatrix}, C^* X C \begin{pmatrix} 1 \\ 0 \end{pmatrix} \rangle < 2a_{1,1}$$

then for a suitable choice of h and k it holds that

$$A_K^* X + X A_K + C^* X C \leq -\delta I \quad \text{for some } \delta > 0$$

then the claim follows (see equation (2.4)).
Now just notice that if $\chi = -c_{1,1}c_{2,1}^{-1}$ and $\zeta = \chi^2 + \epsilon$ then

$$\langle \begin{pmatrix} 1 \\ 0 \end{pmatrix}, C^* X C \begin{pmatrix} 1 \\ 0 \end{pmatrix} \rangle = c_{2,1}^2 \epsilon$$

therefore if ϵ is small enough our claim holds. □

References

[1] A. BESOUSSAN, G. DA PRATO, M. C. DELFOUR, S. K. MITTER, *Representation and Control of Infinite Dimensional Systems*, .

[2] G. DA PRATO AND A. ICHIKAWA, *Stability and quadratic control for linear stochastic equations with unbounded coefficients*, Bollettino U.M.I. **6**, (1985), pp. 987-1001.

[3] M. L. J. HAUTUS, *Stabilization Controllability and Observability of a Linear Autonomous System*, Indagationes Math. **32**, (1970), pp. 448-455.

[4] A. ICHIKAWA, *Equivalence of L_p stability and exponential stability for a class of nonlinear semigroups*, Nonlinear Anal. **8**, (1984), pp. 805-815.

[5] G. TESSITORE, *Some Remarks on the Riccati Equation Arising in an Optimal Control Problem with State- and Control-Dependent Noise*, SIAM J. Contr. Optim. **30**, (1992), pp. 717-744.

[6] G. TESSITORE, *Some Remarks on the Mean Square Stabilizability of a Linear SPDE*, Dynamic Systems and Applications **2**, (1993), pp. 251-266.

[7] G. TESSITORE, *A Note on a Parameter Depending Datko Theorem Applied to Stochastic Systems*, Preprint S.N.S..

[8] G. TESSITORE, *Existence, Uniqueness and Space Regularity of the Adapted Solutions of a Backward SPDE*, to appear in "Stochastic Analysis and Applications".

Suboptimal Shape of a Plate Stretched by Planar Forces

P. VILLAGGIO AND J.P. ZOLÉSIO

Università di Pisa via Diotisalvi,2 56126 PISA,Italy

CNRS-Institut Non-Linéaire de Nice, Sophia Antipolis F 06560 , France

Abstract.

A typical optimization problem in elasticity is that of finding the shape of a plate, stretched by forces acting in its plane, such that the strain energy is a minimum under the constraint that the area is constant.The loads are applied along a known part of the boundary of the plate whereas the remaining part must be determined. The optimum domain is the solution of a problem of minimization formulated in a weak form and, if no restrictions are imposed on the forms of the admissible domains, is turns out that they may have holes but their boundaries must be regular in the sense that they cannot have cusps.

1. INTRODUCTION

A classical optimization problem in planar elasticity can be formulated as follows. Let Ω be a planar domain representing the middle surface of a plate of constant thickness $2h$ and let $\partial\Omega$ denote the boundary of Ω, which is divided into two parts, Γ and Σ, such that $\partial\Omega = \Gamma \cup \Sigma$, with Γ given and Σ unknown (Fig.1). The shape of Ω is thus partially undetermined, but the area of Ω must not exceed a given value α. The plate is stretched by forces acting on the part of its lateral surface having Γ as trace on the middle plane while the remaining part of the lateralsurface is free. The actual distribution of these forces is not of pratical importance and they can be represented by their resultants estimated per unit of length of Γ. The only restriction on these resultants is that they act in the middle plane and are globally balanced in order to maintain the equilibrium of the plate. The loads on Γ generate a state of stress in the interior of the plate, but, if the plate is thin and the plane faces are free, the average values of the stress components taken over the thickness represent a useful approximation of the actual state of stress at each point. The average stress components which are different from zero are those of the type $\sigma_x, \sigma_y, \tau_{xy}$, (x, y) being Cartesian axes placed in the plane of Ω. States of stresses

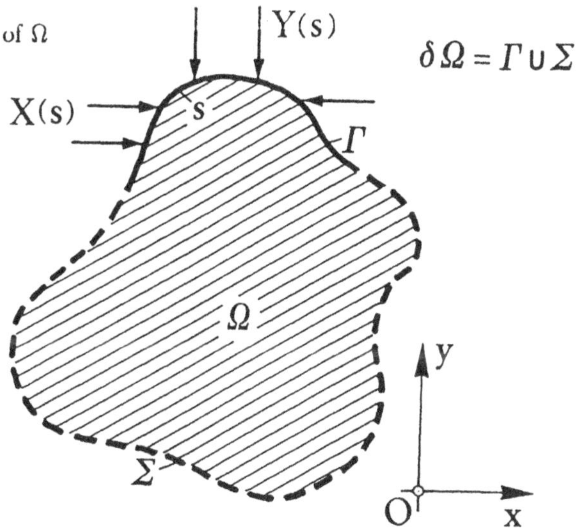

of this type are termed "generalized plane stresses", and can be treated in a simplified way by introducing a stress function $F(x, y)$ such that

(1.1) $$\bar{\sigma}_x = F_{,yy} \ , \ \bar{\sigma}_y = F_{,xx} \ , \ \bar{\tau}_{x,y} = -F_{,xy}.$$

$F(x, y)$ is known as Airy's "stress function". Assuming now that the material composing the plate is elastic, homogeneous, and isotropic, characterized by Young's modulus E $(E > 0)$ and Poisson's ratio σ $(-1 < \sigma < \frac{1}{2})$, the average strains $\bar{\varepsilon}_x, \bar{\varepsilon}_y, \bar{\varepsilon}_{xy}$ are connected with average stress components by the equations

(1.2) $$\bar{\varepsilon}_x = \frac{1}{E}(\bar{\sigma}_x - \sigma\bar{\sigma}_y), \quad \bar{\varepsilon}_y = \frac{1}{E}(\bar{\sigma}_y - \sigma\bar{\sigma}_x),$$

(1.3) $$\bar{\varepsilon}_{xy} = \frac{1+\sigma}{E}\bar{\tau}_{xy}.$$

These strains must satisfy the so called "Beltrami's relations" (LOVE [1927]), which require the quantity $\Delta = \bar{\varepsilon}_{xx} + \bar{\varepsilon}_{yy}$ to be a harmonic function, and therefore F to satisfy the equation

(1.4) $$\nabla^4 F = \frac{\partial^4 F}{\partial x^4} + 2\frac{\partial^4 F}{\partial x^2\,\partial y^2} + \frac{\partial^4 F}{\partial y^4} = 0,$$

hene F to be a planar biharmonic function. The stress function $F(x, y)$ is determined by its boundary conditions along $\partial\Omega$, expressing the balance between the applied tractions and the stress-resultants at the edge. But, as Michell[1900] proved, these conditions assume the surprisingly simple form

(1.5) $$F = g, \frac{\partial F}{\partial n} = h \text{ on } \partial\Omega,$$

where g and h are two functions determined by the components $X(s), Y(s)$ of the edge forces along $\partial\Omega$. However, the detailed expressions of g and h are not necessary, the only required information being a certain regularity and that their support be contained in Γ.
If Σ were known, the state of (generalized) stress inside the plate would be determined by solving a bi-harmonic boundary value problem for the function $F(x, y)$ and, in particular, the solution would permit also to evaluate the strain energy stored in the plate

(1.6) $$\mathcal{E}(\Omega, F) = \frac{1}{2E}\int\int_{\Omega}\{(\nabla^2 F)^2 + 2(1-\sigma)(F_y^2 - F_{xx}F_{yy})\}dxdy.$$

The strain energy depends of course on the shape of Ω, thus the question arises of finding the domain Ω for which $\mathcal{E}$ is a minimum under the constraint that the area remain fixed

$$meas(\Omega) = \alpha,$$

where α is a given constant $(\alpha > 0)$.
Formulated in this way, the problem is not classical because the shape itself of Ω is, besides F , an unknown. There is, however, an additional complication. In a first formulation of the problem it is natural to think that $\Gamma \cup \Sigma$ constitute the boundary of a simply connected domain; but there is no reason to exclude that the optimal domain admit some holes or, in the limit, infinitely many holes, so that the minimum of $\mathcal{E}$ is achieved by a sort of beehive where the material is condensed along several thin ribs. This fact is not pathological since it occurs in other optimization problems in which Ω is fixed, but the thickness h may vary, as KOHN and VOGELIUS [1985] have shown. With the appearence of holes the boundary conditions (1.5) must also be changed since now Σ may be constituted of several connected components, a piece of curve C , prolongation of Γ, and other closed curves Σ_i bounding the

holes. In this case it is possible to adjust the stress function to make F and $\frac{\partial F}{\partial n}$ vanish on $\mathcal{C}$, but not on the other components Σ_i, where they depend on some undetermined constants according to an extension of Michell's theorem. These constants can be evaluated requiring the displacements in Ω be single- valued.
It may seem that the natural way to treat the problem is to reduce it to a variational inequality. But the attempt is unsuccessful for three reasons at least. The first is that the constraint $meas(\Omega) = \alpha$ is implicitly defined through F and cannot be expressed in a tractable form. The second is that bi-harmonic functions don't enjoy a maximum principle and therefore the domain Ω cannot be identified by an inequality constraint on F or its gradient ∇F. A third difficulty arises in the very definition of the function space in which the solution must be sought, for Ω may be so irregular as to prevent the introduction of a classical Sobolev space. The way to overcome these difficulties is two-fold. First of all it is necessary to introduce a larger domain B where Ω must be strictly included ; the boundary of B can be very smooth, for instance piecewise C^∞, so that the Sobolev space $H^2(B)$ is classically defined and so is the subspace $H_0^2(B)$. For any smooth subdomain Ω of B the traces of F and $\frac{\partial F}{\partial n}$ on $\partial\Omega$ have a precise meaning so that $H_0^2(\Omega)$ can be canonically identified wih a closed subspace of $H_0^2(B)$. Unfortunately, $H_0^2(\Omega)$ is not the right space for the optimum problem since F and $\frac{\partial F}{\partial n}$ are not zero on $\partial\Omega$. It is thus necessary to introduce a specific space $H(\Omega)$, but its definition is not immediate and can be done by exploiting the unexpected property that the boundary conditions of F are reducible to $F = constant$, $\frac{\partial F}{\partial n} = 0$. Once the appropriate spaces have been introduced, it is possible to prove an existence theorem , but not uniqueness, for an optimal domain Ω for a "sub-optimal problem" by using standard arguments of compactness, where the term sub-optimal means that we shall solve , instead of the classical optimization problem (1.7) an new one introduced at (1.9) which includs a relaxed form of the optimality necessary conditions of (1.7) associated to possibily smooth optimal holes.
The practical consequences of the abstract treatment of the problem are that in principle, holes are unavoidable, but the question of their number, their shapes, and their extent, remains open so that the optimization problem

$$(1.7) \qquad Min \ \{ \ \mathcal{E}(\Omega), \ \ | \ \Omega \in \mathcal{O}, meas(\Omega) = \alpha \ \}$$

with

$$\mathcal{E}(\Omega) = Min\{ \ \mathcal{E}(\Omega, \phi) \ \ | \ \phi \in H^2(\Omega), \ \phi = g, \ \frac{\partial \phi}{\partial n} = h \text{ on } \Gamma, \ \}$$

may be not well posed for any non trivial family of open sets $\mathcal{O}$. We formaly consider the optimality necessary condition associated to problem (1.7), i.e. when there is an optimal domain Ω^* with smooth boundary verifying the necessary conditions since F is biharmonic, from these optimality conditions and having a smooth enough boundary. From the biharmonicity of F , these optimality conditions and smoothness assumption we derive that if such an open set exists it should be sought in a smaller Hilbert space than $H^2(\Omega)$, namely the subspace of functions having constant value and zero normal derivative on the the boundary of each hole in the domain. Using arguments of capacity this definition of subspace can be relaxed or extended to non smooth sets Ω leading to a specific linear space $H(\Omega)$, defined in Section 3, and then we formulate the "suboptimal problem"

$$(1.8) \qquad Min \ \{ \ \mathcal{E}^s(\Omega) \ | \ \Omega \in \mathcal{O}, meas(\Omega) = \alpha \ \}$$

where the sub-optimal energy functional $\mathcal{E}^s$ satisfies the conditions :

$$(1.9) \quad \mathcal{E}^s(\Omega) = Min \ \{ \ \mathcal{E}(\Omega, \phi) \ \ | \ \phi \in H(\Omega), \ \phi = g, \ \frac{\partial \phi}{\partial n} = h \text{ on } \Gamma, \ \} \leq \mathcal{E}(\Omega)$$

We shall prove that, for a specific family $\mathcal{O}$ large enough, the problem (1.8) has optimal solutions Ω^* (at least one) .Then if problem (1.7) has smooth solutions, by the necessary conditions they should be sought in the linear space $H(\Omega)$ and then these smooth solutions are the solutions Ω^* of the sub-optimal problem (1.8). More precisely we have:

Proposition 1.1. *i) The problem (1.8) has optimal solutions.*
ii) Any smooth solution Ω^* to problem (1.7) is a solution to problem (1.8).

This result is a more precise version of the one in [1991]Villaggio-Zolésio.

2. NECESSARY CONDITIONS

The search of an optimal domain Ω^* leads to a free boundary condition for the solution F associated to Ω^*, which, in the present case, reduces to a constraint on the square of $\frac{\partial^2 F}{\partial n^2}$ along $\partial\Omega^*$. There are many ways to derive this necessary condition : one is, for instance, that described by COURANT and HILBERT [1953,chap. IV , 11] ; another, formally simpler, was proposed by Zolésio [1981],[1981]. In order to obtain the condition in question it is convenient to compute the derivative of the mapping $t \to \mathcal{E}(\Omega_t^*)$ at $t = 0$, where Ω_t^* is a perturbation of Ω^* having the form $\Omega_t^* = T_t(\vec{V})(\Omega^*)$ where T_t is the flow mapping of an admissible smooth vector field $\vec{V}$ verifying $\vec{V} = 0$ on ∂B. The first variation of $\mathcal{E}(\Omega_t^*)$ with respect to t is also called the "domain derivative" and is given by

$$(2.1)\quad d\mathcal{E}(\Omega^* ;\vec{V}) = \left(\frac{d}{dt}\mathcal{E}(\Omega_t^*, FoT_t^{-1} \right)_{t=0} = \frac{\partial}{\partial t}\mathcal{E}(\Omega_t^*, F)\ )_{t=0} + \frac{\partial \mathcal{E}}{\partial F}.(\ -\nabla F.\vec{V}\ ),$$

where ∇F is the gradient of F and $\frac{\partial \mathcal{E}}{\partial F}$ stands for the partial derivative of $\mathcal{E}$ with respect to F. On recalling (1.6), the derivative of $\mathcal{E}(\Omega_t^*)$, at $t = 0$, assumes the form

$$(2.2)\qquad \frac{1}{\bar{E}}\int\int_{\partial\Omega^*} \{\ (\nabla^2 F)^2 + 2(1-\sigma)(\ F_{,xy}^2 - F_{,xx}F_{,yy}\ )\ \}\ \vec{V}.n\ ds$$

$$+\frac{1}{\bar{E}}\int\int_{\Omega^*} \{\nabla^2 F.\nabla^2(-\nabla F.\vec{V}) + 2(1-\sigma)\ (\ 2F_{,xy}(-\nabla F.\vec{V}\ )_{,xy} - F_{,xx}(-\nabla F.\vec{V}\ )_{,yy}$$

$$- F_{,yy}(\ -\nabla F.\vec{V}\ )_{,xx}\ \}dxdy.$$

But in this expression the integral through Ω^*, after two subsequent integrations by parts and use of the formulae

$$\frac{\partial}{\partial s} = cos\vartheta\frac{\partial}{\partial y} - sin\vartheta\frac{\partial}{\partial x}, \quad \frac{\partial}{\partial n} = cos\vartheta\frac{\partial}{\partial x} + sin\vartheta\frac{\partial}{\partial y}, \quad \frac{\partial\vartheta}{\partial n} = 0\ , \quad \frac{\partial\vartheta}{\partial s} = -\frac{1}{\rho},$$

can be transformed into (cf. LOVE [1924])

(2.3)

$$\frac{1}{\bar{E}}\int\int_{\Omega^*} \nabla^4 F(-\nabla F.V\ )dxdy - \int_{\partial\Omega^*} Q(\ -\nabla F.V\ \ )\ ds + \int_{\partial\Omega^*} G\frac{\partial}{\partial n}\ (-\nabla F.\vec{V}\ )\ ds,$$

where G and Q have the following expressions

$$G = -\frac{1}{\bar{E}}\left\{\ \frac{\partial^2 F}{\partial n^2} + \sigma\left(\ \frac{\partial^2 F}{\partial s^2} + \frac{1}{\rho}\frac{\partial F}{\partial n}\ \right)\ \right\}$$

$$Q = -\frac{1}{\bar{E}}\left\{\ (1-\sigma)\frac{\partial^3 F}{\partial n\partial s^2} + \frac{\partial}{\partial n}(\ \Delta F\ )\ \right\}$$

Here the first term vanishes as F is biharmonic in Ω^*, and the boundary terms reduce to two integrals extended on Σ as $\vec{V}$ is zero along Γ which is fixed. The consequence is that the domain derivative of the strain energy assumes the form

(2.4)
$$\int_\Sigma \{ \frac{1}{\bar{E}} (\Delta F)^2 + 2(1-\sigma)(F_{,xy}^2 - F_{,xx}F_{,yy} \ \vec{V}.n + Q(\nabla F.\vec{V}) - G\frac{\partial}{\partial n} (\nabla F.\vec{V}) \}ds$$

The integrand can be further simplified by exploiting the fact that, along Σ, the partial derivatives of F satisfy the relations

$$\Delta F = \frac{\partial^2 F}{\partial n^2} \ f, \quad F_{,xy}^2 - F_{,xx}F_{,yy} = 0,$$

$$Q = -\frac{1}{\bar{E}}\frac{\partial^3 F}{\partial n^3} \ , \ G = -\frac{1}{\bar{E}}\frac{\partial^2 F}{\partial s\, \partial n},$$

since, on the boundary, which is free, $\frac{\partial^2 F}{\partial s^2} = \frac{\partial^2 F}{\partial s \partial n} = 0$. The integral (2.4) can thus be written in the form

$$\int_\Sigma \{ \frac{1}{\bar{E}} (\frac{\partial^2 F}{\partial n^2})^2 \vec{V}.n - \frac{1}{\bar{E}}\frac{\partial^3 F}{\partial n^3}\frac{\partial \nabla F.\vec{V}}{\partial n}) \}ds \ ; \qquad (2.5)$$

where the term $\frac{\partial}{\partial n}$ ($\nabla F.\vec{V}$) is decomposable into the sum

$$\frac{\partial}{\partial n} (\nabla F.\vec{V}) = \frac{\partial}{\partial n} (\nabla F).\vec{V} + \nabla F.\frac{\partial \vec{V}}{\partial n}$$

But, as Ω^* is an optimal domain, its domain derivative $d\mathcal{E}(\Omega^*, V)$ must vanish for all admissible fields $\vec{V}$ preserving the constant measure of the perturbed domains Ω_t. This conditions implies that $div\vec{V} = 0$ in B and $\vec{V}$ is a field of the type $\vec{V} = v(s)n$ where $v(s)$ is a scalar function representing a perturbation of the variable part Σ of the boundary $\partial\Omega^*$ (Γ is fixed) in the direction of the normal. With this choice of $\vec{V}$, $\frac{\partial}{\partial n}$ ($\nabla F.\vec{V}$) becomes

$$\frac{\partial}{\partial n} (\nabla F.\vec{V}),= \frac{\partial}{\partial n} (\nabla F.(vn)),= \frac{\partial^2}{\partial n^2}F \ v + \frac{\partial F}{\partial n}\frac{\partial v}{\partial n} \qquad (2.6)$$

From $\nabla F = 0$ as $F \in K(\Omega)$ (see 3.2) we get $\frac{\partial F}{\partial n} = 0$ (as we assume here the boundary to be smooth) and hence the vanishing of $d\mathcal{E}(\Omega^*; \vec{V})$ with the constraint

$$\int_\Sigma vds = 0, \qquad (2.7)$$

which ensures the constancy of the measure of perturbed domains . This yields as necessary conditions of optimality, consequences of the arbitrariness of v,

$$(\frac{\partial^2 F}{\partial n^2}) = \lambda = \text{ constant }, \ \frac{\partial F}{\partial n}, = 0 , \qquad (2.8)$$

along the free boundary Σ. Since the energy is not zero but finite, this implies that $\lambda \neq 0$, $\lambda \neq \infty$. When the boundary Σ_i of a hole is smooth the last condition may be further refined because $\frac{\partial^2 F}{\partial n^2}$ (which is ΔF) is continuous along Σ_i and thus $\frac{\partial^2 F}{\partial n^2} = \lambda' =$ constant on Σ_i.

This condition generalizes a property already known in problems of optimal design of plates having only one boundary, because, in this case, it is possible to adjust the constants a_1, b_1, c_1 so that $\frac{\partial F}{\partial n} = 0$ on Σ. If the solution admits holes, $\frac{\partial F}{\partial n} = 0$ along the boundary of each hole (cf. NEUBER [1968], BANICHUK [1983.Chap.4]). In this last case it is possible to show that the optimum curve is smooth in the sense that it does not have either external (pointing outwards) or interior (pointing inwards) sharp corners. In fact, as NEUBER [1937] proved that if the curve had an exterior corner the stress would vanish at the vertex and, consequently, also ($\frac{\partial^2 F}{\partial n^2}$) would vanish violating the condition that the constant is not zero ; if the cusp were interior, the stress would tend to infinity there, thus infringing on the condition that the constant is finite. It remains to prove the previously exploited property that $F =$ constant at the boundary Σ_i of each hole. But this result is an immediate consequence of equation $\frac{\partial F}{\partial n} = 0$ on Σ, compared with the second part of (3.5). Since the equation must hold for arbitrary values of n_x, n_y it follows that necessarily $a_i = b_i = 0$ and thus $F = c_i$ on Σ_i. However, the problem of predicting the shape of Σ, or better of each connected component of Σ, and their regularity is still unsolved.

3. THE ADMISSIBLE FAMILY $\mathcal{O}$ OF DOMAINS

The simplest way to characterize the admissible domains is that of considering a family of plane measurable sets $\mathcal{O}$ defined as follows. In the x, y-plane (Fig.2) let E be a smooth bounded domain the points of which satisfy the condition $\{(x, y)|y \geq 0\}$ and let Γ be the boundary of E in the half-plane $y > 0$, that is $\Gamma = E \cap \{(x, y)|y > 0\}$; let then D be a smooth bounded domain contained in the half-plane $y < 0$ such that $D \cap \{(x, y)|y = 0\}$ contains $E \cap \{(x, y)|y = 0\}$. An element Ω of the family $\mathcal{O}$ is a domain of the type

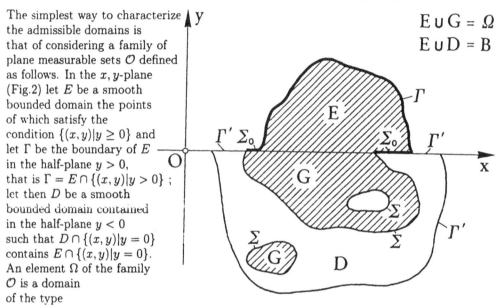

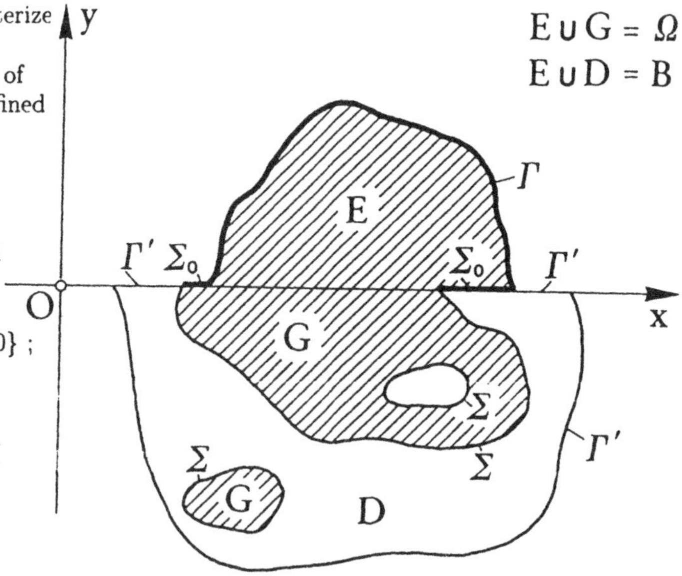

$$\Omega = E \cup G, \tag{3.1}$$

where G is a measurable subset of D, not necessarily connected, sketched in the figure. The boundary Σ of G can be defined as

$$\Sigma = c\ell(G) \cap c\ell(D\ G) \cap \{(x, y)|y < 0\} \tag{3.2}$$

so that the boundary of Ω is

$$\partial\Omega = \Gamma \cup \Sigma \cup c\ell(\Omega) \cap \{(x, y)|y = 0\}^* = \Gamma \cup \Sigma \cup \Sigma_0. \tag{3.3}$$

In a problem of optimal shape the domain E is given and G must be determined . As shown in Fig.2, it has been admitted that G has some disjoint components, although this case will be proved to be impossible ; on the contrary, G may admit holes since the edge forces act only along Γ and are in equilibrium; according to Michell's transformation the function F satisfy the boundary conditions

$$F = g \text{ on } \Gamma, \quad \frac{\partial F}{\partial n} = h \text{ on } \Gamma \tag{3.4}$$

where g is given in $H^{m-\frac{1}{2}}(\Gamma)$ and h in $H^{m-\frac{3}{2}}(\Gamma)$ for some $m \geq 2$; in addition, the support of g and h is contained in $\Gamma \cap \{(x,y)|y > 0\}$, that is g and h are identically zero in some neighborhood of $\{(x,y)|y = 0\}$. As far as the boundary conditions on Σ_0 and Σ are concerned, Michell's theorem states that they are of the type

$$F = a_i x + b_i + c_i \ , \quad \frac{\partial F}{\partial n} = a_i + b_i \quad \text{on } \Gamma \tag{3.5}$$

where a_i, b_i, c_i are constant on each connected component of $\Sigma \cup \Sigma_0$. These constants can be taken all zero on one connected component of $\Sigma \cup \Sigma_0$, for instance the component C which is "close" to ∂D in the sense that $\partial(D - G) = \partial D \cup C$. The other constants are instead determined by imposing the displacements to be single-valued around each curve surrounding any hole or a fragment of G disjoint from E. But, as a consequence of necessary conditions of optimality discussed later, it is possible to show that constants a_i, b_i are equal to zero on each boundary Σ_i of possible holes. In this case, denoting by $D_1, ..., D_i, ...$the parts of $B - \Omega$ interior to each hole, and by D_0 the exterior component of $B - \Omega$ which is not a hole, the function F can be taken $F \equiv 0$ in D_0 and $F \equiv c_i$ on each D_i.

This surprising property of F permits one to seek the solutions in the Sobolev space $H^2(\Omega)$, and, more precisely, in the convex set $K(\Omega)$ of $H^2(\Omega)$ satisfying the non-homogeneous boundary conditions (3.4) imposed on Γ. In fact, if $\mathcal{E}(\Omega, \phi)$ is the strain energy defined by (1.6), corresponding to a generic smooth set Ω of the family $\mathcal{O}$, the stress function F is the unique minimizer of the variational problem

$$min\{\mathcal{E}(\Omega, \phi) | \phi \in K(\Omega)\}. \tag{3.6}$$

The existence of such a minimizer derives from the property that, for each $\sigma > -1$, the inequality

$$\mathcal{E}(\Omega, \phi) \geq \frac{1}{4E} \int\int_\Omega (\phi_{,xx}^2 + \phi_{,xy}^2 + \phi_{,yy}^2) dxdy \tag{3.7}$$

holds which proves the coerciveness of $\mathcal{E}(\Omega, \phi)$. The shape optimization problem can thus be reduced to the following :

$$Inf \ \{ \ \mathcal{E}(\Omega) \ \ |\Omega \in \mathcal{O}, \ meas(\Omega) = \alpha\}, \tag{3.8}$$

where

$$\mathcal{E}(\Omega) = Min \ \{\mathcal{E}(\Omega, \phi) \ | \ \phi \in K(\Omega) \ \}, \tag{3.9}$$

and α is a positive number selected so that

$$\alpha < meas(E) + meas(D). \tag{3.10}$$

However, a difficulty arises in defining the convex set $K(\Omega)$ for Ω is not known in advance except that it is a measurable subset of $B = E \cup D$ and, therefore, the definition of the convex subset $K(\Omega)$ in the Sobolev space $H^2(\Omega)$ is not obvious as Ω is not smooth. It is thus convenient to introduce the convex set

$$K(B) = \{\phi \in H^2(B) | \phi = g, \frac{\partial \phi}{\partial n} = h \text{ on } \partial B\}.$$

Since g and h have supports on Γ, the elements of $K(B)$ have null traces on the part $\Gamma' = \partial B \cap \Gamma$ of the boundary. The closed subspaces of $H^2(B)$ and $H^1(B)$with null traces on Γ' are commonly denoted by $H^2_{\Gamma'}(B)$ and $H^1_{\Gamma'}(B)$, and therefore $K(B)$ is a closed subset of $H^2_{\Gamma'}(B)$.

4. RELAXATION OF THE CONSTRAINT

In order to render the definition of $K(\Omega)$ meaningful it is convenient to start from the convex $K(B)$ which is well defined, provided that both E and D are smooth. Moreover, if F belongs to $H^2(B)$, its gradient ∇F is an element of $H^1(B)^2$, so it is pointwise determined up to a set ω with zero capacity ($cap_B(\omega) = 0$), where, for any compact subset $\bar{\omega} \subset B$,

$$cap_B(\bar{\omega}) = Min\{\int\int_B |\nabla\phi|^2 dxdy \quad | \quad \phi \in C^1_0(B), \quad \forall x \in \omega, \quad \phi(x) \geq 1 \quad \}.$$

Now, if A is a measurable subset of B and ϕ an element of $H^1(B)$ such that $\phi = 0$ *q.e.* in A, that is up to subsets of zero capacity, then $\phi = 0$ *a.e.* in A and it has been proved by STAMPACCHIA [1965] that $\nabla\phi = 0$ *a.e.* in A. It turns out that, if F belong to $H^2(B)$ and each component of ∇F is equal to zero *q.e.* on A , then $\nabla^2 F, F_{,xx}, F_{,yy}, F_{,xy}$ are zero *a.e.* in A. If Ω is smooth, the convex set $K(\Omega)$, considered as a closed subset of $K(B)$, can be defined by making use of the canonical "constant extension' of any element ϕ of $K(\Omega)$, which means that $K(\Omega)$ should be given by :

(4.1) $K(\Omega) = \{\phi \in K(B)|\phi = \text{constant on each simply connected component of } D \}$.

This definition, however, is too restrictive for the present problem since it is not known in advance whether the components of D are smooth domains in D. It is thus necessary to define $K(\Omega)$ as

$$K(\Omega) = \{\phi \in K(B)|\nabla\phi = 0 \text{ q.e.in } D - G \}. \tag{4.2}$$

Since $\nabla\phi$ is in $H^1_0(B)^2$, it is pointwise defined *q.e.*. It turns out that, when Ω is only a measurable set of the family $\mathcal{O}$, definition (4.1) makes no sense while does (4.2). The convex set $K(\Omega)$ defined by (3.2) is also closed in $K(B)$, for, by using a result from potential theory (cf. DENY and LYONS [1957]), from any converging sequence in $H^1_0(B)$ it is possible to extract a subsequence converging *q.e.* to the limit. This property would not hold if, in defining $K(\Omega)$, the condition $\nabla\phi = 0$ *q.e.*in $D - G$ were replaced by $\nabla\phi = 0$ *a.e.*in $D - G$. $K(\Omega)$ is the appropriate set in which the existence of a solution to problem (3.9) can be proved through the following

Theorem 4.1. *Let α be given such that $meas(E) + meas(D) > \alpha$ and m ($m \geq 2$) a real number, then there is at least one $\Omega^* \in \mathcal{O}$ such that $meas(\Omega^*) = \alpha$ and*

$$\mathcal{E}^s(\Omega^*) \leq \mathcal{E}^s(\Omega), \quad \forall\, \Omega \text{ with } meas(\Omega) = \alpha, \tag{4.3}$$

where

$$\mathcal{E}^s(\Omega) = Min\{\mathcal{E}(\Omega, \phi) \mid \phi \in K(\Omega) \} \tag{4.4}$$

Proof. Let Ω_n be a minimizing sequence and let $F_n \in K(\Omega_n)$ be the solution to (3.3), then

$$\frac{1}{2E}\int\int_{\Omega_n} \{ ||\nabla^2 F_n||^2 + 2(1-\sigma)(\; |(F_n)_{,xy}|^2 - (F_n)_{,xx}\,(F_n)_{,yy} \;) \} \, dxdy = \tag{4.5}$$

$$= \frac{1}{2E}\int\int_B \{ ||\nabla^2 F_n||^2 + 2(1-\sigma)(\; |(F_n)_{,xy}|^2 - (F_n)_{,xx}\,(F_n)_{,yy} \;) \} \, dxdy$$

for the integrand is zero $a.e.$ in $B - \Omega_n$. From (4.2) and (4.4) it ensues that the sequence F_n remains bounded in $H^2_{\Gamma'}(B)$, and, therefore, there is a subsequence, still denoted by F_n, weakly convergent to an element F. Then we proceed as in Zolésio[1994]: the sequence χ_Ω of characteristic functions also remains bounded in $L(B)$, it thus weakly converges in $L^2(B)$ to an element $\lambda, 0 \leq \lambda \leq 1, a.e.$ in B. The set

(4.6) $\quad \Omega^* = \{(x,y) \in D \mid \mid \nabla F(x,y) \neq 0 \ \} \cup E \cup \{ \ \bar{E} \cap (\ (x,y) \mid y = 0 \ ) \ \},$

is defined up to a subset $\omega \subset D$ with $cap_B(\omega) = 0$, ∇F being an element of $H^1_{\Gamma'}(B)^2$. Now, as for each n the equation $(1 - \chi_\Omega)\nabla F_n = 0$ holds $a.e.$ in B, and ∇F_n converges strongly in $H^1(B)^2$ to ∇F by the compact embedding of $H^2(B)$ in $H^1(B)$, in the limit the above equation becomes

$$(1 - \lambda)\nabla F = 0 \ a.e. \text{ in } B,$$

so that at the points where $\nabla F \neq 0$, λ is equal to one, that is

(4.7) $$\lambda \geq \chi_{\Omega^*} \ a.e. \text{ in } B.$$

This inequality is essential to prove the theorem. In fact, inequality (3.7) garantees that the integrand in $\mathcal{E}(\Omega, \phi)$ is positive, so, in the limit, by (4.6) the following inequality

$$\mathcal{E}(\Omega^*, F) \leq \frac{1}{2\bar{E}} \int\int_B \ (\nabla^2 F)^2 + 2(1-\sigma) \ (\ (F_{,xy})^2 - F_{,xx}F_{,yy})dxdy \ \leq$$

$$\leq \ lim \ inf \frac{1}{2\bar{E}} \int\int_B \ (\nabla^2 F)^2 + 2(1-\sigma) \ (\ (F_{n,xy})^2 - F_{n,xx}F_{n,yy}dxdy$$

$$= lim \ inf_{\, n \to \infty} \ \mathcal{E}(\Omega_n)$$

must hold. And, since $\mathcal{E}(\Omega_n)$ is a minimizing sequence, this inequality implies that

$$\mathcal{E}(\Omega^*, F) \leq Inf \ \{ \ \mathcal{E}(\Omega) \mid \Omega \in \mathcal{O}, \ meas(\Omega) = \alpha \ \}.$$

On the other hand, the set Ω^* satisfies the two conditions $\Omega^* \in \mathcal{O}, meas(\Omega) \leq \alpha$. Hence, if $meas(\Omega^*) = \alpha$, the theorem is proved. If, on the contrary, $meas(\Omega)$ were less than α, the fact that the mapping $\Omega \to \mathcal{E}(\Omega)$ is decreasing (as $\Omega_1 \subset \Omega_2$ implies$H^1_{\Gamma'}(\Omega_1) \subset H^1_{\Gamma'}(\Omega_2)$) allows us to choose $\Omega^*_1 \supset \Omega^*$ with meas$(\Omega^*) = \alpha$, such that

$$\mathcal{E}(\Omega^*_1) \leq \ \mathcal{E}(\Omega^*) \ \leq \ Inf \ \{\mathcal{E}(\Omega)|\Omega \in \mathcal{O}, \ \text{meas}(\Omega) = \alpha \ \}$$

and the infimum is attained for Ω^*_1.

5. PROPERTIES OF THE OPTIMAL DOMAINS Ω^*

The proof that at least an optimal domain Ω^* exists does not answer the demand for knowing the geometrical properties of the solutions. Though the problem is still unsolved, some of these properties are derivable from the variational formulation and confirm intuition. A first property is that the optimal domain Ω^* cannot be decomposed into the sum $\Omega^* = \Omega_1 \cup A$, with $\Omega_1 \in \mathcal{O}$ and A a measurable subset in D such that

(5.1) $$cap_B(\Omega_1 \cap A) = 0.$$

An Ω^* satisfying this requirement is called "weakly connected ", but Ω^* cannot be weakly connected. In fact, let $G(A)$ be the set

(5.2) $$G(A) = \{ \ \xi \in H^2_{\Gamma'}(B) \mid \nabla\xi = 0 \ q.e. \text{ in } B - A \ \}.$$

Any element ϕ of $K(\Omega)$ can be written as the sum

(5.3) $$\phi = \psi + \xi,$$

with $\psi \in K(\Omega_1)$ and $\xi \in G(A)$. Since ψ and ξ range independently in $K(\Omega_1)$ and $G(A)$, $\mathcal{E}(\Omega, \phi)$ also admits the decomposition

$$\mathcal{E}(\Omega, \phi) = \mathcal{E}(\Omega_1, \psi) + \mathcal{E}(A, \xi),$$

which, considering the minimum, yields

$$\mathcal{E}(\Omega) = \mathcal{E}(\Omega_1) + Min\{\mathcal{E}(A, \xi) | \xi \in G(A)\}$$

Now, as zero belongs to $G(A)$ and $\mathcal{E}(A, \xi) \geq 0$, the minimum of $\mathcal{E}(A, \xi)$ is just zero and $\mathcal{E}(\Omega) = \mathcal{E}(\Omega_1)$.

In order to choose Ω_1 as a solution, it is necessary to consider its measure. If meas$(A) = 0$, then $meas(\Omega_1) = \alpha$ and Ω_1 is an optimal domain. If, conversely, meas$(A) = d > 0$, then meas$(\Omega_1) = \alpha - d < \alpha$. In this case it is always possible to replace Ω_1, composed by the union $\Omega_1 = E \cup G_1$, with another domain $\Omega^* = E \cup G^*$ (fig.3), which is still weakly connected and where G^* is obtained from G_1 through the homothetic transformation

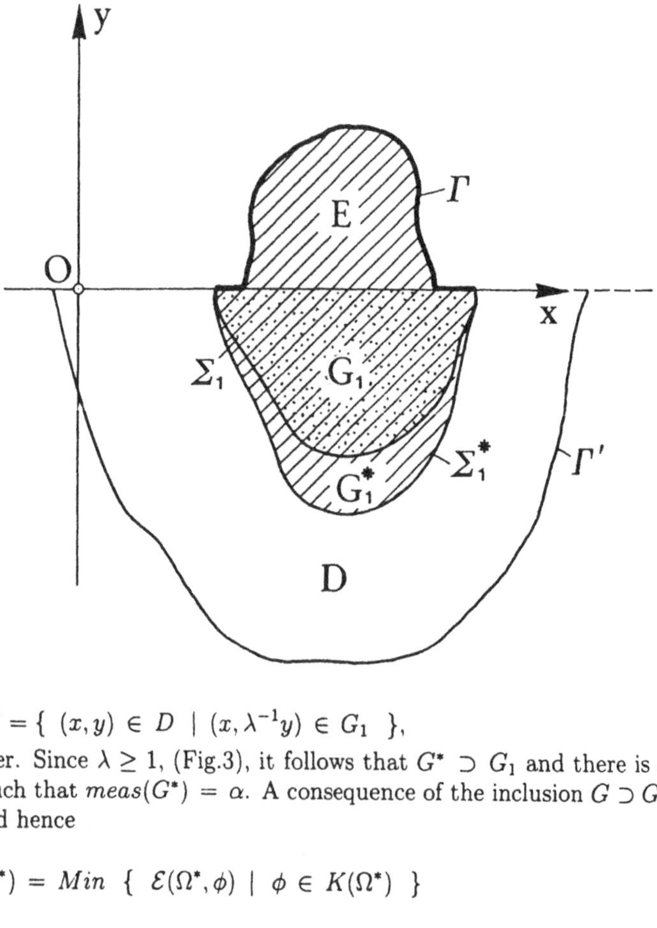

$$G^* = \{ (x, y) \in D \mid (x, \lambda^{-1}y) \in G_1 \},$$

λ $(\lambda \geq 1)$ being a parameter. Since $\lambda \geq 1$, (Fig.3), it follows that $G^* \supset G_1$ and there is a particular value of $\lambda > 1$ such that $meas(G^*) = \alpha$. A consequence of the inclusion $G \supset G_1$ is that $K(\Omega^*) \supset K(\Omega_1)$ and hence

$$\mathcal{E}(\Omega^*) = Min \{ \mathcal{E}(\Omega^*, \phi) \mid \phi \in K(\Omega^*) \}$$

$$\leq min \{ \mathcal{E}(\Omega_1, \phi) \mid \phi \in K(\Omega_1) \} = \mathcal{E}(\Omega_1),$$

which proves that by replacing Ω_1 with Ω^* the strain energy does not increase.

A second property of the optimal solution Ω^* is that it cannot have fractures, in the sense that if $\gamma \subset \partial\Omega$ is a subset with $cap_B(\gamma) > 0$ then $meas(\gamma) > 0$. If indeed $meas(\gamma)$ were equal to zero then the set $\Omega \cup \gamma$ would be admissible since its measure is still α. Obviously, as $\Omega \cup \gamma$ is larger than Ω, also $K(\Omega) \subset K(\Omega \cup \gamma)$ and the minimum of $\mathcal{E}(\Omega \cup \gamma, \phi)$ in $K(\Omega \cup \gamma)$ is lower than the minimum of $\mathcal{E}(\Omega, \phi)$ in $K(\Omega)$. Therefore, $\Omega^* = \Omega \cup \gamma$ is optimal.

6. REFERENCES

[1900] MICHELL, J.H. : "On the direct determination of stresses in an elastic solid, with applications to the theory of plates". Proc. London Math.Soc. V 31, pp. 100-124.
[1924] LOVE, A.H.E. : A Treatise on the Mathematical Theory of Elasticity. Cambridge : The Univ. Press.
[1937] NEUBER, H. : Kerbspannungslehre Berlin/Gottingen/Heidelberg : Springer.
[1953] COURANT, R. and HILBERT, D. : Methods of Mathematical Physics Vol.I, Interscience : New York.
[1957] DENY, H. and LIONS, J.L. : "Les espaces du type Beppo Levi". Ann.Inst.Fourier. V 5, pp.305-370.
[1965] STAMPACCHIA, G. : "Le probleme de Dirichlet pour les equations elliptiques du second ordre avec coefficients discontinus". Ann. Inst.Fourier. V.15, pp.189-258.
[1972] NEUBER, H. : "Zur Optimierung der Spannungskonzentration". In Continuum Mechanics and Related Problems of Analysis. Moskov : Naukq, pp. 375-378.
[1972] GURTIN, M. : "The Classical Theory of Elasticity". In Handbuch der Physics.Vol.VI/A. Berlin/Gottingen/Heidelberg : Springer.
[1981] ZOLÉSIO, J.P. : "The Material Derivative (or Speed) Method for Shape Optimization". In Optimization of Distributed Parameter Structures, (Haug and Cea Ed.), Vol.II, pp.1089-1151,Nordhoff and Sijhoff, Alphen aan Rijn.
[1981] ZOLÉSIO, J.P. : "Domain Variational Formulation of Free Boundary Problems". In Optimization of Distributed Parameter Structure (Haug and Cea Ed.), Vol.II, pp.1152-1194,Nordhoff and Sijhoff, Alphen aan Rijn.
[1983] BANICHUK, N.V. Problems and Methods of Optimal Structural Design. New York and London : Plenum.
[1985] KOHN, R. and VOGELIUS, M. : "A New Model for Thin Plates with Rapidly Varying Thickn ess". Int.J.Solids Structures V.20,No 4, pp.333-350.
[1991] VILLAGGIO, P. and ZOLÉSIO, J.P. : New Results in Shape Optimization, in "Boundary Control and Boundary Variation", L.N.C.I.S.,vol.178, Springer Verlag.
[1994] ZOLÉSIO, J.P. : Weak Shape Formulation of Free Boundary Problems, Annali della Scuola Normale Superiore di Pisa, Serie IV, vol. XI,1, 11-44.

Printed and bound by CPI Group (UK) Ltd, Croydon, CR0 4YY
21/10/2024
01777098-0009